METHODS IN MOLECULAR BIOLOGY

Series Editor
John M. Walker
School of Life and Medical Sciences
University of Hertfordshire
Hatfield, Hertfordshire, AL10 9AB, UK

For further volumes:
http://www.springer.com/series/7651

Split Inteins

Methods and Protocols

Edited by

Henning D. Mootz

*Institute of Biochemistry, Department of Chemistry and Pharmacy,
University of Muenster, Münster, Germany*

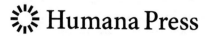 Humana Press

Editor
Henning D. Mootz
Institute of Biochemistry
Department of Chemistry and Pharmacy
University of Muenster
Münster, Germany

ISSN 1064-3745   ISSN 1940-6029   (electronic)
Methods in Molecular Biology
ISBN 978-1-4939-8204-2   ISBN 978-1-4939-6451-2   (eBook)
DOI 10.1007/978-1-4939-6451-2

Printed on acid-free paper

This Humana Press imprint is published by Springer Nature
The registered company is Springer Science+Business Media LLC New York

Preface

Protein splicing performed by inteins is an exceptional rearrangement of the polypeptide backbone of a precursor protein, in which the intein domain removes itself under concomitant linkage of the flanking extein sequences. The reaction follows a remarkable reaction mechanism consisting of typically three covalent intermediates with reactive thioester or ester linkages. First discovered in 1990 [1, 2], countless protein-based applications of protein splicing have been reported until today in the fields of protein biochemistry, biotechnology, chemical biology, and cell biology. These applications are based on both the changes in structure of the backbone connectivity during splicing and the interception of the reactive intermediates in targeted cleavage reactions.

Split inteins represent a subclass of inteins with the intein domain being split into two parts, the N-terminal and C-terminal fragments (Int^N and Int^C). The fact that the two flanking peptide or protein extein segments (Ex^N and Ex^C) reside on two separate polypeptides at the beginning of the protein *trans*-splicing reaction opens up a number of exciting new possibilities compared to regular protein splicing of one-piece inteins. First of all, the two fragments can be prepared separately and thereby by different means. Several technologies exploit this fact, e.g., those for obtaining modified proteins by reconstitution from two purified parts. Examples include segmental isotope labeling, selective chemical labeling, and protein semi-synthesis from synthetic and recombinantly produced split intein fusion proteins, as well as processes to exploit protein reconstitution inside living cells or organisms that can only take place when both complementary intein-extein polypeptides are expressed in the same cell or the same compartment. Second, the split nature allows very simple control over the start of the protein *trans*-splicing reaction. In contrast to a *cis*-splicing intein, which will typically undergo splicing after its synthesis was completed on the ribosome and folding had occurred, protein *trans*-splicing will only occur when both complementary intein fragments are present, for example by simple mixing in an experiment. Owing to the self-processive and usually unregulated nature of both *cis*- and *trans*-splicing inteins, it has been a long-standing goal to engineer inteins with a conditional activity, which can be controlled in the experiment. For example, small-molecule or light-dependent inteins open many exciting ways for artificial regulation of protein activity on the post-translational level. Such tools are very attractive for the manipulation of protein networks inside living cells.

Split inteins are found in nature [3] and can be generated artificially in the laboratory. Native inteins come in three different versions, as maxi-inteins with an inserted homing endonuclease, as mini-inteins without such insertion, and as split inteins [4]. All natively split inteins known until very recently are split at the position where maxi-inteins carry their homing endonuclease, dividing the primary sequence in fragments of about 100 aa (Int^N) and 40 aa (Int^C). Thus, a maxi-intein with a homing endonuclease domain can be regarded as an intramolecular version of a split intein and it may therefore not be too surprising that these two split fragments often also work *in trans*. The AceL-TerL intein identified in metagenomic databases was the first atypically split intein with an Int^N fragment of 25 aa (Int^N) and 104 aa (Int^C) [5]. Artificially split inteins were also created at even more positions [6]. For example, the shortest characterized intein fragments are only 11 aa (Int^N) and 6 aa (Int^C). Such short fragments are of particular interest for chemical synthesis, but may also be advantageous as shorter fusion tags with respect to expression levels and solubility properties. Even three-piece inteins have been reported that contain two split sites and hence a middle fragment without any covalent connection to an extein; however, their performance and usefulness for biotechnological applications still needs to be better evaluated [7, 8]. Finally, another artificial version of a split intein is the inverted intramolecular arrangement, with the Int^N fragment at the C terminus and the Int^C fragment at the N terminus of a polypeptide chain. Intramolecular splicing leads to a cyclic peptide or protein in this case [9].

This book *Split Inteins: Methods and Protocols* focuses on applications of split inteins. It is important to note that many of these technical developments are based on a better fundamental understanding of inteins and split inteins, an active field of research that cannot be focused on in this series. New developments and improvements can be expected for the future. While an enormous progress has been made in the last 5 years or so regarding discovery and engineering of faster and more efficient split inteins, some potential problems still need to be addressed, for example improving solubility of split intein fragments and increasing the tolerance towards foreign sequence contexts.

The first two chapters deal with new twists to use split inteins for affinity purification of overproduced proteins. Affinity chromatography with a tag that can be removed without the requirement for a protease was one of the first applications of inteins [10]. The introduction of mutations at one of the splice junctions blocks protein splicing but allows the induction of a cleavage reaction at the other splice junction, either by the addition of a nucleophile or a shift in temperature or pH. The drawbacks of this method are that some intein fusion proteins display premature cleavage during overexpression in the bacterial cells and that in some cases the cleavage reaction is not efficient enough. In Chapter 1, Zhilei Chen and Dongli Guan describe the use of an affinity chromatography based on the high inherent affinity of the split *Npu* DnaE intein pieces. Since just one half is fused to the protein of interest for overproduction, the problem of premature cleavage in cells is entirely circumvented. Furthermore, premature cleavage during purification with the other intein half linked to the beads is inhibited by Zn^{2+} ions. In Chapter 2, David W. Wood and colleagues report on their protocol to use elastin-like peptide tags (ELP) in combination with split inteins. ELP tags allow inexpensive protein purification because ELP fusions can be selectively precipitated using pH shifts. Thereby an expensive affinity matrix is not required. By employing split inteins, this protocol also efficiently avoids premature cleavage in cells.

Chapters 3–6 report on split intein-based technologies to prepare cyclic peptides. Peptide macrocycles are promising for inhibitor development and are considered of particular value for the inhibition of protein-protein interactions. Intramolecular splicing leads to circularization of genetically encoded peptides; thus large libraries of different peptide sequences can be easily created by randomization of the encoding gene on the DNA level and then subjected to a selection for the best binders. In Chapter 3, Ali Tavassoli and Eliot L. Osher describe the generation of such large cyclic peptide libraries for intracellular selections. Julio A. Camarero and Krishnappa Jagadish show in Chapter 4 how the approach can be used to prepare a bioactive cyclotide. Cyclotides are small globular microproteins of 28–37 amino acids with a cystine-knot motif to increase rigidity. They are highly promising scaffolds for the design of new inhibitors. In Chapter 5, Rudi Fasan and Nina Bionda report on their new protocol to introduce an additional intramolecular crosslink into the cyclic peptide backbone. The increased rigidification in this bicyclic peptide was made possible by the incorporation of an unnatural amino acid with unique electrophilic reactivity. Henning D. Mootz and Shubhendu Palei describe in Chapter 6 how a semisynthetic peptide macrocycle can be obtained by linking a synthetic and a genetically encoded peptide fragment. Here protein *trans*-splicing with the M86 DnaB mutant intein is performed to link the fragments with a peptide bond while the second bond is formed by a bioorthogonal, intramolecular oxime ligation involving an unnatural ketone amino acid. This protocol allows for the incorporation of synthetic, not genetically encodable, moieties into the macrocycle.

Semisynthetic protein *trans*-splicing using one synthetic intein piece is also employed in Chapter 7 by Christian F. W. Becker and coworkers to attach a lipidated peptide as a membrane anchor to the prion protein. To this end, a polypeptide containing the 35 aa Int^C fragment of the *Ssp* DnaE or *Npu* DnaE split inteins is prepared by solid-phase peptide synthesis. The hydrophobicity of the lipid modification requires the use of liposomes and nicely underlines the range of conditions under which these natively split inteins are active. The synthetic intein-extein pieces can also be used to deliver other cargos like synthetic fluorophores for protein labeling, as described previously in this book series [11] and also in Chapter 8. Here, Julio A. Camarero and Radhika Borra even take synthetic protein labeling by protein *trans*-splicing to proteins in living mammalian cells. While the targeted protein is expressed in fusion with an Int^N fragment for endogenous expression, the cells are transfected with the synthetic complementary intein piece. The access to synthetic fluorophores and minimal tags of only a few amino acids by this approach is potentially advantageous over more photolabile fluorescent proteins or other much larger fusion tags. To avoid the high background signal of unreacted Int^C-fluorophore, an elegant quencher strategy was devised.

The next two chapters describe state-of-the-art protocols for segmental isotopic labeling of proteins for NMR studies. In Chapter 9, David Cowburn and Dongsheng Liu show how two protein parts, one of which expressed in medium with enriched isotopes, can be ligated to reduce the complexity of NMR resonances in a soluble protein. Strictly speaking, their protocol does not involve protein *trans*-splicing, but uses the one-piece intein to generate a protein with a C-terminal thioester, that is then ligated with the other protein part exhibiting an N-terminal cysteine residue. This special form of native chemical ligation (NCL) is referred to as expressed protein ligation (EPL). In Chapter 10, Christiane Ritter and coworkers report on segmentally labeled proteins for solid-state NMR spectroscopy. They exploit the high activity of the split *Npu* DnaE intein in the presence of chaotropic reagents and the possibility to denature and refold the corresponding intein fusion constructs. Their protocol gives aggregation-prone, amyloidogenic proteins as example but is certainly useful also for other proteins.

In Chapter 11, Zhilei Chen and Miguel A. Ramirez show how protein block copolymers can be generated by protein *trans*-splicing to form protein hydrogels. This application at the interface to materials science is made possible by the defined start point of protein polymerization through mixing of the two split intein constructs.

In Chapter 12, Mario Gils takes advantage of protein *trans*-splicing in transgenic plants. Two split intein fusions that generate the toxic enzyme barnase upon splicing are encoded on two homologous chromosomes in wheat. Their allelic expression allows the selection of male sterility as an agriculturally important trait. The chapter details optimization steps of this technique for improved split-transgene expression, including intron-mediated enhancement and the design of flexible polypeptide linkers flanking the intein.

Chapters 13–15 deal with conditional inteins that can be regulated in artificial ways by small molecules or light. These tools are highly desirable, because the large structural changes associated with protein splicing may make the design of a switch to control protein function potentially much easier and more general than the incorporation of artificial switches into the protein itself. The trick for the intein design is to turn off its activity in the absence of the artificial trigger. Along these lines, Perry L. Howard and coworkers describe in Chapter 13 the design of a conditional version of the toxic enzyme barnase by inserting a previously reported split intein that can be activated with the small molecule rapamycin. This system is used for conditional killing of mammalian cells. In this case, the split *Sce* VMA intein fragment has only negligible activity without the proximity induced by the small molecule binding to the two rapamycin-binding domains FKBP and FRB [12]. In Chapter 14, Hui-Wang Ai and Wei Ren show how a photoactivatable version of the *Npu* DnaE intein is generated by unnatural amino acid mutagenesis. By inserting the intein into the Src protein kinase, the light-dependent regulation of this enzyme involved in signaling pathways can be achieved in mammalian cell culture. In Chapter 15, Kevin Truong and coworkers describe the regulation of the *Npu* DnaE intein by the LOV2 photoreceptor. The chapter details considerations on the design of the blue-light-dependent LOV2 photoswitches and how to insert the intein into the protein of interest. An advantage of the Flavin-dependent LOV2 photoreceptor from plants is that it does not require any further genetic manipulation in the host cell or addition of synthetic compounds.

The selection of the right insertion site of an intein into another protein, as discussed in many of the chapters in this book, is one of the principal hurdles to develop a functional intein fusion construct. Tim Sonntag describes in Chapter 16 a cassette-based approach to quickly test many positions by simplifying the cloning effort usually associated with these steps. The method is used to insert the above-mentioned rapamycin-dependent *Sce* VMA intein into TEV protease and the ultrafast naturally split NrdJ-1 intein into the cAMP response element-binding protein (CREB).

The generation of new split sites into inteins has been a major goal of many studies in recent years. When splitting *cis*-inteins the choice of the split position can be crucial for the activity in the *trans*-splicing reaction. In Chapter 17, Shih-Che Sue and coworkers describe a computational approach to predict new intein split sites (the approach also works for other proteins). Their protocol takes not only loop structures on the surface of the intein into account but also the nature of the amino acids adjacent to the new split position.

For the selection of chapters in *Split Inteins: Methods and Protocols* I have attempted to include what I consider are the most important, most current, and also the typical types of methodologies involving split inteins. Together, these chapters should provide guidance to the possibilities of split intein applications, provide proven and detailed protocols adaptable to various research projects as well as inspire new method developments.

I would like to thank all authors who have contributed to this edition of *Methods in Molecular Biology*. With the considerably more detailed description of their protocols compared to the original papers, this book will hopefully contribute to the growing awareness and dissemination of intein techniques in research and production laboratories. I also like to thank Prof. John Walker, the Series Editor at Springer Publishing for *Methods in Molecular Biology*, for guiding me through all stages of this book project.

Münster, Germany   *Henning D. Mootz*

References

1. Hirata R, Ohsumk Y, Nakano A, Kawasaki H, Suzuki K, Anraku Y (1990) Molecular structure of a gene, VMA1, encoding the catalytic subunit of H(+)-translocating adenosine triphosphatase from vacuolar membranes of Saccharomyces cerevisiae. J Biol Chem 265(12):6726

2. Kane PM, Yamashiro CT, Wolczyk DF, Neff N, Goebl M, Stevens TH (1990) Protein splicing converts the yeast TFP1 gene product to the 69-kD subunit of the vacuolar H(+)-adenosine triphosphatase. Science 250(4981):651

3. Wu H, Hu Z, Liu XQ (1998) Protein trans-splicing by a split intein encoded in a split DnaE gene of Synechocystis sp. PCC6803. Proc Natl Acad Sci U S A 95(16):9226

4. Saleh L, Perler FB (2006) Protein splicing in cis and in trans. Chem Rec 6(4):183. doi:10.1002/tcr.20082

5. Thiel IV, Volkmann G, Pietrokovski S, Mootz HD (2014) An atypical naturally split intein engineered for highly efficient protein labeling. Angew Chem Int Ed Engl 53(5):1306. doi:10.1002/anie.201307969

6. Aranko AS, Wlodawer A, Iwai H (2014) Nature's recipe for splitting inteins. Protein Eng Des Sel 27(8):263. doi:10.1093/protein/gzu028

7. Sun W, Yang J, Liu XQ (2004) Synthetic two-piece and three-piece split inteins for protein trans-splicing. J Biol Chem 279(34):35281. doi:10.1074/jbc.M405491200

8. Shah NH, Eryilmaz E, Cowburn D, Muir TW (2013) Naturally split inteins assemble through a "capture and collapse" mechanism. J Am Chem Soc 135(49):18673. doi:10.1021/ja4104364

9. Scott CP, Abel-Santos E, Wall M, Wahnon DC, Benkovic SJ (1999) Production of cyclic peptides and proteins in vivo. Proc Natl Acad Sci U S A 96(24):13638

10. Chong S, Mersha FB, Comb DG, Scott ME, Landry D, Vence LM, Perler FB, Benner J, Kucera RB, Hirvonen CA, Pelletier JJ, Paulus H, Xu MQ (1997) Single-column purification of free recombinant proteins using a self-cleavable affinity tag derived from a protein splicing element. Gene 192(2):271

11. Matern JC, Bachmann AL, Thiel IV, Volkmann G, Wasmuth A, Binschik J, Mootz HD (2015) Ligation of synthetic peptides to proteins using semisynthetic protein trans-splicing. Methods Mol Biol 1266:129. doi:10.1007/978-1-4939-2272-7_9

12. Mootz HD, Blum ES, Tyszkiewicz AB, Muir TW (2003) Conditional protein splicing: a new tool to control protein structure and function in vitro and in vivo. J Am Chem Soc 125(35):10561. doi:10.1021/ja0362813

Contents

Contributors

HUI-WANG AI • *Department of Chemistry, University of California-Riverside, Riverside, CA, USA*

SPENCER C. ALFORD • *Department of Bioengineering, Stanford University, Stanford, CA, USA; Department of Biochemistry and Microbiology, University of Victoria, Victoria, BC, Canada*

CHRISTIAN F.W. BECKER • *Department of Chemistry, Institute of Biological Chemistry, University of Vienna, Vienna, Austria*

NINA BIONDA • *Department of Chemistry, University of Rochester, Rochester, NY, USA*

RADHIKA BORRA • *Department of Pharmacology and Pharmaceutical Sciences, University of Southern California, Los Angeles, CA, USA*

JULIO A. CAMARERO • *Department of Pharmacology and Pharmaceutical Sciences, University of Southern California, Los Angeles, CA, USA; Department of Chemistry, University of Southern California, Los Angeles, CA, USA*

ZHILEI CHEN • *Department of Chemical Engineering, Texas A&M University, College Station, TX, USA; Department of Microbial and Molecular Pathogenesis, Texas A&M University Health Science Center, College Station, TX, USA*

DAVID COWBURN • *Department of Biochemistry, Albert Einstein College of Medicine, Bronx, NY, USA*

RUDI FASAN • *Department of Chemistry, University of Rochester, Rochester, NY, USA*

MARIO GILS • *Nordsaat Saatzucht GmbH, Langenstein, Germany*

DONGLI GUAN • *Department of Chemical Engineering, Texas A&M University, College Station, TX, USA*

STEFANIE HACKL • *Department of Chemistry, Institute of Biological Chemistry, University of Vienna, Vienna, Austria*

TZU-CHIANG HAN • *Department of Chemical and Biomolecular Engineering, Ohio State University, Columbus, OH, USA*

PERRY L. HOWARD • *Department of Biochemistry and Microbiology, University of Victoria, Victoria, BC, Canada*

KRISHNAPPA JAGADISH • *Department of Pharmacology and Pharmaceutical Sciences, University of Southern California, Los Angeles, CA, USA*

YI-ZONG LEE • *Institute of Bioinformatics and Structural Biology, National Tsing Hua University, Hsinchu, Taiwan*

DONGSHENG LIU • *iHuman Institute, ShanghaiTech University, Pudong, Shanghai, China*

WEI-CHENG LO • *Department of Biological Science and Technology, Institute of Bioinformatics and Systems Biology, National Chiao Tung University, Hsinchu, Taiwan*

HENNING D. MOOTZ • *Institute of Biochemistry, Department of Chemistry and Pharmacy, University of Muenster, Münster, Germany*

ABDULLAH MOSABBIR • *Institute of Biomaterials and Biomedical Engineering, University of Toronto, Toronto, ON, Canada*

MADHU NAGARAJ • *Macromolecular Interactions, Helmholtz Centre for Infection Research, Braunschweig, Germany; Leibniz-Institut für Molekulare Pharmakologie, Berlin, Germany*

CONNOR O'SULLIVAN • *Department of Biochemistry and Microbiology, University of Victoria, Victoria, BC, Canada*

ELIOT L. OSHER • *Department of Chemistry, University of Southampton, Southampton, UK*

SHUBHENDU PALEI • *Department of Chemistry and Pharmacy, Institute of Biochemistry, University of Muenster, Münster, Germany*

ANAM QUDRAT • *Institute of Biomaterials and Biomedical Engineering, University of Toronto, Toronto, ON, Canada*

MIGUEL A. RAMIREZ • *Department of Chemical Engineering, Texas A&M University, College Station, TX, USA*

WEI REN • *Department of Chemistry, University of California-Riverside, Riverside, CA, USA*

CHRISTIANE RITTER • *Macromolecular Interactions, Helmholtz Centre for Infection Research, Braunschweig, Germany; SeNostic, Hannover, Germany*

ALANCA SCHMID • *Department of Chemistry, Institute of Biological Chemistry, University of Vienna, Vienna, Austria*

TOBIAS SCHUBEIS • *Macromolecular Interactions, Helmholtz Centre for Infection Research, Braunschweig, Germany; Centre de RMN à Très Hauts Champs, Institut des Sciences Analytiques, UMR 5280 CNRS/Ecole Normale Supérieure de Lyon/UCBL, University of Lyon, Villeurbanne, France*

CHANGHUA SHI • *Biodesign Institute at Arizona State University, Tempe, AZ, USA*

TIM SONNTAG • *Clayton Foundation Laboratories for Peptide Biology, The Salk Institute for Biological Studies, La Jolla, California, USA*

SHIH-CHE SUE • *Department of Life Sciences, Institute of Bioinformatics and Structural Biology, National Tsing Hua University, Hsinchu, Taiwan*

ALI TAVASSOLI • *Department of Chemistry, University of Southampton, Southampton, UK*

KEVIN TRUONG • *Institute of Biomaterials and Biomedical Engineering, University of Toronto, Toronto, ON, Canada; Department of Electrical and Computer Engineering, University of Toronto, Toronto, ON, Canada*

DAVID W. WOOD • *Department of Chemical and Biomolecular Engineering, Ohio State University, Columbus, OH, USA*

Chapter 1

Affinity Purification of Proteins in Tag-Free Form: Split Intein-Mediated Ultrarapid Purification (SIRP)

Dongli Guan and Zhilei Chen

Abstract

Proteins purified using affinity-based chromatography often exploit a recombinant affinity tag. Existing methods for the removal of the extraneous tag, needed for many applications, suffer from poor efficiency and/or high cost. Here we describe a simple, efficient, and potentially low-cost approach—split intein-mediated ultrarapid purification (SIRP)—for both the purification of the desired tagged protein from *Escherichia coli* lysate and removal of the tag in less than 1 h. The N- and C-fragment of a self-cleaving variant of a naturally split DnaE intein from *Nostoc punctiforme* are genetically fused to the N-terminus of an affinity tag and a protein of interest (POI), respectively. The N-intein/affinity tag is used to functionalize an affinity resin. The high affinity between the N- and C-fragment of DnaE intein enables the POI to be purified from the lysate via affinity to the resin, and the intein-mediated C-terminal cleavage reaction causes tagless POI to be released into the flow-through. The intein cleavage reaction is strongly inhibited by divalent ions (e.g., Zn^{2+}) under non-reducing conditions and is significantly enhanced by reducing conditions. The POI is cleaved efficiently regardless of the identity of the N-terminal amino acid except in the cases of threonine and proline, and the N-intein-functionalized affinity resin can be regenerated for multiple cycles of use.

Key words Protein purification, Chromatography, Chitin-binding domain, Tag removal, Tagless

1 Introduction

Affinity tags have greatly simplified the process of recombinant protein purification [1], but many applications require native proteins free of foreign amino acids. The removal of affinity tags from purified recombinant proteins remains cumbersome and expensive. The most widely use method for removing appended affinity tags from a protein of interest (POI) employs an array of precision proteases that often suffer from low activity, low specificity and/or high cost and requires additional purification step(s) to remove the cleaved affinity tag and proteases from the final product [2–5]. Consequently, while popular in academic laboratories, proteases are not used for large-scale purification.

Henning D. Mootz (ed.), *Split Inteins: Methods and Protocols*, Methods in Molecular Biology, vol. 1495,
DOI 10.1007/978-1-4939-6451-2_1, © Springer Science+Business Media New York 2017

Inteins are protein splicing elements that can excise themselves from precursor proteins and ligate the surrounding sequences (exteins) [6–8]. Many inteins have been engineered to undergo a single N- or C-terminal cleavage reaction under acidic or reducing conditions and these have been employed for cost-effective removal of affinity tags following recombinant protein purification [9–11]. In these cases, the N- and C-exteins are replaced with an affinity tag and the POI. However, most of these engineered inteins exhibit low catalytic efficiency, requiring overnight incubation to achieve significant cleavage. Furthermore, incomplete regulation of the intein cleavage reaction can cause premature cleavage of the affinity tag from the POI in vivo during protein expression, thus significantly impacting purification yield [12, 13].

Recently, our lab engineered a naturally split DnaE intein from *Nostoc punctiforme* (*Npu*DnaE) that undergoes a rapid C-terminal cleavage reaction upon exposure to reducing conditions [14]. Unlike the conventional continuous inteins, the catalytic residues of a split intein are located on two separate protein fragments. Intein cleavage occurs only upon reconstitution of the two intein fragments. Thus, premature cleavage of POI during expression in vivo is avoided by separate expression of the two fragments in different hosts, potentially increasing the yield of the purified POI. With the N- and C-exteins substituted with a chitin-binding domain (CBD) affinity tag and POI, respectively, we observed >80% cleavage completion within 3 h incubation at room temperature. In follow-up work, we further enhanced the cleavage kinetics by positioning the affinity tag in the split intein junction at the C-terminus of the intein N-fragment (Fig. 1), and achieved >90% POI release within 30 min at 22 °C and within 3 h at 6 °C [16]. The method employing this new configuration—split intein-mediated ultrarapid purification

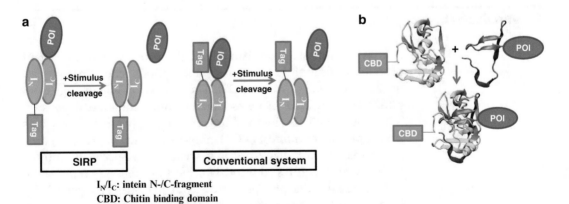

I$_N$/I$_C$: intein N-/C-fragment
CBD: Chitin binding domain
POI: Protein of Interest

Fig. 1 Schematic of the engineered intein pair. (**a**) Schematic of the configuration of the current intein design for tag removal vs. the "conventional" design. (**b**) Cartoon representation of fusion proteins before and after intein association. The intein N-fragment (*yellow*) and C-fragment (*brown*) are adapted from the NMR structure of *Npu*DnaE (PDB code: 2keq) [15] (Reproduced from ref. [16] with permission from John Wiley & Sons)

(SIRP)—exhibits dramatically increased intein cleavage efficiency, likely due to alleviation of the steric hindrance between the N- and C-exteins. The cleavage reaction is significantly accelerated under reducing conditions and can be efficiently suppressed by divalent ions (e.g., Zn^{2+}) under non-reducing conditions [17, 18]. Moreover, the cleavage efficiency is largely independent of the identity of the first residue of the POI except for threonine and proline, making it possible to purify most natural proteins in unmodified form. Using CBD as the affinity tag, we demonstrated the utility of SIRP by purifying three model proteins, GFP, phosphite dehydrogenase (PTDH) [19], and β-galactosidase (β-Gal) [14] with monomeric, dimeric, and tetrameric tertiary structure, respectively, to 80–90% purity from a single column with a yield of purified protein of ~10 mg per mL of chitin resin. We also demonstrated the efficient regeneration of the intein-functionalized column via conditional dissociation of the intein N- and C-fragments for at least four cycles, enabling recycling of the column and potentially further reducing the purification cost. In this chapter we provide protocols for the expression of the split intein fusion proteins (Subheading 3.1), the preparation of a split intein affinity column (Subheading 3.2), its application in purifying a protein of interest (Subheading 3.3) and its regeneration for multiple rounds of protein purification (Subheading 3.4).

2 Materials

Prepare culture media using deionized water (diH$_2$O) and all other solutions using double deionized water (ddH$_2$O, prepared by purifying diH$_2$O through a MilliQ purification system). Prepare and store all reagents at room temperature unless indicated otherwise.

2.1 Protein Expression

1. Expression host: *E. coli* BL21 (DE3), chemically competent cells in 50 μL aliquots stored at –80 °C.

2. Plasmids: pNpuN$_{C1A}$-CBD, pNpuC$_{D118G}$-PTDH, pNpuC$_{D118G}$-GFP, pNpuC$_{D118G}$-β-Gal, purified plasmid DNA stored at –20 °C.

3. Luria-Bertani (LB) broth: 25 g LB powder per liter. Sterilize by autoclaving at 121 °C, 20 psi for 20 min.

4. 50 mg/mL Kanamycin (Kan): Filter through a 0.45 μm sterile filter, aliquot into 1 mL per tube, and store at –20 °C.

5. LB agar plate: 25 g LB powder and 15 g agar per liter. Autoclave at 121 °C, 20 psi for 20 min. When the medium is cooled to ~55 °C (*see* **Note 1**), add 1 mL of 50 mg/mL Kan per liter of medium, mix well, and immediately transfer 20 mL aliquots into each petri dish (100 mm diameter) using sterile technique. After solidification, store plates at 4 °C.

6. 1 M Isopropyl β-D-1-thiogalactopyranoside (IPTG): Filter through a 0.45 μm sterile filter, aliquot to 1 mL per tube, and store at −20 °C.

2.2 SIRP

1. Plastic columns.

2. Chitin resin (New England Biolabs), in 20 % (v/v) ethanol and stored at 4 °C.

3. **Buffer A**: 0.5 M NaCl, 10 mM Tris–HCl, pH 8.0, filter through a 0.2 μm filter, and store at 4 °C.

4. **Buffer B**: 0.5 M NaCl, 50 mM sodium phosphate buffer, pH 6.0, filter through a 0.2 μm filter, and store at 4 °C (*see* **Note 2**).

5. 20 mM $ZnCl_2$: Filter through a 0.2 μm filter.

6. 1 M Dithiothreitol (DTT): Filter through a 0.2 μm filter, aliquot to 100 μL per tube, and store at −20 °C (*see* **Note 3**).

7. **Buffer C**: 1.5 M NaCl, 0.5 mM $ZnCl_2$, 50 mM Na_2HPO_4–NaOH, pH 11.4, filter through a 0.2 μm filter.

8. **Storage buffer**: 0.5 M NaCl, 10 mM Tris–HCl, 1 mM EDTA, 0.15 % (w/v) NaN_3, pH 8.0, filter through a 0.2 μm filter.

3 Methods

Carry out all procedures at room temperature unless otherwise specified.

3.1 Protein Expression

1. Transform *E. coli* BL21 (DE3) with pNpuN$_{C1A}$-CBD and pNpuC$_{D118G}$-GFP/PTDH/β-Gal, and plate the cells on LB agar plates containing 50 μg/mL Kan.

2. Incubate plates at 37 °C for 12–15 h (*see* **Note 4**).

3. Pick a single colony from each plate and grow in 4 mL of LB broth containing 50 μg/mL Kan (LB/Kan50) at 37 °C, 250 rpm until $OD_{600} \sim 0.6$.

4. Transfer each 4 mL culture into 1 L LB broth containing 50 μg/mL Kan and continue to grow the culture until OD_{600} 0.6–1.1 (*see* **Note 5**).

5. Cool down each culture to ~18 °C in an ice–water bath for 6–8 min with swirling every 2 min (*see* **Note 6**), and then add IPTG to a final concentration of 1 mM (*see* **Note 7**).

6. Incubate the cultures at 18 °C, 250 rpm for 14–15 h (*see* **Note 8**).

7. Harvest cells by centrifugation at $8000 \times g$ and 4 °C for 15 min. The cell pellets can be used directly for the next step or stored at −80 °C until use (*see* **Note 9**).

3.2 Preparation of N$_{C1A}$-CBD-Chitin Affinity Column

1. Resuspend cell pellet harboring pNpuN$_{C1A}$-CBD with 10 mL Buffer A per gram of wet pellet on ice.

2. Lyse cells in ice–water bath with sonication (*see* **Note 10**), and then centrifuge the lysate at $16,000 \times g$ and 4 °C for 15 min.

3. Transfer supernatant (clarified lysate) to a clean tube and keep the lysate on ice (*see* **Note 11**).

4. Equilibrate an appropriate volume of chitin resin (*see* **Note 12**) in a plastic column with 10× column volume (10CV) of Buffer A.

5. Resuspend the resin with the soluble lysate, and then incubate the suspension for ~10 min with gentle rocking or rotation.

6. Wash the column with 10CV of Buffer B at least four times (*see* **Note 13**), and add 0.5 mM ZnCl$_2$ to the final wash.

3.3 Purification of Protein-of-Interest in Sample via SIRP

A schematic of SIRP is shown in Fig. 2.

1. Resuspend cell pellet harboring pNpuC$_{D118G}$-GFP/PTDH/β-Gal (*see* **Note 14**) in 10 mL Buffer B (*see* **Note 15**) per gram wet pellet on ice.

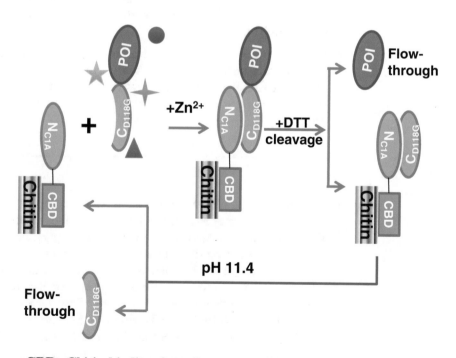

CBD: Chitin binding domain
POI: Protein of Interest

Fig. 2 Schematic of protein purification using SIRP. Lysate of **C$_{D118G}$**-POI is applied to a column containing prebound **N$_{C1A}$**-CBD in the presence of 0.5 mM ZnCl$_2$. After washing, the intein C-terminal cleavage reaction is induced by the addition of DTT and purified POI is collected in the flow-through. The column is then regenerated by washing with pH 11.4 buffer to dissociate the intein complex (Reproduced from ref. [16] with permission from John Wiley & Sons)

2. Lyse cells in ice–water bath with sonication (*see* **Note 10**), and then centrifuge at $16,000 \times g$ and 4 °C for 15 min.

3. Transfer supernatant (clarified lysate, Fig. 3, lane 2) to a clean tube prechilled in ice, and add 0.5 mM $ZnCl_2$ (*see* **Note 16**).

4. Resuspend the N_{C1A}-CBD-chitin resin in an appropriate volume of soluble lysate (*see* **Note 17**), and then incubate for ~10 min with gentle rocking or rotation.

5. Wash the column with 10CV of Buffer A plus 0.5 mM $ZnCl_2$ at least four times (Fig. 3, lane 4).

6. Incubate the column with 4CV of Buffer A plus 50 mM DTT (*see* **Note 18**) with gentle rocking or rotation for 30 min at 22 °C or for 3 h at 6 °C (Fig. 3, lane 6 and 7) (*see* **Note 19**).

7. Collect the flow-through as purified protein-of-interest (Fig. 3, lane 8 and 9 for GFP/PTDH/β-Gal) (*see* **Note 20**).

3.4 Regeneration of N_{C1A}-CBD-Chitin Affinity Column

1. High pH condition is able to dissociate the N- and C-inteins without affecting the binding of CBD to chitin resin. N_{C1A}-functionalized chitin beads can be regenerated by resuspending the chitin beads in 15CV of Buffer C and incubating the suspension at RT for ~10 min with gentle rocking or rotation (*see* **Note 21**).

2. Wash the beads with 10CV of storage buffer. The regenerated column can be used directly for purification as described above starting from the last wash in **step 6** in Subheading 3.2 or be stored in Storage Buffer at 4 °C for at least a week (*see* **Note 22**).

4   Notes

1. Agar starts to solidify at ~50 °C and most antibiotics are heat labile, meaning they should not be added at high temperatures. Therefore, Kan should be added to LB agar medium at ~55 °C. This can be done by cooling the medium in a 55 °C water bath for ≥ 1 h. Alternatively, the medium can be rapidly cooled in an ice–water bath for 5–10 min with swirling every minute. If the bottle is warm to the touch, allowing placement of the hand on the bottle for 5–10 s without experiencing an intense burning sensation, the concentrated Kan stock may be added.

2. We prepared 0.2 M sodium phosphate buffer at pH 6.0 by mixing 87.7 mL 0.2 M NaH_2PO_4 and 12.3 mL 0.2 M Na_2HPO_4, and then dilute this solution fourfold in 0.5 M NaCl solution to make Buffer B. The pH of Buffer B needs to be verified by a pH meter before each use. We observed a decrease of the solution pH over time during storage at RT. The solution pH is adjusted by the addition of 1 M NaOH.

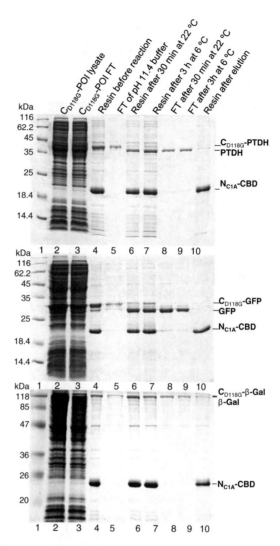

Fig. 3 Purification of PTDH, GFP and β-Gal using N_{C1A}-CBD-chitin resin. SDS-PAGE analysis of the purification of PTDH, GFP and β-Gal. *Lane 1*, EZ-Run Protein Ladder; *lane 2*, supernatant of lysate containing C_{D118G}-POI; *lane 3*, flow-through after applying lysate to column; *lane 4*, chitin resin after loading lysate and washing with buffer containing 0.5 mM $ZnCl_2$; *lane 5*, elution of C_{D118G}-POI in pH 11.4 buffer; *lane 6*, chitin resin in buffer containing 50 mM DTT incubated at 22 °C for 30 min; *lane 7*, chitin resin in buffer containing 50 mM DTT incubated at 6 °C for 3 h; *lane 8*, flow-through after incubation at 22 °C for 30 min; *lane 9*, flow-through after incubation at 6 °C for 3 h; *lane 10*, chitin resin after elution of POI. Samples were taken at different time points, immediately frozen in liquid N_2, boiled in 2× SDS sample buffer at 95 °C for 10 min and analyzed on 12 % SDS-PAGE gels. The gels were stained with Coomassie brilliant blue R250. Band intensities corresponding to reactants and products were quantified using the Trace Quantity module in Quantity One software (Bio-Rad) (reproduced from ref. [16] with permission from John Wiley & Sons)

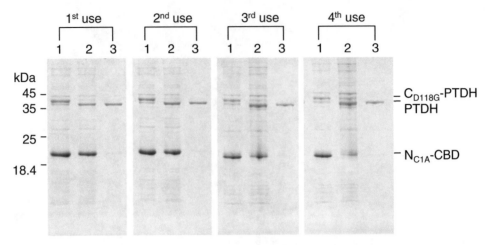

Fig. 4 Chitin-bound N_{C1A}-CBD can be regenerated after purification. SDS-gels of samples collected during the purification of PTDH using regenerated column. *Lane 1*, chitin resin before cleavage; *lane 2*, chitin resin in buffer containing 50 mM DTT incubated at 22 °C for 30 min; *lane 3*, flow through containing purified PTDH. The cleaved C_{D118G} is 4 kDa and is not visible from the gel (reproduced from ref. [16] with permission from John Wiley & Sons)

3. DTT is volatile so the solution concentration decreases over time. Storing it frozen at –20 °C and in small quantities can minimize concentration variations.

4. There should be <100 colonies on each 10 cm plate. Incubation for >16 h may reduce the yield of expressed protein.

5. Cultures should be stopped at a defined optical density range OD_{600} (e.g., 0.8–1.0) to maximize protein yield. Do not allow OD_{600} to exceed 1.1 as older cultures tend to yield a lower level of protein expression.

6. For large (≥1 L) cultures, cooling in air will take more than 20 min to lower the temperature from 37 °C to 18 °C. Since *E. coli* BL21(DE3) cells at exponential growth phase double every ~20 min at 37 °C, OD_{600} will increase significantly during the cooling period. Therefore, the cooling process should be done in an ice–water bath (usually take 6–8 min) prior to induction.

7. 0.4–1 mM IPTG can be used without affecting expression level. Typically, a stock comprising 1000× concentrated IPTG is added to achieve the final desired concentration.

8. Expression time longer than 16 h may result in proteolytic degradation of NpuC fusions by cellular proteases.

9. Expressed cells can be stored at –80 °C for 2–3 months or at –20 °C for 2–3 weeks. Longer storage time may cause degradation of NpuC fusions by cellular proteases.

10. For efficient cell lysis, the sonication conditions need to be optimized for the specific instrument. We used QSonica Misonix 200 at Amp 20, with 1 s pulse 4 s pause cycles with sample immersed in an ice–water bath. For small amounts of

cells (e.g., ≤1 g wet cell pellet), we use 1 min total pulse time. For large amounts of cells (e.g., >2 g wet cell pellet), we use 2 min total pulse time.

11. After cell lysis and centrifugation, the clarified lysate should be kept on ice for no longer than 2 h. It is recommended that the lysate be used immediately to minimize degradation by cellular proteases.

12. It is critical to saturate the chitin column with N_{C1A}-CBD to minimize the nonspecific binding of other cellular proteins. For example, for 2 mL clarified lysate, we typically use ~150 μL of chitin resin (This is the dry bed volume excluding any liquid, corresponding to ~200 μL of chitin slurry). The binding capacity of N_{C1A}-CBD is ~5 mg/mL of chitin resin and the crude yield of N_{C1A}-CBD in the clarified cell lysate is typically 0.62–0.75 mg/mL.

13. We observed that the 4× wash ensured removal of most non-specific proteins bound to the column. Removal of nonspecific proteins can be monitored using a "quick Bradford" assay by mixing 5 μL Bradford reagent with 5 μL flow-through after each wash on a piece of Parafilm. A clear solution without any blue color suggests the absence of protein.

14. The three C_{D118G} fusion proteins used here contain a cysteine at the +1 position (the first residue of C-extein) which is required for the intein *trans*-splicing reactions [20–22], but is not needed for the C-terminal cleavage reaction. We have tested the cleavage efficiency of POI constructs harboring each of the 20 amino acids as the +1 residue [16]. Eighteen constructs (all except those with Pro and Thr as the +1 residue) yielded >70% completion of C-terminal cleavage following 1 h incubation at room temperature. Fifteen constructs exhibited >80% cleavage completion within 30 min at room temperature (all except those with Lys, Ser, Gly, Thr, Pro as the +1 residue) [16]. The cleavage efficiency of proteins with Thr and Pro as the +1 residue were 50% and 7%, respectively, following 1 h at room temperature. The particularly low cleavage efficiency of the Pro+1 protein likely derives from the steric hindrance caused by unusual backbone conformation of Proline.

15. Buffer B (pH 6.0) instead of Buffer A (pH 8.0) is used for the purification of all NpuC-fusion proteins. The lower pH is intended to reduce the proteolytic degradation of NpuC fusions by cellular proteases.

16. The reaction between N_{C1A}-CBD and C_{D118G}-POI is greatly accelerated by reducing conditions. However, a significant amount of cleavage takes place even under non-reducing conditions. We observed about 50% C-terminal cleavage after 30 min incubation at 22 °C under non-reducing conditions [16]. The basal C-terminal cleavage reaction can be efficiently

suppressed by divalent ions, with only ~14% cleavage after 3 h incubation at 22 °C in the presence of $ZnCl_2$ at 0.5 mM for POI with a Cys+1 residue. A higher concentration of Zn^{2+} may be used to more efficiently inhibit the C-terminal cleavage reaction, if needed. However, ≥1 mM Zn^{2+} can cause precipitation of some cellular proteins and thus should not be applied directly to the cell lysate. Additional divalent metal ions such as Cu^{2+}, Ca^{2+}, Mg^{2+}, Mn^{2+}, and Fe^{2+} can also inhibit the intein C-terminal cleavage under non-reducing conditions albeit with varying efficiencies [16].

17. It is critical to saturate the N_{C1A}-CBD-chitin column with C_{D118G}-POI to minimize nonspecific binding of cellular proteins to the column. For example, for 150 μL N_{C1A}-CBD-chitin resin, we typically use ~2 mL of clarified C_{D118G}-POI lysate (may vary depending on the expression levels of different NpuC fusions). The binding capacity and yield of purified protein is in general inversely related to the size of the POI, likely due to steric hindrance caused by larger proteins.

18. 5–50 mM DTT can be used without affecting cleavage efficiency. For proteins that cannot tolerate reducing conditions, such as those with surface-exposed disulfide bonds, EDTA can be used as an alternative to induce intein cleavage by chelating the Zn^{2+} ions that suppress basal cleavage. The cleavage kinetics induced by EDTA are slightly reduced relative to that obtained via DTT induction. In our experience, 5 mM EDTA is sufficient to induce nearly 100% intein cleavage after 1 h incubation at room temperature [16].

19. The intein C-terminal cleavage efficiency for all the three proteins we processed—GFP, PTDH, and β-Gal—were comparable on-column and in-solution, with >80% cleavage completion in 30 min at 22 °C and in 3 h at 6 °C. However, in some cases, the intein cleavage rate is affected not only by the first residue, but the entire C-extein structure [14]. Thus, the actual cleavage efficiency is POI-dependent.

20. The yield of purified POI was determined by absorbance at 280 nm using a NanoDrop 1000 (Thermo Fisher Scientific). In most cases, the yield of purified protein was ~10 mg per mL of chitin resin. The purity was measured by analyzing SDS-PAGE gels using the Trace Quantity module in Quantity One software (Bio-Rad). For all three of the POI we processed, the final purified protein sample contained 80–90% of the tagless POI and 10–15% N_{C1A}-CBD that dissociated from the chitin beads. N_{C1A}-CBD dissociated from the chitin resin is usually the major impurity and can be removed by passing the sample through a fresh chitin column. The use of affinity tags with higher affinity towards their respective resins relative to chitin, such as 6xHistidine in conjunction with Ni^{2+}-functionalized resin, may enable further enhanced sample purity.

21. The calculated isoelectric point for C_{D118G} and N_{C1A} are 10.4 and 4.2, respectively. Incubation in Buffer C (pH 11.4) significantly weakens the intein association, and enables efficient removal of the cleaved C_{D118G} and uncleaved C_{D118G}-POI from N_{C1A} and chitin beads, enabling regeneration of chitin-bound N_{C1A}-CBD for multiple cycles of purification.

22. We carried out the purification of PTDH using the same chitin column four times (Fig. 4). The yields of purified PTDH were comparable for each use, confirming the ability of N_{C1A}-CBD-chitin affinity matrix to be regenerated for at least four cycles. We believe that cleaved C_{D118G} dissociates more readily from N_{C1A}-CBD at pH 11.4 than full-length C_{D118G}-PTDH. The same N_{C1A}-CBD-chitin column was stored at 4 °C for a week without significant loss of activity.

Acknowledgements

Funding for this work was provided by the Norman Hackerman Advanced Research Program, National Science Foundation and the Chemical Engineering Department at the Texas A&M University.

References

1. Waugh DS (2005) Making the most of affinity tags. Trends Biotechnol 23(6):316–320. doi:10.1016/j.tibtech.2005.03.012

2. Isetti G, Maurer MC (2007) Employing mutants to study thrombin residues responsible for factor XIII activation peptide recognition: a kinetic study. Biochemistry 46(9):2444–2452. doi:10.1021/bi0622120

3. Kapust RB, Tozser J, Fox JD, Anderson DE, Cherry S, Copeland TD, Waugh DS (2001) Tobacco etch virus protease: mechanism of autolysis and rational design of stable mutants with wild-type catalytic proficiency. Protein Eng 14(12):993–1000

4. Gasparian ME, Ostapchenko VG, Schulga AA, Dolgikh DA, Kirpichnikov MP (2003) Expression, purification, and characterization of human enteropeptidase catalytic subunit in Escherichia coli. Protein Expr Purif 31(1):133–139

5. Waugh DS (2011) An overview of enzymatic reagents for the removal of affinity tags. Protein Expr Purif 80:283–293

6. Saleh L, Perler FB (2006) Protein splicing in cis and in trans. Chem Rec 6(4):183–193. doi:10.1002/tcr.20082

7. Evans TC, Xu MQ (1999) Intein-mediated protein ligation: harnessing natures' escape artists. Biopolymers 51(5):333–342

8. Shi J, Muir TW (2005) Development of a tandem protein trans-splicing system based on native and engineered split inteins. J Am Chem Soc 127(17):6198–6206

9. Lew BM, Mills KV, Paulus H (1999) Characteristics of protein splicing in trans mediated by a semisynthetic split intein. Biopolymers 51(5):355–362. doi:10.1002/(SICI)1097-0282(1999)51:5<355::AID-BIP5>3.0.CO;2-M

10. Southworth MW, Amaya K, Evans TC, Xu MQ, Perler FB (1999) Purification of proteins fused to either the amino or carboxy terminus of the Mycobacterium xenopi gyrase A intein. Biotechniques 27(1):110–114, 116, 118–120

11. Mathys S, Evans TC, Chute IC, Wu H, Chong S, Benner J, Liu XQ, Xu MQ (1999) Characterization of a self-splicing mini-intein and its conversion into autocatalytic N- and C-terminal cleavage elements: facile production of protein building blocks for protein ligation. Gene 231(1–2):1–13

12. Li Y (2011) Self-cleaving fusion tags for recombinant protein production. Biotechnol Lett 33(5):869–881. doi:10.1007/s10529-011-0533-8

13. Cui C, Zhao W, Chen J, Wang J, Li Q (2006) Elimination of in vivo cleavage between target protein and intein in the intein-mediated protein purification systems. Protein Expr Purif 50(1):74–81. doi:10.1016/j.pep.2006.05.019

14. Ramirez M, Valdes N, Guan D, Chen Z (2013) Engineering split intein DnaE from Nostoc punctiforme for rapid protein purification. Protein Eng Des Sel 26(3):215–223. doi:10.1093/protein/gzs097

15. Oeemig JS, Aranko AS, Djupsjobacka J, Heinamaki K, Iwai H (2009) Solution structure of DnaE intein from Nostoc punctiforme: structural basis for the design of a new split intein suitable for site-specific chemical modification. FEBS Lett 583(9):1451–1456. doi:10.1016/j.febslet.2009.03.058, S0014-5793(09)00248-8 [pii]

16. Guan D, Ramirez M, Chen Z (2013) Split intein mediated ultra-rapid purification of tagless protein (SIRP). Biotechnol Bioeng 110(9):2471–2481. doi:10.1002/bit.24913

17. Nichols NM, Benner JS, Martin DD, Evans TC Jr (2003) Zinc ion effects on individual Ssp DnaE intein splicing steps: regulating pathway progression. Biochemistry 42(18):5301–5311. doi:10.1021/bi020679e

18. Sun P, Ye S, Ferrandon S, Evans TC, Xu MQ, Rao Z (2005) Crystal structures of an intein from the split dnaE gene of Synechocystis sp. PCC6803 reveal the catalytic model without the penultimate histidine and the mechanism of zinc ion inhibition of protein splicing. J Mol Biol 353(5):1093–1105. doi:10.1016/j.jmb.2005.09.039

19. Johannes TW, Woodyer RD, Zhao H (2005) Directed evolution of a thermostable phosphite dehydrogenase for NAD(P)H regeneration. Appl Environ Microbiol 71(10):5728–5734. doi:10.1128/AEM.71.10.5728-5734.2005

20. Xu MQ, Perler FB (1996) The mechanism of protein splicing and its modulation by mutation. EMBO J 15(19):5146–5153

21. Nichols NM, Evans TC Jr (2004) Mutational analysis of protein splicing, cleavage, and self-association reactions mediated by the naturally split Ssp DnaE intein. Biochemistry 43(31):10265–10276. doi:10.1021/bi0494065

22. Zettler J, Schutz V, Mootz HD (2009) The naturally split Npu DnaE intein exhibits an extraordinarily high rate in the protein trans-splicing reaction. FEBS Lett 583(5):909–914. doi:10.1016/j.febslet.2009.02.003

Chapter 2

Purification of Microbially Expressed Recombinant Proteins via a Dual ELP Split Intein System

Changhua Shi, Tzu-Chiang Han, and David W. Wood

Abstract

segments then restores self-cleaving activity to deliver the tagless target protein.

Key words Intein, Elastin like protein, Trans cleavage, Split intein, Non-chromatographic purification, Recombinant protein purification

1 Introduction

Inteins are self-splicing protein elements that excise themselves from their native host proteins while ligating their flanking sequences by a peptide bond [1]. This ability has enabled new tools in biotechnology, medicine and pure research in molecular biology [2]. In one of these applications, modified inteins have been generated where the native splicing reaction is replaced by a self-cleaving activity at either the N- or C-terminus of the intein, thus providing self-cleaving affinity tags for recombinant protein purification [3–5]. In practice, the intein cleaving reaction can be triggered by the addition of thiol compounds, or through changes in pH and/or temperature, allowing on-column tag cleavage during purification. The significant potential of inteins to simplify and streamline the purification of tagless recombinant target proteins resulted in the first commercial intein system released by New England Biolabs nearly 20 years ago. Despite this promise, however, pH-inducible intein systems typically suffer from premature cleaving of the tag during expression, while thiol-inducible inteins are inconvenient for proteins that contain disulfide bonds.

In our previous work, we developed an engineered variant of the *Mtu* RecA intein with C-terminal cleaving activity that can be triggered by changes in pH and temperature [6, 7]. This intein, referred to as ΔI-CM, has been combined with a number of conventional affinity tags, and we have demonstrated that it can be used effectively in microbial hosts for a variety of target proteins. Notably, this intein has been combined with an elastin like polypeptide (ELP) tag [8], allowing the simple purification of tagless proteins via cycles of selective salt precipitation followed by tag self-cleavage [9, 10]. Thus, the ELP-intein combination can provide a highly attractive purification platform that lacks the requirement for conventional packed bed columns or affinity chemistries. As with many pH-inducible inteins, however, this intein exhibits significant premature cleavage during protein expression in vivo, which is dependent on both the target protein and expression temperature [11].

Recently, issues associated with premature intein cleaving have been resolved through the development of split-intein systems, where a self-cleaving intein domain is split into two inactive segments and expressed separately [12–14]. Once recombined, the segments associate to form the active cleaving complex, allowing release of the tagless target protein with little or no premature cleaving during expression. In the method described here, we have split our ΔI-CM intein into two inactive segments and joined each to an N-terminal ELP precipitation tag (Fig. 1a). The target protein is then fused to the C-terminus of the C-terminal intein segment, expressed from the plasmid pET/EI$_{0C}$-TP (TP = Target Protein). The N-terminal intein segment is separately expressed from the plasmid pET/EI$_{0N}$, and acts as a "switch protein" that can control the cleavage activity of the C-terminal segment by spontaneous association. The purification can then be performed as with the continuous ELP-intein fusion, using salt precipitation followed by self-cleaving of the associated intein segments. We have implemented this strategy using two configurations. In the "two step" method, the intein segments are expressed and purified separately, and then mixed to induce the cleaving reaction. In the alternate "one step" method, the segments are expressed separately, but the expressing cells are mixed before lysis, allowing the segments to associate in the cell lysate before purification. Herein, we will demonstrate these methods using β-galactosidase as a model target protein for each of these methods.

2   Materials

Prepare all solutions using ultrapure water (prepared by purifying deionized water to attain a sensitivity of 18 MΩ/cm at 25 °C) and analytical grade reagents. Prepare and store all reagents at room temperature unless indicated otherwise. Diligently follow all waste

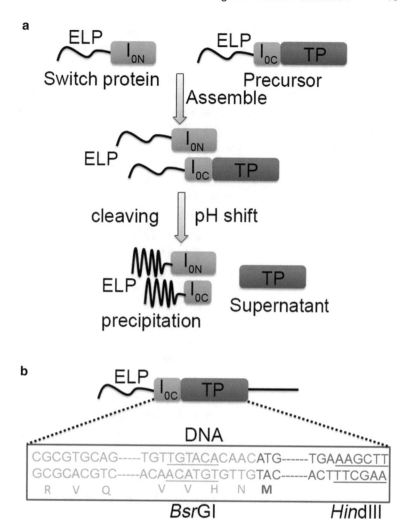

Fig. 1 Illustration of the dual-ELP system and the plasmid pET/EI$_{oc}$-TP construction. (**a**) Schematic representation of the dual-ELP purification system. The switch protein and tagged target are purified separately by ELP precipitation and then mixed together to restore cleaving activity. The cleaving incubation is triggered by a pH shift, releasing the soluble target protein from the ELP-intein segment tags. An additional ELP precipitation allows simple removal of the cleaved tags from the target protein. (**b**) Construction of the tagged target protein fusion plasmid. The target protein is fused two the C-terminus of the C-terminal intein segment (I$_{oc}$) via conventional forced cloning using an internal *Bsr*GI and additional *Hind*III site. To place the target protein in frame with the intein and retain the final critical intein amino acids, the internal *Bsr*GI (TGTACA) and the last four nucleic acids of the intein sequence (CAAC) must be added to the 5′ end of the target protein DNA sequence. The downstream cloning primer can use a variety of sites, where a *Hind*III site is shown here for illustration. Plasmid sequences can be requested from the authors

disposal regulations when disposing waste materials. Specific vendors listed below can be substituted in cases where reagents and equipment for general procedures are described (e.g., restriction digests for cloning or gel electrophoresis apparatus).

2.1 Construction of Recombinant pET/EI$_{0C}$-TP Plasmid

1. Cell strains: cloning strain DH5α (F⁻ Φ80*lac*ZΔM15 Δ(*lac*ZYA-*arg*F) U169 *rec*A1 *end*A1 *hsd*R17 (rK⁻, mK⁺) *pho*A *sup*E44 λ– *thi*-1 *gyr*A96 *rel*A1) and expression strain BLR(DE3) (F⁻ *ompT hsd*S$_B$(r$_B$⁻ m$_B$⁻) *gal dcm* (DE3) Δ(srl-recA)306::Tn10 (*Tet*^{R})) (EMD Millipore) are from lab stock. The cells were made competent using the Z-competent cell™ buffer kit (Zymo Research) according to the manufacturer's instructions (*see* **Note 1**).

2. Plasmids pET/EI$_{0C}$-X and pET/EI$_{0N}$ available from the authors [14].

3. Ampicillin stock solution, 100 μg/mL ampicillin: Dissolve 1 g ampicillin in 10 mL water and sterilize by filtration through a 0.22 μm filter. Store at –20 °C.

4. LB (Luria Broth) medium: Dissolve 10 g tryptone, 10 g NaCl, 5 g yeast extract in 600 mL distilled water, adjust volume to 1 L and autoclave at 121 °C for 30 min.

5. 50× TAE buffer: Dissolve 242 g Tris base, 57.1 mL acetic acid, 100 mL of 0.5 M EDTA (pH 8.0) stock solution into 600 mL water and adjust into 1 L.

6. Phusion DNA polymerase and buffer concentrate (New England Biolabs).

7. Taq DNA polymerase and buffer concentrate.

8. 10 mM dNTP stock solution.

9. 10 μM DNA primers: PCR primers must be designed to place the target protein in frame with the intein and retain the final critical intein amino acids (Fig. 1b). Thus, the internal *Bsr*G I restriction site (TGTACA) and the last four nucleic acids of the intein sequence (CAAC) must be added to the 5′ end of the target protein DNA sequence as described in previously published work [15] (*see* **Note 2**).

10. Qiagen gel extraction kit (Qiagen).

11. Restriction enzymes *Hin*d III and *Bsr*G I.

12. T4 DNA ligase.

13. Electrophoresis Grade Agarose.

2.2 Protein Expression

1. TB (Terrific Broth) medium: dissolve 12 g tryptone, 24 g yeast extract in 800 mL water, adjust to 900 mL with water and autoclave at 121 °C for 30 min. Separately dissolve 2.31 g of KH$_2$PO$_4$ and 12.54 g of K$_2$HPO$_4$ in 90 mL of H$_2$O, adjust the volume of

the solution to 100 mL with water and sterilize by autoclaving for 20 min at 121 °C. When cool, add this solution to the tryptone/yeast extract solution above to get 1 L total volume.

2. Two 500 mL baffled flasks for each target protein to be purified.

2.3 Protein Purification

1. Lysis buffer, 10 mM Tris–HCl, pH 8.5, 1 mM EDTA.

2. Ammonium sulfate stock solution, 1.6 M $(NH_4)_2SO_4$. Dissolve 211.4 g $(NH_4)_2SO_4$ in 700 mL distilled water and then adjust into 1 L.

3. Cleaving buffer, 10 mM PBS, 40 mM bis-tris, pH 6.2, 1 mM EDTA.

4. Protein gel chemicals: 30 % acrylamide–bis stock solution, 10 % ammonium persulfate stock solution and N,N,N,N'-tetramethyl-ethylenediamine (TEMED). Store at 4 °C.

5. 12 % SDS-PAGE: Prepare protein gel as described previously [16], or according to manufacturer's instructions.

6. 1× protein gel running buffer, 25 mM Tris base, 192 mM Glycine, 0.1 % sodium dodecyl sulfate (SDS).

7. 2× SDS-PAGE loading buffer, 4 % (w/v) SDS, 0.2 % (w/v) bromophenol blue, 20 % (v/v) glycerol, 200 mM dithiothreitol (DTT).

3 Method

3.1 Construction of the Recombinant pET/EI₀C-TP Expression Plasmid

1. Set up a 50 μL PCR reaction containing Phusion polymerase buffer (5 μL of 10× buffer concentrate), 0.5 μM concentration of each primer (2.5 μL of each 10 μM stock solution), 200 μM dNTP mix (2.5 μL of 10 mM stock solution), 2–10 ng of template DNA and 1.0 unit of Phusion-HF polymerase. The PCR reaction should start with a hold 98 °C for 60s, followed by 25 cycles of 15 s melting at 98 °C, 30 s annealing (at the primer melting temperature plus 5 °C), and extension at 72 °C for 20 s/kb of target protein DNA sequence. After 25 cycles, the reaction should hold at 72 °C for 10 min.

2. Resolve the PCR products via 1 % agarose gel electrophoresis (*see* **Note 3**). The expected PCR product is cut from the gel and extracted using a Qiagen gel extraction kit according to manufacturer's instructions.

3. Digest 1 μg of the purified PCR products along with 1 μg of the vector pET/EI₀C-X with *Bsr*G I and *Hin*d III in NEB buffer 2 at 37 °C for 1 h.

4. Heat the digested vector and PCR products at 80 °C for 20 min to kill the enzymes and then resolve via a 1 % agarose

gel. Excise the vector backbone and PCR product bands and extract the DNA from the agarose gel using the Qiagen gel extraction kit according to manufacturer's instructions.

5. Ligate the digested PCR and vector backbone products at a molar ratio of 3:1 respectively, according to manufacturer's instructions for the ligase. The molar ratios can be determined by an additional agarose gel, or by absorbance at 280 nm. We also recommend a control ligation reaction without the PCR product insert (*see* **Note 4**).

6. Inactivate the ligation product at 65 °C for 15 min (*see* **Note 5**).

7. Transform 5 μL of each ligation product (with and without PCR product insert in separate transformation reactions) into separate vials of 100 μL each of Z-competent DH5α cells, prepared and transformed according to manufacturer's instructions.

8. Spread the cells on pre-warmed LB agar plates (*see* **Note 6**) supplemented with 100 μg/mL ampicillin and incubate at 37 °C overnight (*see* **Note 7**).

9. Check the transformation agar plates including the control plate (*see* **Note 8**).

10. Screen the transformation colonies by colony PCR using the cloning primers used in **step 1** (above). Briefly, a single colony from the transformed plate is first patched onto a fresh pre-warmed LB agar plate with a sharp 10 μL tip and then the same tip is dipped into a 20 μL PCR reaction with Taq polymerase. As above, the PCR reaction contains 1× polymerase buffer, 0.5 μM of each primer, 200 μM dNTP and 0.5 units of Taq polymerase. Positive and negative control reactions are recommended here as well (*see* **Note 9**).

11. The PCR reaction is performed with the following program: 95 °C for 5 min, followed by 25 cycles of 95 °C for 10s, 55 °C for 10s, 72 °C for 1 min/kb of insert DNA, then an additional incubation at 72 °C for 5 min (*see* **Note 10**).

12. Analyze 5 μL of each colony PCR reaction using 1% agarose gel electrophoresis visualized by a UV imager with an appropriate dye (e.g., Gel Red or ethidium bromide).

13. Prepare overnight cultures using three to five positive colonies from the patch plate in 5 mL liquid LB medium supplemented with 100 μg/mL ampicillin. Incubate each at 37 °C and 200 rpm agitation in a temperature-controlled shaker for 12–16 h.

14. Extract the plasmid from each overnight culture using a Qiagen mini prep kit according to the manufacturer's instructions. The extracted plasmids should then be confirmed by DNA sequencing (*see* **Note 11**).

**3.2 Expression
of ELP-Intein Segment
Fusion Proteins**

1. Transform the recombinant plasmids which express the switch protein (pET/EI$_{0N}$) and tagged target protein (pET/EI$_{0C}$-TP, prepared above) into Z-competent BLR (DE3) cells separately according to manufacturer's instructions using pre-warmed agar plates supplemented with 100 µg/mL ampicillin, and incubate at 37 °C for 12–16 h.

2. Pick a single colony from each plate and grow in 3 mL LB medium supplemented with 100 µg/mL ampicillin in a culture tube overnight at 37 °C.

3. Dilute 1 mL of the overnight cultures into 100 mL fresh TB medium in 500 mL baffle flasks and grow at 37 °C with an agitation speed of 160 rpm until the optical density at 600 nm (OD$_{600nm}$) reaches ~0.8. This normally takes 2.5–3 h (*see* **Note 12**).

4. Cool each 100 mL culture by placing in a room temperature water bath for a half hour with 100 rpm agitation speed.

5. Add 100 µL of the 0.8 M IPTG into each flask (final IPTG concentration should be 0.8 mM) to induce protein expression. Continue to grow at 16 °C (or room temperature) for 20 h (*see* **Note 13**).

6. Cool the expressed cell culture on ice for 20 min and divide the cells from each flask into two 50 mL falcon tubes and pellet by centrifugation at 5000×*g* for 10 min. The harvested cell pellets can be stored at –80 °C until purification.

**3.3 Purification
and Cleavage
of the Target Protein**

In this method, the switch protein and the tagged target protein are purified separately by ELP precipitation, and then the two purified fusion proteins are mixed to trigger the cleaving reaction (Fig. 2a).

*3.3.1 Standard Two-Step
Purification Method*

1. Suspend cell pellets from 50 mL expression cultures (tagged target and switch protein) in separate 50 mL falcon tubes using 6 mL lysis buffer (*see* **Note 14**). Lyse by sonication using ten rounds of 30s pulses at 7 W RMS power with 30 s pauses between pulses in an ice-water bath. After sonication, take a whole lysate sample for SDS-PAGE analysis by mixing 10 µL of the lysate with 10 µL of 2× loading dye (this is referred to as the WL sample).

2. Clarify the cell lysate by centrifugation at 10,000×*g* for 25 min at 4 °C.

3. Transfer the clarified cell lysates (6 mL) into fresh 50 mL tubes and take clarified cell lysate samples for SDS-PAGE by mixing 10 µL of each sample with 10 µL of the 2× loading dye (this is referred to as the CL sample).

4. Precipitate the ELP fusion proteins by adding 2 mL of the 1.6 M (NH$_4$)$_2$SO$_4$ stock solution into each 50 mL falcon tube and mix (the final (NH$_4$)$_2$SO$_4$ concentration should be 0.4 M). Then incubate at 37 °C for 5 min to allow ELP precipitation (*see* **Note 15**).

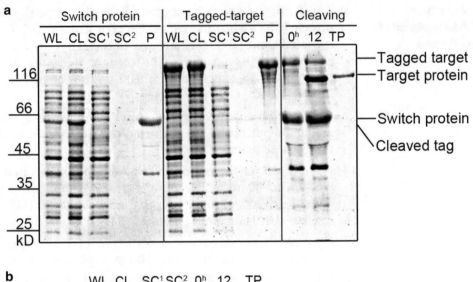

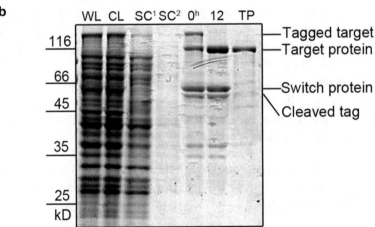

Fig. 2 Purification of model protein β-galactosidase with the dual-ELP purification system. (**a**) The conventional two-step purification method, and (**b**) the alternate one-step purification method. All the samples are as described in the text. Note: Because of the ELP tag is rich in glycine, which is not sensitive to Coomassie blue dye, the staining of the cleaved tags is not as sensitive as with the other proteins on the gel [17]

5. Pellet the precipitated ELP fusion proteins by centrifugation at $10,000 \times g$ for 10 min at room temperature.

6. Decant the supernatant into a fresh 15 mL falcon tube for SDS-PAGE analysis. Prepare a soluble contaminants sample by mixing 10 μL of supernatant sample with 10 μL 2× loading dye (this is referred to as the SC[1] sample).

7. Redissolve the ELP pellets in 3 mL of cold lysis buffer (*see* **Note 16**). We recommend using a pipette to suspend the ELP pellets, and sonication may also be used here for more difficult proteins (*see* **Note 17**). The pellets should go completely into solution.

8. Centrifuge the ELP fusion protein solutions at $5000 \times g$ and room temperature for 1 min to get rid of any residual insoluble contaminants (*see* **Note 18**).

9. Transfer the supernatants from **step 8** to fresh falcon tubes and add 1 mL of 1.6 M $(NH_4)_2SO_4$ (to a final concentration of 0.4 M), mix well and incubate at room temperature for 5 min to precipitate the ELP fusion protein again (*see* **Note 19**).

10. Centrifuge the mixture at $10,000 \times g$ and room temperature for 5 min to collect the ELP fusion protein.

11. Decant the supernatant into a fresh falcon tube and prepare a soluble contaminants sample for the second ELP precipitation by mixing 10 μL of this supernatant with 10 μL of 2× SDS-PAGE loading dye (this sample is referred to as SC^2).

12. Dissolve the precipitated ELP fusion proteins in 3 mL cleaving buffer (*see* **Note 20**). Prepare samples of the purified switch protein and tagged target protein for SDS-PAGE by mixing 10 μL of each redissolved ELP pellet with 10 μL of 2× SDS-PAGE loading dye (these samples are respectively referred to as P, for Purified protein).

13. Initiate the cleaving reaction by mixing 2 mL of switch protein (expressed from pET/EI_{0N}) with 1 mL of the tagged target (expressed from pET/EI_{0C}-TP). We recommend immediately taking a 0 h incubation sample, prepared by mixing 10 μL of the mixture and mixing with 10 μL of 2× SDS-PAGE loading dye (*see* **Note 21**).

14. Incubate the mixture at room temperature for 12 h to allow cleavage to take place (*see* **Note 22**).

15. Take 10 μL of cleaving reaction and mix with 10 μL of 2× SDS-PAGE loading dye (12 h sample; *see* **Note 23**).

16. Add 1 mL of 1.6 M $(NH_4)_2SO_4$ to the cleaving reaction and incubate at room temperature for 5 min to precipitate the ELP fusion proteins.

17. Spin the mixture at $10,000 \times g$ for 5 min and transfer the supernatant into a fresh tube. This is the purified target protein. Prepare a sample of the purified target by mixing 10 μL of the purified protein with 10 μL of 2× loading buffer (this sample is referred to as TP). Analyze by SDS-PAGE.

18. Heat all protein samples at 95 °C for 10 min and resolve by 12% SDS-PAGE to analyze the protein purification process. The purified protein can then be used.

3.3.2 Alternate One-Step Purification Method

The separate expression of the switch protein (pET/EI_{0N}) and tagged target protein (pET/EI_{0C}-TP) are the same for this method as with the two-step method described above. In the one-step method described here, however, the cell pellets are before lysis (Fig. 2b).

This leads to an increase in cleaving efficiency, but can also lead to some losses due to premature cleaving during the lysis step. The cell pellets are prepared as in Subheading 3.2 for use in this method.

1. Mix the expressed switch protein and tagged target protein cells at a weight ratio of 2:1. In practice, the cell pellets for each protein are resuspended in ten times their mass in lysis buffer (see **Note 14**) and then mixed at a ratio of two volumes of switch protein per volume of tagged target protein.

2. Lyse the cells by ten rounds of sonication using a of 30s pulse at 7 W RMS with a 30 s pause at in an ice-water bath for each round. Take 10 μL of the whole cell lysate (WL sample) sample and mix with 10 μL of 2× SDS-PAGE loading dye.

3. Clarify the cell lysate at $10,000 \times g$ for 25 min at 4 °C.

4. Transfer the clarified cell lysate (15 mL) into a 50 mL falcon tube and take a 10 μL clarified cell lysate sample (CL sample) and mix with 10 μL of 2× SDS-PAGE loading dye.

5. Precipitate the ELP fusion proteins by adding 5 mL of 1.6 M $(NH_4)_2SO_4$ (final concentration is 0.4 M) and incubate the mixture at 37 °C for 5 min.

6. Pellet the ELP fusion protein by centrifugation at $10,000 \times g$ for 10 min at room temperature.

7. Decant the supernatant into a 15 mL Falcon tube and take a 10 μL soluble contaminants sample from the supernatant and mix with 10 μL of 2× SDS-PAGE loading dye (SC^1 sample).

8. Redissolve the ELP pellet from **step 6** in 7.5 mL of ice cold lysis buffer (see **Notes 16** and **17**).

9. Centrifuge the redissolved ELP pellet at $5000 \times g$ for 1 min (see **Note 18**).

10. Transfer the supernatant into a fresh tube and add 2.5 mL of 1.6 M $(NH_4)_2SO_4$ (final concentration is 0.4 M) and incubate at room temperature for 5 min to precipitate the ELP fusions.

11. Centrifuge the mixture at $10,000 \times g$ for 5 min to pellet the ELP fusion proteins.

12. Decent the supernatant into 15 mL tube and take a 10 μL soluble contaminants sample from the supernatant and mix with 10 μL of 2× SDS-PAGE loading dye (SC^2 sample).

13. Redissolve the ELP fusions pellet in 7.5 mL of cleaving buffer, and take a 10 μL sample and mix with 10 μL of 2× SDS-PAGE loading dye (0 h sample).

14. Incubate the cleaving reaction mixture at room temperature for 12 h for intein to cleave the target from the ELP-intein fusion.

15. Take a 10 μL sample and mix with 10 μL of 2× SDS-PAGE loading dye (12 h sample).

16. Add 2.5 mL of 1.6 M $(NH_4)_2SO_4$ to the cleaving reaction mixture and incubate at room temperature for 5 min to precipitate the ELP fusion proteins.

17. Centrifuge the mixture at $10,000 \times g$ for 5 min and then transfer the supernatant into a fresh tube. Take a 10 µL sample and mix with 10 µL of 2× SDS-PAGE loading dye (TP sample).

18. Analyze all samples via 12 % SDS-PAGE as shown in Fig. 2b.

4   Notes

1. These are the cloning and expression strains typically used in our laboratories. It is likely that any other cloning and expression strains will also work.

2. *Critical*: The forward PCR primer must incorporate a *BsrG* I restriction site (TGTACA) followed by the nucleotide sequence "CAAC" before the start codon of the Target Protein sequence (*see* Fig. 1b). The specific sequence and reading frame is shown in Fig. 1b. The downstream (reverse) primer must include the stop codon for the Target Protein, typically followed by a *Hind* III restriction site. Plasmid maps and plasmids are available from the authors.

3. The gel concentration varies on the target gene size. Normally a 1 % agarose gel can efficiently separate 500–5000 base pair gene fragments. Larger fragments may need lower concentration gel and smaller fragments may require a higher concentration gel.

4. The control reaction serves to indicate the number of background colonies resulting from incomplete digestion of the vector backbone, and can be useful for troubleshooting in cases where the correct insert cannot be isolated from the transformed colonies.

5. We have found that the inactivation step can increase the transformation efficiency in most cases.

6. We have found that pre-warming the plates at 37 °C for 30–60 min before transformation will typically result in at least a tenfold higher transformation efficiency.

7. Normally colonies show up after 12–16 h incubation. Longer incubation should be avoided due to the formation of untransformed satellite colonies.

8. Ideally there will be no colony formation on the control plate (with no PCR product insert) and up to several hundred colonies on the ligation plates. If similar numbers of colonies appear on both plates, then we recommend repeating the digestion and ligation procedure with greater care given to the digestion conditions.

9. The insert gene template DNA can be used as a positive control for the PCR reaction, while the original vector can be used as the negative control in the colony PCR screen.

10. In colony PCR, the initial denaturing step needs to be longer than normal PCR to break the cell walls and release the plasmid DNA in the reaction.

11. Typically one sequencing reaction can read around 800 bp, so larger targets may need several reactions to read through the whole gene. Alternate mini-prep methods can also be used.

12. Normally the diluted cell OD_{600} is around 0.1, so the dilution factor can be varied according to the overnight cell density.

13. We initially induce at room temperature for uncharacterized target proteins to maximize the likely solubility of the target protein. In cases where the target protein is known to express and fold well, the time and temperature of induction can be optimized for efficiency.

14. The lysis buffer volume is typically ten times of the wet cell weight (v/m). For example, 1 g of wet cell pellet requires 10 mL of lysis buffer.

15. The incubation time may vary based on the target proteins. Large and highly soluble targets may need slightly longer incubation times, but our experience has been that most target proteins can be precipitated in 5–10 min.

16. Because the precipitation of ELP fusion protein is based on the salt and temperature, using cold lysis buffer can help solubilize the precipitated ELP fusions.

17. Sonication in an ice water bath can also help dissolve the aggregated ELP fusion protein, and normally 30s at 3 W RMS can efficiently suspend ELP fusion proteins.

18. This step is to remove contaminant proteins that either make it through the lysate clarification step or irreversibly aggregate during the first precipitation step. Because the ELP fusion proteins fully dissolve at low salt concentrations, this step can significantly increase the purity of ELP fusion proteins.

19. Normally it is not necessary to incubate this precipitation reaction at 37°, and higher concentrations of ELP fusion proteins can generally be precipitated at lower temperatures (unpublished results).

20. The volume of this resuspension step can be adjusted to increase the final concentration of the target protein as desired.

21. We recommend that the switch protein be supplied at a 2:1 molar excess to the tagged target protein. This can be quantified using additional SDS-PAGE or a conventional protein concentration measurement (e.g., Bradford assay).

22. The cleaving reaction can usually be finished in 12 h, and we typically incubate overnight, but there is some variation relative to different target proteins. For a specific target, the incubation time can be optimized if desired by performing a cleaving kinetics time course experiment.

23. We occasionally see some precipitation of ELP fusion at this stage, and therefore suggest mixing the samples by vortex before taking the 12 h cleaving samples.

References

1. Liu XQ (2000) Protein-splicing intein: genetic mobility, origin, and evolution. Annu Rev Genet 34:61–76. doi:10.1146/annurev.genet.34.1.61

2. Perler FB, Adam E (2000) Protein splicing and its applications. Curr Opin Biotechnol 11(4):377–383

3. Chong S, Mersha FB, Comb DG et al (1997) Single-column purification of free recombinant proteins using a self-cleavable affinity tag derived from a protein splicing element. Gene 192(2):271–281

4. Chong S, Montello GE, Zhang A et al (1998) Utilizing the C-terminal cleavage activity of a protein splicing element to purify recombinant proteins in a single chromatographic step. Nucleic Acids Res 26(22):5109–5115

5. Southworth MW, Amaya K, Evans TC et al (1999) Purification of proteins fused to either the amino or carboxy terminus of the Mycobacterium xenopi gyrase A intein. Biotechniques 27(1):110–114

6. Wood DW, Wu W, Belfort G et al (1999) A genetic system yields self-cleaving inteins for bioseparations. Nat Biotechnol 17(9):889–892

7. Derbyshire V, Wood DW, Wu W et al (1997) Genetic definition of a protein-splicing domain: functional mini-inteins support structure predictions and a model for intein evolution. Proc Natl Acad Sci U S A 94(21):11466–11471

8. Meyer DE, Chilkoti A (1999) Purification of recombinant proteins by fusion with thermally-responsive polypeptides. Nat Biotechnol 17(11):1112–1115

9. Banki MR, Feng L, Wood DW (2005) Simple bioseparations using self-cleaving elastin-like polypeptide tags. Nat Methods 2(9):659–661

10. Fong BA, Wu WY, Wood DW (2009) Optimization of ELP-intein mediated protein purification by salt substitution. Protein Expr Purif 66(2):198–202. doi:10.1016/j.pep.2009.03.009, S1046-5928(09)00078-3 [pii]

11. Wood DW, Derbyshire V, Wu W et al (2000) Optimized single-step affinity purification with a self-cleaving intein applied to human acidic fibroblast growth factor. Biotechnol Prog 16(6):1055–1063. doi:10.1021/bp0000858

12. Ramirez M, Valdes N, Guan D et al (2013) Engineering split intein DnaE from Nostoc punctiforme for rapid protein purification. Protein Eng Des Sel 26(3):215–223. doi:10.1093/protein/gzs097

13. Lu W, Sun Z, Tang Y et al (2011) Split intein facilitated tag affinity purification for recombinant proteins with controllable tag removal by inducible auto-cleavage. J Chromatogr A 1218(18):2553–2560. doi:10.1016/j.chroma.2011.02.053

14. Shi C, Meng Q, Wood DW (2013) A dual ELP-tagged split intein system for non-chromatographic recombinant protein purification. Appl Microbiol Biotechnol 97(2):829–835. doi:10.1007/s00253-012-4601-3

15. Wu WY, Mee C, Califano F et al (2006) Recombinant protein purification by self-cleaving aggregation tag. Nat Protoc 1(5):2257–2262. doi:10.1038/nprot.2006.314

16. Sambrook J, Russell DW (2001) Molecular cloning: a laboratory manual, 3rd edn. Cold Spring Harbor Laboratory Press, New York

17. de Moreno MR, Smith JF, Smith RV (1986) Mechanism studies of Coomassie blue and silver staining of proteins. J Pharm Sci 75(9):907–911

Chapter 3

Intracellular Production of Cyclic Peptide Libraries with SICLOPPS

Eliot L. Osher and Ali Tavassoli

Abstract

Cyclic peptides are an important class of molecules that are increasingly viewed as an ideal scaffold for inhibition of protein–protein interactions (PPI). Here we detail an approach that enables the intracellular synthesis of cyclic peptide libraries of around 10^8 members. The method utilizes split intein mediated circular ligation of peptides and proteins (SICLOPPS), taking advantage of split intein splicing to cyclize a library of peptide sequences. SICLOPPS allows the ring size, set residues and number of random residues within a library to be predetermined by the user. SICLOPPS libraries have been combined with a variety of cell-based screens to identify cyclic peptide inhibitors of a variety of enzymes and protein–protein interactions.

Key words SICLOPPS, Cyclic peptide, High-throughput screening, Protein–protein interaction, Split intein

1 Introduction

Over the past two decades high-throughput screening has become the cornerstone of drug discovery [1, 2]. Combinatorial small molecule library based approaches in combination with HTS have proven to be invaluable and over the past decade have been extensively employed in both industrial and academic settings [3–6]. While small molecule libraries provide a fruitful source of functional group diversity, however, it has become increasingly evident that a peptidic backbone is ideal for inhibitors of protein–protein interactions [7]. The prominent rise of cyclic peptides for this purpose has strong roots in academia, with several techniques for the generation of cyclic peptide libraries having been developed there [8–13]. As these libraries are genetically encoded, deconvolution of hits from libraries of hundreds of millions of compounds is relatively straightforward. The majority of these techniques produce cyclic peptides in vitro and therefore are only compatible with in vitro screening methods.

Henning D. Mootz (ed.), *Split Inteins: Methods and Protocols*, Methods in Molecular Biology, vol. 1495,
DOI 10.1007/978-1-4939-6451-2_3, © Springer Science+Business Media New York 2017

Split-intein circular ligation of proteins and peptides (SICLOPPS) in contrast enables the intracellular generation of genetically encoded cyclic peptide libraries of around a hundred million members. The techniques utilizes the *Synechocystis sp.* (*Ssp*) PCC6803 DnaE trans inteins [11] rearranged so that the C-intein (I$_C$) precedes the N-intein (I$_N$) flanking an extein sequence (in the form I$_C$:extein:I$_N$), which upon splicing cyclizes the extein (Fig. 1) [14]. Backbone cyclization of the peptides not only results in enhanced resistances towards cellular catabolism, but also restricts the conformational freedom of the peptide, potentially improving the binding affinity of the inhibitor towards its binding partner [15, 16]. By introducing a randomized sequence in to a SICLOPPS vector between the intein elements through the incorporation of a PCR primer, a library of SICLOPPS plasmids is generated, that will encode a library of cyclic peptides when transformed into cells (Fig. 2). The randomized segment of these primers (*see* Subheading 2.1) is encoded in the form NNS where N is any of the four DNA bases (A, T, C, and G) and S represents a C or G; by using an NNS codon library (rather than the fully randomized NNN) all amino acids are still covered, but two of the three stop codons (TGA and TAA) are eliminated from the library [14]. The randomization at the amino acid level is introduced via a degenerate oligonucleotide by PCR, however a second PCR is necessary to ensure that all DNA sequences are correctly annealed to their complimentary strands. The libraries are typically restricted to six

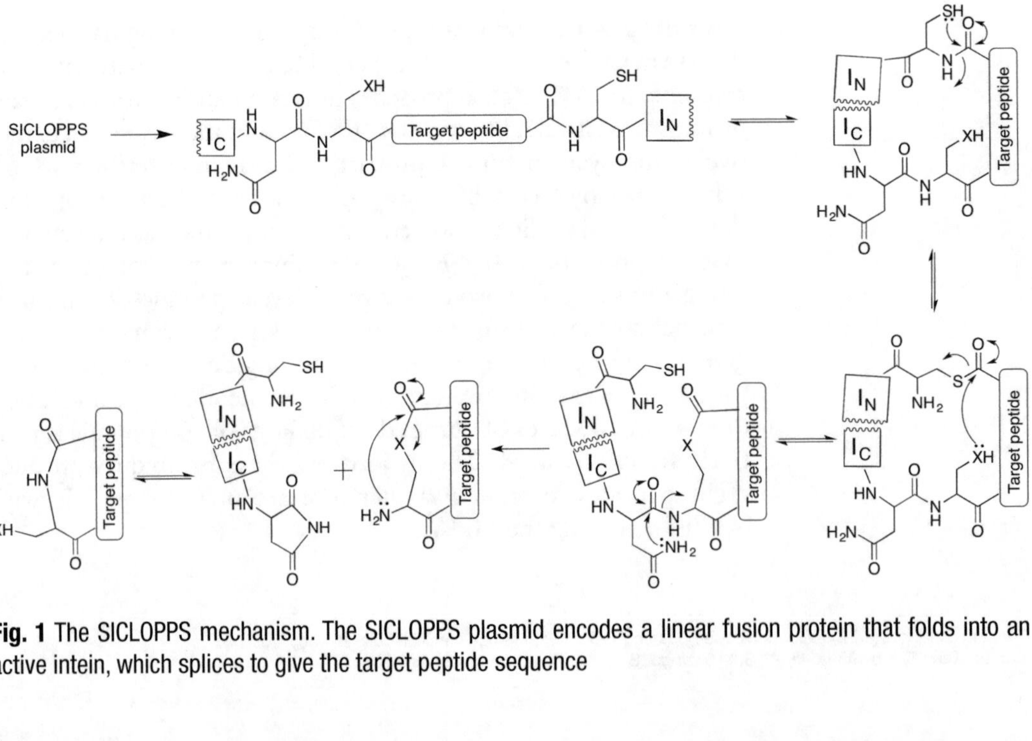

Fig. 1 The SICLOPPS mechanism. The SICLOPPS plasmid encodes a linear fusion protein that folds into an active intein, which splices to give the target peptide sequence

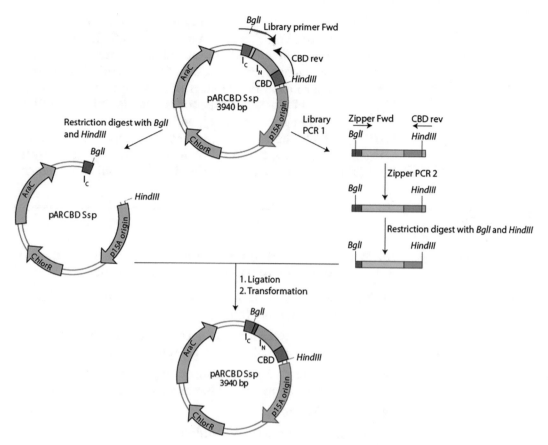

Fig. 2 Graphical representation of SICLOPPS library construction. The approach uses standard molecular biology techniques and the pARCBD plasmid as template

randomized positions (6.4×10^7 potential peptides) as this number of transformants is comfortably covered by electroporation into the *E. coli* host. It should be noted, however, that larger libraries may be generated and used in screens, with the knowledge that the whole library is not present in the screen. The only limitation on the library is requirement for a nucleophilic cysteine or serine residue in peptide position 1 to initiate the splicing of the inteins.

The power of SICLOPPS libraries and the ease with which they may be generated is demonstrated in large number of reported applications [7]. SICLOPPS libraries have been generated in a variety of prokaryotic and several eukaryotic host cells [17, 18]. A frequently successful implementation of this technique has been coupled with a bacterial reverse two-hybrid system (RTHS), allowing for the screening of inhibitors of challenging PPIs [19]. In an early example, SICLOPPS was used to identify a cyclic peptide inhibitor (*cyclo*-CRYFNV) of the homodimerization of AICAR transformylase, an enzyme with a homodimeric interaction interface of ~5000 Å² [12]. The active motif of this inhibitor was identified by alanine scanning and derivatized into a more potent small molecule that is active in vivo [20, 21]. More recently this approach has been used to identify inhibitors of transcription factor assembly, with first in class cyclic peptides that inhibit the PPI of C-terminal binding protein [22], and hypoxia inducible factor-1 (HIF-1) [23]. In this chapter, we detail the precise method for the construction of these SICLOPPS libraries; once generated, these SICLOPPS libraries may be interfaced with any cell-based screen.

2   Materials

2.1   Reagents

All molecular biology reagents were autoclaved with deionized water (15 MΩ) prior to use to prevent contamination.

1. Electrocompetent *E. coli* cells (e.g., ElectroMAX DH5α-E from Thermo Fisher Scientific).

2. pARCBD-p expression vector containing the *Synechocystis sp.* PCC6803 C- and N- (I_C and I_N) split inteins.

3. Luria–Bertani (LB) broth, according to the protocol supplemented with 35 μg/mL chloramphenicol, 6.5 μM arabinose.

4. LB Agar Vegitone.

5. 6.5 mM arabinose stock solution.

6. Deionized water.

7. 1× TAE buffer, 2 M Tris base, 1 M glacial acetic acid, 50 nM EDTA, pH 8.0 in deionized water. In a final volume of 1 L ddH₂O, mix 242 g Tris base, 57.1 mL glacial acetic acid, and 100 mL of 0.5 M EDTA to give a 50× TAE buffer stock. Adjust the solution to pH 8.0 with sodium hydroxide solution. A 1× TAE buffer can then be prepared by dissolving 20 mL of 50× TAE in 1 L of autoclaved ddH₂O.

8. Agarose powder, molecular biology grade.

9. Ethidium bromide (Caution, known carcinogen).

10. DNA polymerase (e.g., GoTaq DNA polymerase and GoTaq 5× reaction buffer).

11. 10 mM dNTP mix, dATP, dCTP, dGTP, and dTTP, 10 mM each.

12. Miniprep, PCR and gel purification kits. We use GenJET kits.

13. HindIII-HF restriction endonuclease.

14. BglI restriction endonuclease.

15. Suitable buffer for restriction endonucleases (e.g., 10× NEB buffer 2.0 or 10× CutSmart buffer).

16. Shrimp Alkaline Phosphatase (rSAP).

17. T4 DNA ligase and 10× ligase buffer.

18. Electrocompetent DH5α or RTHS cells. These should be prepared as detailed in the literature [24] or according to favorite protocol.

19. Chitin buffer, 20 mM Tris–HCl, 1 mM TCEP, 0.5 mM NaCl, pH 7.8.

20. Chitin beads (e.g., NEB).

21. 2× Laemmli buffer, 100 mM Tris–HCl, 4% (v/v) SDS, 20% (v/v) glycerol, 50 mM DTT, trace bromophenol blue, pH 6.8.

22. PCR primers for library construction synthesized either "In house" if a synthesizer is available or ordered unless otherwise stated.

23. CX_5-F primer: 5'-GGA ATT CGC CAA TGG GGC GAT CGC CCA CAA TTGC NNS NNS NNS NNS NNS TGC TTA AGT TTT GGC-3'.

24. CBD-R primer: 5'-GGA ATT CAA GCT TTC ATT GAA GCT GCC ACA AGG-3'.

25. Zipper-F primer: 5'-GGA ATT CGC CAA TGG GGC GAT CGC C-3'.

2.2 Equipment

Below is a list of essential equipment required for library construction.

1. Shaking incubator capable of 37 °C temperatures.

2. Water bath capable of 37 °C temperatures.

3. Thermal cycler.

4. UV spectrometer.

5. Oligonucleotide electrophoresis equipment.

6. Gel imaging and UV transilluminator.

7. 1 mm Electroporation cuvettes.

8. Electroporator (e.g., BIO-RAD MicroPulser electroporator).

9. Dialysis filter paper, 0.025 μm (e.g., Merck Millipore).

10. Scraper to scrape off *E. coli* colonies from agar plates.

11. 0.22 μM sterile filter (e.g., Merck Millipore).

Construction of an expression vector for the I_C and I_N genes should be carried out as previously described [14]. In our laboratory we use the pARCBD plasmid for expression of the SICLOPPS construct from an arabinose inducible expression vector pAR3. This vector contains a chloramphenicol resistance gene, araC repressor and araB promoter-operator placing it under direct control of arabinose. The SICLOPPS (I_C-Peptide-I_N) construct was cloned downstream of the araB promoter proceeded by a chitin binding domain (CBD) for characterization and purification purposes (Fig. 1) [11]. The pARCBD plasmid contains a p15A origin of replication (~10 copies per cell). This origin maintains a low copy number thus minimizing potential toxicity upon expression of the SICLOPPS construct.

3   Methods

3.1   SICLOPPS Library Construction

For the purpose of this protocol a CX_5 (X = any proteinogenic amino acid) library primer has been used. The length of the randomized region of the primer, as well as the amino acids present anywhere in the extein can be varied to generate different libraries, but a cysteine or serine residue must be present in position 1 for splicing (Fig. 1). This part of the method takes about 2 weeks.

1. Grow overnight cultures of pARCBD, or another SICLOPPS plasmid, supplemented with 35 μg/mL chloramphenicol in freshly autoclaved LB broth and incubate at 37 °C with shaking for ~16 h.

2. Isolate and purify the plasmid using a plasmid purification kit, according to the manufacturer's instructions. Determine the concentration of isolated plasmid DNA using a UV spectrophotometer.

3. Set up a 50 μL PCR reaction in a 200 μL PCR tube as detailed in Table 1.

4. Run the PCR using the program in Table 2.

5. **Optional but recommended step**: Make a 1% agarose gel with 1× TAE buffer and run ~5 μL of PCR product by electrophoresis. A positive result should yield a DNA band of ~560 bp.

Table 1
PCR mixture for generation of SICLOPPS library

Reagent	Amount (μL)	Stock concentration	Final concentration
GoTaq polymerase buffer (5×)	10	5×	1×
dNTP mix	1	10 mM each	200 μM
CX_5-F	1	10 μM	0.2 μM
CBD-R	1	10 μM	0.2 μM
Template plasmid (pARCBD)	–	–	100 ng
GoTaq polymerase	0.25	5 U/μL	0.025 U/μL
ddH₂O	35.75	–	–

Table 2
PCR program for generation of SICLOPPS library

Cycle	Denaturation	Annealing	Extension
1	95 °C for 3 min	55 °C for 30 s	72 °C for 1 min
2–30	95 °C for 30 s	55 °C for 30 s	72 °C for 1 min
31	/	/	72 °C for 5 min

6. Purify the PCR products with a PCR purification kit, according to the manufacturer's instructions. Determine the concentration of the purified DNA using a UV spectrophotometer.

7. Set up a second round of PCR using the zipper-F primer and the PCR product from **step 4** as a template. This PCR will ensure the correct alignment of any mismatched base pairs. The zipper PCR reaction mixture is set up as detailed in Table 3.

8. Run the same PCR reaction with the same program as detailed in **step 4**.

9. **Optional but recommended step**: Make a 1% agarose gel with 1× TAE buffer and run ~5 μL of PCR product by electrophoresis. A positive result should yield a DNA band of ~560 bp.

10. Purify the PCR products with a PCR purification kit, according to the manufacturer's instructions. Determine the concentration of the purified DNA using a UV spectrophotometer.

11. Digest the purified product of the zipper PCR with BglI and HindIII-HF as detailed in Table 4. We have found this digestion to be most efficient by digesting ~2500 ng of insert in a 37 °C water bath for ~16 h, as detailed for **step 12**.

12. Digest the SICLOPPS pARCBD plasmid with BglI and HindIII-HF in a 50 μL reaction as detailed in Table 4.

*Warning if you do not have access to NEB buffer 2.0 we recommend a sequential digestion in CutSmart (HindIII-HF) and NEB buffer 3.1 (BglI) as detailed by NEB (*see* **Note 1**).

13. Inactivate the restriction enzymes by incubating at 80 °C for 20 min. Treat the pARCBD digested vector with shrimp alkaline phosphatase (rSAP) by incubating at 37 °C for 1 h.

14. Purify the digested library PCR product using a PCR purification kit and determine the concentration of the subsequent DNA using a UV spectrophotometer.

15. Purify the digested linear pARCBD vector by gel extraction from a 1% agarose gel in 1× TAE buffer using a gel purification kit (following manufacturer's guidelines) and determine the concentration of DNA.

Table 3
PCR mixture for zipper PCR

Reagent	Amount (μL)	Stock concentration	Final concentration
GoTaq polymerase buffer (5×)	10	5×	1×
dNTP mix	1	10 mM each	200 μM
Zipper-F	1	10 μM	0.2 μM
CBD-R	1	10 μM	0.2 μM
PCR product 1	–	–	100 ng
GoTaq polymerase	0.25	5 U/μL	0.025 U/μL
ddH$_2$O	35.75	–	–

Table 4
Restriction digestion mixture

Reagent	Amount (μL)	Stock concentration	Final concentration
NEB buffer 2.0	5	10×	1×
BglI	2	10 U/μL	20 U
HindIII-HF	1	20 U/μL	20 U
pARCBD vector/Zipper PCR product	/	/	2500 ng
dH$_2$O	Up to 50 μL	/	/

16. Set up two 10 μL ligation reactions in 200 μL PCR tubes as detailed in Table 5 and incubate at 4 °C for ~16 h (*see* **Note 2**). A vector-only control ligation (lacking the insert) should also be set up.

17. Inactivate the ligase enzyme by incubating at 70 °C for 10 min.

18. Combine the library ligation mixtures and dialyze the mixture by placing a 20 μL spot (by micropipette) on dialysis filter paper (*see* Subheading 2.2) suspended on 1 L of dH$_2$O for 3–4 h. The vector-only control should also be dialyzed at the same time. Carefully remove each spot (by micropipette) and place in a PCR tube on ice in preparation for the electroporation process. *Warning, this step is critical (*see* **Note 3**).

19. Thaw 100 μL aliquots of electrocompetent cells for 5 min (on ice) and then add 5 μL of ligation mixture or vector-only control. Transfer the mixtures to an ice-cold electroporation cuvette (*see* **Note 4**). Apply an electrical field with an electroporator according to the manufacturer's instructions.

Table 5
Ligation reaction conditions

Reagent	Amount (µL)	Stock concentration	Final concentration
Digested pARCBD vector	Dependent on concentration	/	/
Digested library insert	Dependent on size and concentration	/	/
T4 DNA ligase buffer	1	10×	1×
T4 DNA ligase	1	3 U/µL	3 U
dH₂O	Up to 10 µL	/	/

Following the pulse, immediately add 895 µL of SOC medium to the cuvette, mix once (gently by micropipette) and transfer to a 10 mL aerated recovery culture tube. Repeat as appropriate for the vector-only control. Allow the cells to recover at 37 °C with shaking for 1 h (*see* **Note 5**).

20. Prepare 10× serial dilutions of the recovery mixture (from 10^{-3} to 10^{-7}); this is achieved by taking 20 µL of recovery mixture and mixing it with 180 µL of LB broth and then performing a serial dilution (*see* **Note 6**).

21. Plate 100 µL of each serial dilution onto LB agar plates (100 × 15 mm) supplemented with 35 µg/mL chloramphenicol. Incubate these LB agar plates at 37 °C for ~16 h.

22. If you are conducting a screen (electrocompetent cells of a selection strain were used in **step 21**) plate the remainder (980 µL) of the recovery solution onto your selection plates. Remember to supplement the selection plates with l-arabinose to initiate SICLOPPS expression (in addition to any other compounds and antibiotics required for your selection). If you are preparing a plasmid library using ElectroMAX DH5α-E electrocompetent cells, plate the remainder of the recovery mixture (980 µL) LB agar plates containing 35 µg/mL chloramphenicol. Incubate your plates overnight, or until colonies have grown (for our RTHS this is typically 2–3 days).

23. The next day, determine library size by counting the number of colonies on the 10^{-7} or 10^{-6} dilution plates (*see* **Note 6**). A minimum of 4×10^6 efficiency must be achieved for a CX_5 (3.2×10^6 members) library to ensure complete coverage, however we typically aim for tenfold coverage of each library. The SICLOPPS library has now been constructed.

24. If you wish to isolate the library, scrape all the colonies off the library plate and transfer them to 10 mL of LB broth using a scraper. Pellet the cells by centrifugation (~1600×g, 15 min). Decant the supernatant and resuspend the pellet in 2 mL of LB broth.

25. If the library is required for a biological screen, plasmids should be extracted from 0.5 mL of cells using a GeneJET plasmid miniprep kit. The SICLOPPS library has now been constructed and can be transformed into a selection strain for screening (*see* **Notes 7** and **8**).

3.2 Splicing Assay

Splicing assays can be carried out on isolated hits from the library, or random library members to ascertain whether the resulting peptide is cyclic or remains attached to the intein precursor (thus acting as an aptamer-like scaffold) [25]. This part of the method takes about 2 days.

1. Grow *E. coli containing* a pARCBD Ssp plasmid encoding the peptide of interest for ~16 h at 37 °C with shaking in LB medium supplemented with 35 µg/mL chloramphenicol.

2. The following day make a 1% subculture in 100 mL of sterile LB broth supplemented with 35 µg/mL chloramphenicol and incubate at 37 °C with shaking until an OD_{600} of 0.5–0.7 is reached.

3. Induce the culture with 0.05% (w/v) l-arabinose and incubate at 37 °C for a further 3 h.

4. Harvest the cells by centrifugation (1550×g, 15 min, 4 °C), decant the supernatant and store the pelleted cells at −80 °C until required.

5. Lyse the cell pellet in 4 mL of chitin buffer by sonication three times the following cycle: six repeats of 10 s on, 10 s off. Centrifuge the lysed samples (10,500×g, 45 min, 4 °C) and filter the supernatant using a 0.22 µM sterile filter.

6. Add 1 mL of filtered protein to 200 µL of chitin beads (pre-washed four times with 1 mL of chitin buffer) by centrifugation (1000×g, 5 min, with slow acceleration, 4 °C).

7. Incubate the protein–bead sample on ice with shaking for 1 h, gently inverting the sample every 15 min.

8. Centrifuge the protein–bead suspension (1000×g, 5 min, with slow acceleration, 4 °C) and discard the supernatant. Wash the protein–bead suspension four times with chitin buffer as detailed in **step 31** and then incubate at 4 °C overnight.

9. Add ~15 µL of chitin beads to 2× Laemmli buffer.

10. Denature by heating to 95 °C for 10 min and analyze the sample on a 15% SDS-page gel. The unspliced protein (I_C:peptide:I_N) and post spliced fragments (peptide-I_N and I_C-peptide) should be observed for peptides that correctly splice.

4 Notes

1. The BglI and HindIII-HF restriction enzymes have 75 % and 100 % respective activity in the presence of NEB buffer 2.0. Therefore, an overnight reaction is strongly recommended to ensure complete digestion. We digest pARCBD for ~16 h using the above buffer. If NEB buffer 2.0 cannot be used then we recommend a sequential digestion with CutSmart (HindIII-HF) and NEB buffer 3.1 (BglI).

2. If the ligation/transformation stage of library construction is not working, ensure the vector–insert ratio of your construct has been optimized. We have found a 1:6 ratio at 4 °C for ~16 h to work optimal for the pARCBD construct. Also ensure critical reagents such as the T4 DNA ligase buffer does not undergo constant freeze/thaw cycles (make aliquots) and that the buffer has been thawed and mixed well (by vortex if available) prior to use.

3. **Warning, step 18** is CRITICAL to ensure removal of salts that could either hinder or cause failure (by arcing) of the electroporation process, resulting in significantly reduced transformants. Failure to dialyze the ligation mixture is a very common reason for loss of transformation efficiency. This loss of transformation efficiency will result in the number of transformants being significantly lower than the number of members in the SICLOPPS library. We typically see an improvement in transformation efficiency from 10^4–10^5 to 10^6–10^7 as a result of this step.

4. We have found the electroporation process to be more efficient if all materials and reagents are ice cold before beginning. This includes electroporation cuvettes, SOC, electrocompetent cells and the ligation mixture.

5. If the number of transformants is lower than you might expect, this could be due to the efficiency of your electrocompetent cells. We recommend transforming a positive plasmid control into your cells. Ideally this plasmid should be the construct, which you have used for library construction (in our case pARCBD). Repeat the electroporation process as detailed in **step 19** and plate serial dilutions. Your plasmid control should yield ~10^8 transformants, if this is not the case then your electrocompetent cells need to be optimized. However, if you do achieve transformation efficiency of ~10^8 with your cells, then it is likely that the ligation is failing.

6. When preparing your serial dilutions, mixing each dilution by pipetting can result in the shearing of cells especially at 10^{6-7} dilutions. We have found that gently agitating each 1.7 mL eppendorf tube to mix can result in improved efficiency and only using a pipette to transfer between dilutions.

7. All plasmids should be stored long term at −20 °C, and kept on ice during use to maintain stability.

8. When using the GeneJET gel purification kit for the purification of the pARCBD vector, we have found that incubation of 35 µL of dH$_2$O for ~45 min with the column lid open will give a greater yield and higher purity sample.

References

1. Hughes JP, Rees S, Kalindjian SB, Philpott KL (2011) Principles of early drug discovery. Br J Pharmacol 162(6):1239–1249

2. Drews J (1996) Genomic sciences and the medicine of tomorrow. Nat Biotechnol 14(11):1516–1518

3. Boldt GE, Dickerson TJ, Janda KD (2006) Emerging chemical and biological approaches for the preparation of discovery libraries. Drug Discov Today 11(3–4):143–148

4. Brey DM, Motlekar NA, Diamond SL, Mauck RL, Garino JP, Burdick JA (2011) High-throughput screening of a small molecule library for promoters and inhibitors of mesenchymal stem cell osteogenic differentiation. Biotechnol Bioeng 108(1):163–174

5. Kneller R (2010) The importance of new companies for drug discovery: origins of a decade of new drugs. Nat Rev Drug Discov 9(11):867–882

6. Liu RW, Marik J, Lam KS (2003) Design, synthesis, screening, and decoding of encoded one-bead one-compound peptidomimetic and small molecule combinatorial libraries. Methods Enzymol 369:271–287

7. Lennard KR, Tavassoli A (2014) Peptides Come round: using SICLOPPS libraries for early stage drug discovery. Chemistry 20(34):10608–10614

8. Angelini A, Cendron L, Chen S, Touati J, Winter G, Zanotti G, Heinis C (2012) Bicyclic peptide inhibitor reveals large contact interface with a protease target. ACS Chem Biol 7(5):817–821

9. Goto Y, Katoh T, Suga H (2011) Flexizymes for genetic code reprogramming. Nat Protoc 6(6):779–790

10. Ito K, Passioura T, Suga H (2013) Technologies for the synthesis of mRNA-encoding libraries and discovery of bioactive natural product-inspired non-traditional macrocyclic peptides. Molecules 18(3):3502–3528

11. Scott CP, Abel-Santos E, Wall M, Wahnon DC, Benkovic SJ (1999) Production of cyclic peptides and proteins in vivo. Proc Natl Acad Sci 96(24):13638–13643

12. Tavassoli A, Benkovic SJ (2005) Genetically selected cyclic-peptide inhibitors of AICAR transformylase homodimerization. Angew Chem Int Ed Engl 44(18):2760–2763

13. Timmerman P, Beld J, Puijk WC, Meloen RH (2005) Rapid and quantitative cyclization of multiple peptide loops onto synthetic scaffolds for structural mimicry of protein surfaces. Chembiochem 6(5):821–824

14. Tavassoli A, Benkovic SJ (2007) Split-intein mediated circular ligation used in the synthesis of cyclic peptide libraries in E-coli. Nat Protoc 2(5):1126–1133

15. Liskamp RMJ, Rijkers DTS, Kruijtzer JAW, Kemmink J (2011) Peptides and proteins as a continuing exciting source of inspiration for peptidomimetics. Chembiochem 12(11):1626–1653

16. Menegatti S, Hussain M, Naik AD, Carbonell RG, Rao BM (2013) mRNA display selection and solid-phase synthesis of Fc-binding cyclic peptide affinity ligands. Biotechnol Bioeng 110(3):857–870

17. Kinsella TM, Ohashi CT, Harder AG, Yam GC, Li WQ, Peelle B, Pali ES, Bennett MK, Molineaux SM, Anderson DA, Masuda ES, Payan DG (2002) Retrovirally delivered random cyclic peptide libraries yield inhibitors of interleukin-4 signaling in human B cells. J Biol Chem 277(40):37512–37518

18. Kritzer JA, Hamamichi S, McCaffery JM, Santagata S, Naumann TA, Caldwell KA, Caldwell GA, Lindquist S (2009) Rapid selection of cyclic peptides that reduce [alpha]-synuclein toxicity in yeast and animal models. Nat Chem Biol 5(9):655–663

19. Horswill AR, Savinov SN, Benkovic SJ (2004) A systematic method for identifying small-molecule modulators of protein-protein interactions. Proc Natl Acad Sci U S A 101(44):15591–15596

20. Asby DJ, Cuda F, Beyaert M, Houghton FD, Cagampang FR, Tavassoli A (2015) AMPK activation via modulation of de novo purine biosynthesis with an inhibitor of ATIC homodimerization. Chem Biol 22(7):838–848

21. Spurr IB, Birts CN, Cuda F, Benkovic SJ, Blaydes JP, Tavassoli A (2012) Targeting tumour proliferation with a small-molecule inhibitor of AICAR transformylase homodimerization. Chembiochem 13(11):1628–1634

22. Birts CN, Nijjar SK, Mardle CA, Hoakwie F, Duriez PJ, Blaydes JP, Tavassoli A (2013) A cyclic peptide inhibitor of C-terminal binding protein dimerization links metabolism with mitotic fidelity in breast cancer cells. Chem Sci 4(8):3046–3057. doi:10.1039/C3sc50481f

23. Miranda E, Nordgren IK, Male AL, Lawrence CE, Hoakwie F, Cuda F, Court W, Fox KR, Townsend PA, Packham GK, Eccles SA, Tavassoli A (2013) A cyclic peptide inhibitor of HIF-1 heterodimerization that inhibits hypoxia signaling in cancer cells. J Am Chem Soc 135(28):10418–10425

24. Warren DJ (2011) Preparation of highly efficient electrocompetent Escherichia coli using glycerol/mannitol density step centrifugation. Anal Biochem 413(2):206–207

25. Naumann TA, Savinov SN, Benkovic SJ (2005) Engineering an affinity tag for genetically encoded cyclic peptides. Biotechnol Bioeng 92(7):820–830

Chapter 4

Recombinant Expression of Cyclotides Using Split Inteins

Krishnappa Jagadish and Julio A. Camarero

Abstract

Cyclotides are fascinating microproteins ($\approx$30 residues long) present in several families of plants that share a unique head-to-tail circular knotted topology of three disulfide bridges, with one disulfide penetrating through a macrocycle formed by the two other disulfides and inter-connecting peptide backbones, forming what is called a cystine knot topology. Naturally occurring cyclotides have shown to posses various pharmacologically relevant activities and have been reported to cross cell membranes. Altogether, these features make the cyclotide scaffold an excellent molecular framework for the design of novel peptide-based therapeutics, making them ideal substrates for molecular grafting of biological peptide epitopes. In this chapter we describe how to express a native folded cyclotide using intein-mediated protein trans-splicing in live *Escherichia coli* cells.

Key words Cyclotides, CCK motif, Split-intein, Protein trans-splicing, *Npu* intein

1 Introduction

Cyclotides are small globular microproteins (ranging from 28 to 37 residues) containing a unique head-to-tail cyclized backbone topology that is stabilized by three disulfide bonds to form a cystine-knot (CCK) motif [1, 2] (Fig. 1). This CCK molecular framework provides a very rigid molecular platform [3–5] conferring an exceptional stability towards physical, chemical, and biological degradation [1, 2]. In fact, the use of cyclotide-containing plants in indigenous medicine first highlighted the fact that the peptides are resistant to boiling and are orally bioavailable [6].

Cyclotides can be considered as natural combinatorial peptide libraries structurally constrained by the cystine-knot scaffold and head-to-tail cyclization but in which hypermutation of essentially all residues is permitted with the exception of the strictly conserved cysteines that comprise the knot [7–9]. The main features of cyclotides are a remarkable stability due to the cystine knot, a small size making them readily accessible to chemical synthesis, and an excellent tolerance to sequence variations. Naturally occurring

Henning D. Mootz (ed.), *Split Inteins: Methods and Protocols*, Methods in Molecular Biology, vol. 1495,
DOI 10.1007/978-1-4939-6451-2_4, © Springer Science+Business Media New York 2017

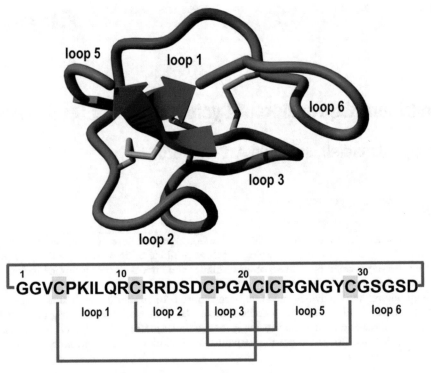

Fig. 1 Tertiary structure of the cyclotide MCoTI-II (PDB code: 1IB9) and primary structures of cyclotides used in this study. The backbone cyclized peptide (connecting bond shown in *green*) is stabilized by the three disulfide bonds (shown in *red*)

cyclotides have shown to posses various pharmacologically relevant activities [1, 10]. Cyclotides have been also engineered to target extracellular [11–13] and intracellular [14] molecular targets in animal models. Some of these novel cyclotides are orally bioavailable [12] and are able to cross cellular membranes efficiently [15, 16]. Cyclotides thus appear as highly promising leads or frameworks for peptide drug design [10, 17].

Naturally occurring cyclotides are ribosomally produced in plants from precursors that comprise between one and three cyclotide domains [18–21]. However, the mechanism of excision of the cyclotide domains and ligation of the free N- and C-termini to produce the circular peptides has not yet been completely elucidated although it has been speculated that asparaginyl endopeptidases are involved in the cyclization process [22–24]. Cyclotides can be also produced recombinantly using standard microbial expression systems by making use of modified protein splicing units [25–28] allowing for the first time the production of biologically generated libraries of these microproteins [26].

We describe in this chapter how to produce cyclotide MCoTI-I (Fig. 1) in *E. coli* cells making use of protein trans-splicing. Cyclotide MCoTI-I is a very potent trypsin inhibitor ($K_i \approx 20$ pM) [27] that has been recently isolated from dormant seeds of

Momordica cochinchinensis, a plant member of the *Cucurbitaceae* family [29]. Trypsin inhibitor cyclotides are interesting candidates for drug design because they can cross mammalian cell membranes [15, 16] and their specificity for inhibition can be altered and their structures can be used as natural scaffolds to generate novel binding activities [11, 14]. Protein trans-splicing is a post-translational modification similar to protein splicing with the difference that the intein self-processing domain is split into N- (I_N) and C-intein (I_C) fragments. The split-intein fragments are not active individually; however, they can bind to each other with high specificity under appropriate conditions to form an active protein splicing or intein domain in *trans* [30]. PTS-mediated backbone cyclization can be accomplished by rearranging the order of the intein fragments. By fusing the I_N and I_C fragments to the C- and N-termini of the polypeptide for cyclization, the trans-splicing reaction yields a backbone-cyclized polypeptide (Fig. 2).

In cell cyclization and folding of cyclotide MCoTI-I will be accomplished using the naturally occurring *Nostoc puntiforme* PCC73102 (*Npu*) DnaE split-intein. This DnaE intein has the highest reported rate of protein trans-splicing ($\tau_{1/2} \approx 60$ s) [31], high splicing yield [31, 32] and has shown high tolerance to the amino acid composition of the intein–extein junctions for efficient protein splicing [25, 33]. To accomplish this, we designed the split-intein construct **1** (Fig. 3). In this construct, the MCoTI-I linear precursor was fused in-frame at the C- and N-termini directly to the *Npu* DnaE I_N and I_C polypeptides. None of the additional native C- or N-extein residues were added in this construct. We used the native Cys residue located at the beginning of loop 6 of MCoTI-I (Fig. 1) to facilitate backbone cyclization. A His-tag was also added at the N-terminus of the construct to facilitate purification. In-cell expression of cyclotide MCoTI-I using PTS-mediated backbone cyclization was achieved by transforming the plasmid encoding the split-precursor **1** into Origami 2(DE3) cells to facilitate folding.

2 Materials

All solutions were prepared using ultrapure water with a resistivity of 18 M$\Omega\times$cm at 25 °C and analytical grade reagents. All reagents and solutions were stored at room temperature unless indicated otherwise.

2.1 Instruments

1. Sonicator for cell lysis (e.g., Sonifier 250 Branson CA, USA).

2. 5 mL polypropylene columns (QIAGEN).

3. Lyophilizer (e.g., Flexi Dry™ μp, Frigeco Inc. USA).

4. HPLC system equipped with gradient capability and UV–Vis detection (e.g., Agilent 1100, Agilent Technologies, USA).

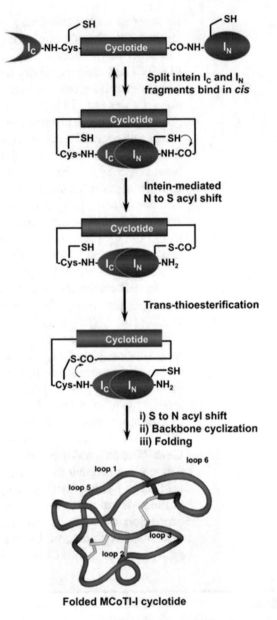

Fig. 2 In-cell expression of native folded cyclotide MCoTI-I using intein-mediated protein trans-splicing

MCoTI-intein construct, 1

Fig. 3 Architecture of the intein precursor used for the expression of cyclotide MCoTI-I described in this protocol

5. C18 reverse phase HPLC columns (e.g., Vydac C18 column 5 μm, 4.6×150 mm and 2.1×100 mm columns, Grace Discovery Sciences, USA).

6. Electrospray mass spectrometer (ES-MS) (e.g., API-3000, Applied Biosystems, USA).

2.2 Cloning of MCoTI-Intein Contruct 1

1. DNA ultramers encoding MCoTI-I, *Npu* DnaE I$_C$ and I$_N$ were ordered as synthetic oligonucleotides (20 nmol scale and purified by PAGE) (Table 1).

2. TE buffer: 10 mM Tris–HCl, 1 mM EDTA, pH 8.0.

Table 1
ssDNA sequences used to generated the dsDNA fragments encoding *Npu* DnaE I$_C$ and I$_N$, and MCoTI-I

Name of Sequence	Nucleotide sequence
P5-I$_C$	C ATG GGC AGC AGC CAT CAT CAT CAT CAT CAC AGC AGC GGC CTG GTG CCG CGC GGC AGC ATG ATC AAA ATA GCC ACA CGT AAA TAT TTA GGC AAA CAA AAT GTC TAT GAC ATT GGA GTT GAG CGC GAC CAT AAT TTT GCA CTC AAA AAT GGC TTC ATA GCT T
P3-I$_C$	CGA AGC TAT GAA GCC ATT TTT GAG TGC AAA ATT ATG GTC GCG CTC AAC TCC AAT GTC ATA GAC ATT TTG TTT GCC TAA ATA TTT ACG TGT GGC TAT TTT GAT CAT GCT GCC GCG CGG CAC CAG GCC GCT GCT GTG ATG ATG ATG ATG ATG GCT GCT GCC C
P5-MCoTI	CGA ACT GCG GTT CTG GTT CTG ACG GTG GTG TTT GCC CGA AAA TCC TGC AGC GTT GCC GTC GTG ACT CTG ACT GCC CGG GTG CTT GCA TCT GCC GTG GTA ACG GTT ACT GTT TAT CA
P3-MCoTI	TAT GAT AAA CAG TAA CCG TTA CCA CGG CAG ATG CAA GCA CCC GGG CAG TCA GAG TCA CGA CGG CAA CGC TGC AGG ATT TTC GGG CAA ACA CCA CCG TCA GAA CCA GAA CCG CAG TT
P5-I$_N$	TAT GAA ACG GAA ATA TTG ACA GTA GAA TAT GGA TTA TTA CCG ATT GGT AAA ATT GTA GAA AAG CGC ATC GAA TGT ACT GTT TAT AGC GTT GAT AAT AAT GGA AAT ATT TAT ACA CAA CCT GTA GCA CAA TGG CAC GAT CGC GGA GAA CAA GAG GTG TTT GAG TAT TGT TTG GAA GAT GGT TCA TTG ATT CGG GCA ACA AAA GAC CAT AAG TTT ATG ACT GTT GAT GGT CAA ATG TTG CCA ATT GAT GAA ATA TTT GAA CGT GAA TTG GAT TTG ATG CGG GTT GAT AAT TTG CCG AAT TA
P3-I$_N$	AGC TTA ATT CGG CAA ATT ATC AAC CCG CAT CAA ATC CAA TTC ACG TTC AAA TAT TTC ATC AAT TGG CAA CAT TTG ACC ATC AAC AGT CAT AAA CTT ATG GTC TTT TGT TGC CCG AAT CAA TGA ACC ATC TTC CAA ACA ATA CTC AAA CAC CTC TTG TTC TCC GCG ATC GTG CCA TTG TGC TAC AGG TTG TGT ATA AAT ATT TCC ATT ATT ATC AAC GCT ATA AAC AGTA CAT TCG ATG CGC TTT TCT ACA ATT TTA CCA ATC GGT AAT AAT CCA TAT TCT ACT GTC AAT ATT TCC GTT TCA

DNA sequences were generated using optimal codons for expression in *E. coli*. Bases in red are as overhangs to facilitate ligation

3. Annealing buffer: phosphate buffer, 20 mM Na_2HPO_4, 300 mM NaCl buffer, pH 7.4.

4. QIAquick PCR Purification Kit (QIAGEN).

5. QIAprep Spin Miniprep Kit (QIAGEN).

6. QIAquick Gel Extraction Kit (QIAGEN).

7. Synthetic DNA primers used to amplify DNA encoding MCoTI-intein construct 1 (20 nmol scale, HPLC purified) (Table 2).

8. Vent DNA polymerase, TaqDNA polymerase, dNTPs solution, 10× ThermoPol PCR buffer, 10× TaqDNA polymerase buffer.

9. Restriction enzymes: *NcoI* and *HindIII*.

10. NEB buffer 2.1, 50 mM NaCl, 10 mM Tris–HCl, 10 mM $MgCl_2$, 100 μg/mL bovine serum albumin (BSA), pH 7.9.

11. Chemical competent DH5α cells.

12. Expression plasmid pET28a (Novagen-EMD Millipore).

13. T4 DNA ligase buffer.

14. LB medium: 25 g of LB broth was dissolved in 1 L of pure H_2O and sterilized by autoclaving at 120 °C for 30 min.

15. LB medium-agar: 3.3 g of LB agar was suspended in 100 mL of pure H_2O and sterilized by autoclaving at 120 °C for 30 min. To prepare plates, allow LB medium-agar to cool to ≈50 °C, then add 0.1 mL of kanamycin stock solution (25 mg kanamycin/mL in H_2O, sterilized by filtration over a 45 μm filter), gently mix and pipet 20 mL into a sterile petri dish (100 mm diameter).

16. SOC Medium: 20 g of tryptone, 5 g yeast extract, 0.5 g NaCl, and 0.186 g KCl was suspended into 980 mL of pure water and sterilized by autoclaving at 120 °C for 30 min. Dissolve 4.8 g $MgSO_4$, 3.603 g dextrose in 20 mL of pure H_2O and filter sterilize over a 45 μm filter and add to the autoclaved medium.

2.3 Cyclotide Expression, Purification, and Characterization

1. Chemical competent Origami2 (DE3) cells (EMD Millipore).

2. Isopropyl-thio-β-D-galactopyranoside (IPTG), analytical grade. Prepare a stock solution of 1 M in H_2O and sterilize by filtration over 45 μm filter. Store at –20 °C.

Table 2
DNA oligonucleotides used to generate the dsDNA encoding the MCoTI-intein precursor construct 1

Name of Sequence	Nucleotide sequence
Forward I_C primer	5′- AAA ACC ATG GGC AGC AGC CAT CAT CAT -3′
Reverse I_N primer	5′- TTT TAA GCT TAA TTC GGC AAA TTA TCA ACC C -3′

3. Sterile conical bottom flasks.

4. Ni-lysis buffer: 10 mM imidazole, 50 mM Na_2HPO_4, 150 mM NaCl, pH 8.

5. Ni-wash buffer: 20 mM imidazole, 50 mM Na_2HPO_4, 150 mM NaCl, pH 8.

6. Ni-elution buffer: 50 mM Na_2HPO_4, 150 mM NaCl, 250 mM imidazole, pH 8.

7. Ni-NTA agarose beads (EMD MilliPore).

8. 100 mM phenylmethylsulfonyl fluoride (PMSF) in EtOH (better to prepare fresh before use).

9. 4× SDS-PAGE sample buffer: 1.5 mL of 1 M Tris–HCl buffer, pH 6.8, 3 mL of 1 M DTT (dithiothreitol) in pure H_2O, 0.6 g of sodium dodecyl sulfate (SDS), 30 mg of bromophenol blue, 2.4 mL of glycerol, bring final volume to 7.5 mL.

10. SDS-PAGE sample buffer: dilute four times 4× SDS-PAGE sample buffer in pure H_2O and add 20% 2-mercaptoethanol (by volume). Prepare fresh.

11. SDS-4–20% PAGE gels, 1× SDS running buffer.

12. Gel stain: Gelcode® Blue (Thermo scientific).

13. N-hydroxy-succinimide ester (NHS)-activated sepharose beads (GE Healthcare life sciences).

14. Porcine pancreatic trypsin type Ix-S (14,000 units/mg) (Sigma Aldrich).

15. Coupling buffer: 200 mM sodium phosphate, 250 mM NaCl, pH 6.0.

16. Washing buffer: 200 mM sodium acetate, 250 mM NaCl, pH 4.5.

17. Column buffer: 0.1 mM ethylenediaminetetraacetic acid (EDTA), 50 mM Na_2HPO_4, 150 mM NaCl, pH 7.4.

18. 100 mM ethanolamine (Eastman Kodak).

19. 8 M guanidinium chloride in pure water for molecular biology.

20. Solid-phase extraction silica-C18 cartridge (820 mg of silica-C18, 55–105 μm particle size) (Sep-Pak C18 Plus Long Cartridge, Waters).

21. HPLC buffers. Buffer A: pure and filtered (over 45 μm filter) H_2O with 0.1% trifluoroacetic acid (TFA) (HPLC grade). Buffer B: 90% acetonitrile (HPLC grade) in pure and filtered (over 45 μm filter) H_2O with 0.1% TFA.

3   Methods

3.1   Construction of Intein-MCoTI-I Construct 1

3.1.1   Annealing of the DNA Fragments Encoding Split Intein Npu DnaE and Cyclotide MCoTI-I

1. Dissolve each ultramer shown in Table 1 in TE buffer to a concentration of 1 µg/µL with TE buffer.

2. The annealing reaction for every DNA fragment (Npu DnaE I_C and I_N, and MCoTI-I) is carried out as follows: 5 µL of solution containing the upper DNA strand (P5) and 5 µL of solution containing the lower strand (P3) are added into a 0.5 mL centrifuge tube containing 2.5 µL of 10× annealing buffer and 12 µL of pure H_2O. This should provide three annealing reactions for the DNA encoding regions corresponding to the DnaE I_N and I_C, and MCoTI-I polypeptides.

3. Incubate the above samples on a preheated water bath to 95 °C for 15 min. Turn off the power of the water bath, and allow the samples to slowly cool down to room temperature. The cooling process should not take less than 60 min.

4. Purify the double strand DNA fragments by using the QIAGEN PCR cleanup kit following the manufacturer instructions.

5. Double strand DNA fragments were obtained in TE buffer and quantified by UV–Vis spectroscopy (for a 1-cm pathlength, an optical density at 260 nm (OD_{260}) of 1.0 equals to a concentration of 50 µg/mL solution of dsDNA).

3.1.2   Ligation and Amplification of the DNA Fragment Encoding Construct 1

1. Mix equimolar amounts ($\approx$20 nmol) of each dsDNA fragment encoding for the DnaE I_C/I_N and MCoTI-I polypeptides in a thin walled 0.5 mL centrifuge tube. Add enough pure sterile water to have a final volume reaction of 50 µL, add 5 µL of 10× T4 DNA ligase buffer, 1 µL of 10 mM ATP, 1 µL of 10 mM dNTP, and then add 1 µL (400 units) of T4 DNA ligase enzyme. Incubate the ligation reaction at 16 °C overnight.

2. Purify the ligated DNA encoding construct 1 using Qiagen PCR cleanup kit according to the manufacturer instructions, and quantify it by UV–Vis spectroscopy.

3. Amplify the construct by PCR using primers shown in Table 2, which introduce $NcoI$ and $HindIII$ restriction sites in the 5′ and 3′ positions of the coding DNA sequence. Carry out the PCR reaction as follows: 40 µL sterile pure H_2O, 1 µL of ligated dsDNA ($\approx$10 ng/µL), 5 µL of 10× ThermoPol reaction buffer, 1.0 µL of dNTP solution (10 mM each), 1 µL of forward I_C primer solution (0.2 µM), 1 µL of I_N reverse primer solution (0.2 µM), and 1 µL Vent DNA polymerase (2 units).

4. PCR cycle conditions used: initial denaturation at 94 °C for 5 min followed by 30 cycles (94 °C denaturation for 30 s, annealing at 56 °C for 45 s, and extension at 72 °C for 60 s) and final extension at 72 °C for 10 min.

5. Purify the PCR amplified fragment encoding construct 1 using the QIAquick PCR purification kit following the manufacturer instructions and quantify it by UV–visible spectroscopy.

3.1.3 Preparation of Expression Plasmid pET28-MCoTI-TS

1. Digest the plasmid pET28a (Novagen-EMD Millipore) and the PCR-amplified gene encoding MCoTI-intein construct with restriction enzymes *NcoI* and *HindIII*. Use a 0.5 mL centrifuge tube and add 5 μL of NEB buffer 2.1 (New England Biolabs), add enough pure sterile water to have a final volume reaction of 50 μL, add ≈10 μg of the corresponding dsDNA to be digested add finally add 1 μL (20 units) of restriction enzyme *NcoI* (New England Biolabs). Incubate at 37 °C for 3 h. Then, add 1 μL (20 units) of restriction enzyme *HindIII* (New England Biolabs) to the same tube and incubate at 37 °C for 1 h.

2. Purify the double digested PCR-product and pET28a plasmid by agarose (0.8 % and 2 % agarose gels for pET28s and PCR product should be used, respectively) gel electrophoresis. The bands corresponding to the double digested DNA are cut and purified using the QIAquick Gel Extraction Kit (QIAGEN), eluted with TE buffer and quantified using UV–visible spectroscopy.

3. Ligate double digested pET28a and PCR-product encoding MCoTI-intein construct 1. Use a 0.5 mL centrifuge tube, add ≈100 ng of *NcoI*, *HindIII*-digested pET28a, ≈50 ng of *NcoI*, *HindIII*-digested PCR-amplified DNA encoding MCoTI-intein construct 1, enough pure sterile H$_2$O to make a final reaction volume of 20 μL, 2 μL of 10× T4 DNA ligase buffer, 1 μL of 10 mM ATP, and 1 μL (400 units) T4 DNA ligase. Incubate at 16 °C overnight.

4. Transform the ligation mixture into DH5α competent cells. ≈100 μL of chemical competent cells are thawed on ice and mixed with the ligation mixture (20 μL) for 30 min. The cells are heat-shocked at 42 °C for 45 s and then kept on ice for an extra 10 min. Add 900 μL of SOC medium and incubate at 37 °C for 1 h in an orbital shaker. Plate 100 μL on LB agar plate containing kanamycin (25 μg/mL) and incubate the plate at 37 °C overnight.

5. Pick up several colonies (most of the times five colonies should be enough) and inoculate into 5 mL of LB medium, 25 μg/mL kanamycin. Incubate tubes at 37 °C overnight in an orbital shaker.

6. Pellet down cells and extract DNA using the QIAprep Spin Miniprep Kit (QIAGEN) following the manufacturer protocol and quantify plasmid using UV–visible spectroscopy.

7. Verify the presence of DNA encoding MCoTI-intein construct in each colony using PCR. Carry out the PCR reaction as follows: 40 μL sterile pure H$_2$O, 1 μL of plasmid DNA (≈50 ng/ μL), 5 μL of 10× TaqDNA polymerase buffer, 1.0 μL of dNTP

solution (10 mM each), 1 μL of forward I_C primer solution (0.2 μM), 1 μL of I_N reverse primer solution (0.2 μM), and 1 μL Taq DNA polymerase (5 units).

8. PCR cycle conditions used: initial denaturation at 94 °C for 5 min followed by 30 cycles (94 °C denaturation for 30 s, annealing at 56 °C for 45 s, and extension at 72 °C for 60 s) and final extension at 72 °C for 10 min.

3.2 Expression and Purification of Cyclotides

3.2.1 Expression of Precursor Protein Encoding the MCoTI-Intein Construct 1

1. Transform chemical competent Origami2(DE3) cells with plasmid containing the DNA encoding MCoTI-intein construct 1 (plasmid pET28-MCoTI-TS) (*see* **Note 1**). Plate transformed cells on LB plate, 2 μg/mL kanamycin, and incubate at 37 °C overnight as described in Subheading 3.1.3 (*see* **Note 2**).

2. Resuspend the colonies from two plates in 2 mL of LB medium and use resuspension to inoculate 1 L of LB medium, 25 μg/mL kanamycin, in a 2.5 L flask.

3. Grow cells in an orbital shaker incubator at 37 °C for 2–3 h to reach mid-log phase (OD at 600 nm≈0.5). Add IPTG to reach a final concentration of 0.3 mM. Adjust the temperature of the incubator to 25 °C and incubate cells in shaker for 16 h.

4. Pellet cells by centrifugation at 6000×g for 15 min at 4 °C. Discard the supernatant and process the pellet immediately (*see* **Note 3**).

3.2.2 Protein Extraction

1. Resuspend cell pellet with 30 mL of Ni-lysis buffer containing 1 mM PMSF. Lyse cells by sonication on ice using 25 s bursts spaced 30 s each (*see* **Note 4**). Repeat the cycle six times (*see* **Note 5**). Take two 100 μL aliquots. In one of the samples add 33 μL 4× SDS-PAGE sample buffer, 20 % 2-mercaptoethanol, and heat it at 94 °C for 5 min (label it total cell lysate, T). For the other aliquot, separate the insoluble and soluble fractions by centrifugation at 15,000×g in a microcentrifuge at 4 °C for 30 min. Take the supernatant fraction and resuspend the pellet in 100 μL of Ni-lysis buffer. Add 33 μL of 4× SDS-PAGE sample buffer containing 20 % 2-mercaptoethanol to both fractions and heat them at 94 °C for 5 min (label them soluble and insoluble cell lysate samples, P and S, respectively). Save the samples for later SDS-PAGE analysis.

2. Separate the soluble cell lysate fraction by centrifugation at 15,000×g for 20 min at 4 °C. Stored the pellets at −80 °C in case they need to be reprocessed.

3. Transfer the soluble cell lysate fraction (≈30 mL) into a 50 mL centrifuge tube and add 1 mL of pre-equilibrated Ni-NTA-agarose beads. Incubate with gentle rocking for 30 min at 4 °C.

4. Separate the beads from supernatant by centrifugation at 3000×g for 10 min at 4 °C. Take the supernatant and save it at 4 °C for

later analysis (Subheading 3.2.4). Separate the beads and wash them in a 5 mL polypropylene column with no less than 15 column volumes of Ni-wash buffer. Take ≈100 μL of Ni-NTA-agarose beads into a 0.5 mL centrifuge tube, add 33 μL of 4× SDS-PAGE sample buffer, 20% 2-mercaptoethanol, and heat it at 94 °C for 5 min (label it soluble cell lysate bound to Ni-NTA agarose beads, B). Save the sample for later SDS-PAGE analysis.

5. Analyze the expression level of the precursor protein 1 using SDS-PAGE (Fig. 4). Load 25 μL of samples labeled T, P, S, and B (see above) onto an SDS-4–20% PAGE gel. Run the samples at 125 V for about 1 h and 30 min in 1× SDS running buffer. Remove SDS with pure water and stain the gel with 20 mL GelCode® Blue reagent (*see* **Note 6**) using the manufacturer protocol (Fig. 4).

3.2.3 Preparation of Trypsin-Immobilized Agarose Beads for Affinity Chromatography

1. Wash ≈1 mL of NHS-activated sepharose with 15 column volumes of ice-cold 1 mM HCl using a 5 mL polypropylene column.

2. Equilibrate column with 15 volumes of coupling buffer.

3. Dissolve 4 mg of porcine pancreatic trypsin in 500 μL of coupling buffer using gentle rocking.

4. Add the trypsin solution to the equilibrated NHS-activated sepharose beads and incubate for 3 h with gentle rocking at room temperature.

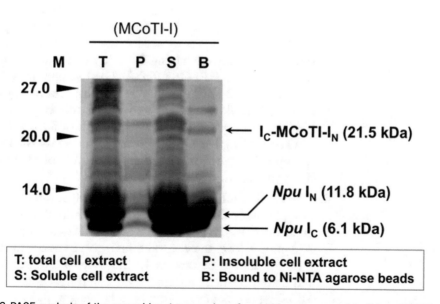

Fig. 4 SDS-PAGE analysis of the recombinant expression of cyclotide precursors **1** in Origami2(DE3) cells for in-cell production of cyclotide MCoTI-I. The bands corresponding to precursor 1, and I_N and I_C polypeptides are marked with *arrows*. Despite that only the I_C polypeptide has a His-tag, the I_N binds the I_C polypeptide with low nM affinity and is co-purified during the Ni-NTA pre-purification step. M, stands for protein markers

5. Wash the sepharose beads with ten volumes of coupling buffer containing 100 mM ethanolamine.

6. Incubate beads with three column volumes of coupling buffer containing 100 mM ethanolamine for 3 h with gentle rocking at room temperature.

7. Wash the sepharose beads with 50 column volumes of washing buffer and store a 4 °C until use (*see* **Note 7**).

3.2.4 Affinity-Purification of Cyclotide MCoTI-I from Bacterial Soluble Cell Lysate

1. Wash ≈500 μL of trypsin-sepharose beads with ten column volumes of column buffer.

2. Incubate the washed beads with the soluble cell lysate flow-through from the Ni-NTA agarose beads purification step (Subheading 3.2.2, **step 4**) for 1 h at room temperature with gentle rocking.

3. Separate the supernatant by centrifugation at $3000 \times g$ for 10 min at 4 °C.

4. Transfer the trypsin-beads to a 5 mL polypropylene column and wash the beads with 50 volumes of column buffer containing 0.1% Tween 20.

5. Wash trypsin-beads with 50 volumes of column buffer with no detergent added.

6. Elute bound cyclotide MCoTI-I with three volumes (3×500 μL) of 8 M guanidinium hydrochloride at room temperature for 15 min by gravity.

7. Desalt the sample using a solid-phase extraction cartridge (SepPak, Waters) by following the manufacturer protocol. Briefly, pre-swell the SepPak cartridge with 20 mL of 50% acetonitrile in water containing 0.1% TFA. Equilibrate cartridge with 50 mL of HPLC buffer A. Load sample into cartridge using a plastic 20 mL syringe slowly with a flow rate of ≈1 mL/min. Collect flow-through and repeat this step for efficient binding. Wash the cartridge with 20 mL of 5% acetonitrile in H_2O containing 0.1% TFA. Elute cyclotide MCoTI-I from the solid-phase extraction cartridge with 5 mL of HPLC buffer-B. Lyophilize to remove solvents.

8. Dissolve in 5 mL of HPLC buffer A. Analyze sample by HPLC using an isocratic of 0% buffer B for 2 min and then a linear gradient of 0–70% buffer B in 30 min. Use detection at 220 and 280 nm. Using these conditions the retention time of the cyclotide should be around 15 min (Fig. 5). Collect the peak and analyze by mass spectrometry to confirm identity of cyclotide MCoTI-I (Fig. 5; expected molecular weight: 3481.0 Da).

9. Quantify the cyclotide using UV–visible spectroscopy and a molar absorptivity at 280 nm of 2240 $M^{-1} \times cm^{-1}$. Around 150 μg of folded cyclotide should be obtained per liter of LB culture.

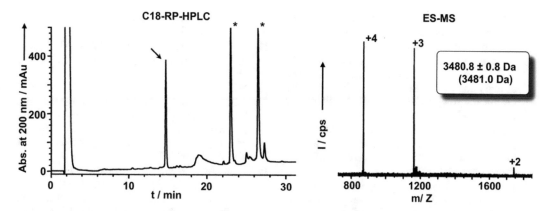

Fig. 5 Analytical HPLC trace (*left panel*) of the soluble cell extract of bacterial cells expressing precursor **1** after purification by affinity chromatography on a trypsin-sepharose column. Folded MCoTI-I is marked with an *arrow*. Endogenous bacterial proteins that bind trypsin are marked with an *asterisk*. Mass spectrum (*right panel*) of affinity purified MCoTI-I. The expected average molecular weight is shown in parentheses

4   Notes

1. We recommend to screen at least 3–4 different colonies for protein expression to make sure that the MCoTI-I intein construct 1 is expressed efficiently.

2. When plating the transformed cells with pET28-MCoTI-TS, it is better to aim for plates containing 200–300 colonies.

3. Cell pellets can be stored at –80 °C for no more than 2–3 weeks before being processed.

4. During sonication, be sure the temperature of the sample does not overheat.

5. A french press can be also used to lyse cells, depending on the availability.

6. Coomassie brilliant blue can be also used for staining PAGE gels.

7. The trypsin-sepharose column should not be stored for more than 2 weeks. The loading of the trypsin-sepharose can be determined by incubating a small aliquot of the beads with a known amount of pure MCoTI-I and determining the amount of cyclotide captured on the beads using HPLC, *see* Subheading 3.2.4).

References

1. Daly NL, Rosengren KJ, Craik DJ (2009) Discovery, structure and biological activities of cyclotides. Adv Drug Deliv Rev 61:918–930

2. Gould A, Ji Y, Aboye TL, Camarero JA (2011) Cyclotides, a novel ultrastable polypeptide scaffold for drug discovery. Curr Pharm Des 17:4294–4307

3. Puttamadappa SS, Jagadish K, Shekhtman A, Camarero JA (2010) Backbone dynamics of cyclotide MCoTI-I free and complexed with trypsin. Angew Chem Int Ed Engl 49:7030–7034

4. Puttamadappa SS, Jagadish K, Shekhtman A, Camarero JA (2011) Erratum in: backbone dynamics of cyclotide MCoTI-I free and complexed with trypsin. Angew Chem Int Ed Engl 50:6948–6949

5. Daly NL, Thorstholm L, Greenwood KP, King GJ, Rosengren KJ, Heras B, Martin JL, Craik

DJ (2013) Structural insights into the role of the cyclic backbone in a squash trypsin inhibitor. J Biol Chem 288:36141–36148

6. Saether O, Craik DJ, Campbell ID, Sletten K, Juul J, Norman DG (1995) Elucidation of the primary and three-dimensional structure of the uterotonic polypeptide kalata B1. Biochemistry 34:4147–4158

7. Austin J, Kimura RH, Woo YH, Camarero JA (2010) In vivo biosynthesis of an Ala-scan library based on the cyclic peptide SFTI-1. Amino Acids 38:1313–1322

8. Huang YH, Colgrave ML, Clark RJ, Kotze AC, Craik DJ (2010) Lysine-scanning mutagenesis reveals an amendable face of the cyclotide kalata B1 for the optimization of nematocidal activity. J Biol Chem 285:10797–10805

9. Simonsen SM, Sando L, Rosengren KJ, Wang CK, Colgrave ML, Daly NL, Craik DJ (2008) Alanine scanning mutagenesis of the prototypic cyclotide reveals a cluster of residues essential for bioactivity. J Biol Chem 283:9805–9813

10. Garcia AE, Camarero JA (2010) Biological activities of natural and engineered cyclotides, a novel molecular scaffold for peptide-based therapeutics. Curr Mol Pharmacol 3:153–163

11. Aboye TL, Ha H, Majumder S, Christ F, Debyser Z, Shekhtman A, Neamati N, Camarero JA (2012) Design of a novel cyclotide-based CXCR4 antagonist with anti-human immunodeficiency virus (HIV)-1 activity. J Med Chem 55:10729–10734

12. Wong CT, Rowlands DK, Wong CH, Lo TW, Nguyen GK, Li HY, Tam JP (2012) Orally active peptidic bradykinin B1 receptor antagonists engineered from a cyclotide scaffold for inflammatory pain treatment. Angew Chem Int Ed Engl 51:5620–5624

13. Chan LY, Gunasekera S, Henriques ST, Worth NF, Le SJ, Clark RJ, Campbell JH, Craik DJ, Daly NL (2011) Engineering pro-angiogenic peptides using stable, disulfide-rich cyclic scaffolds. Blood 118:6709–6717

14. Ji Y, Majumder S, Millard M, Borra R, Bi T, Elnagar AY, Neamati N, Shekhtman A, Camarero JA (2013) In vivo activation of the p53 tumor suppressor pathway by an engineered cyclotide. J Am Chem Soc 135:11623–11633

15. Contreras J, Elnagar AY, Hamm-Alvarez SF, Camarero JA (2011) Cellular uptake of cyclotide MCoTI-I follows multiple endocytic pathways. J Control Release 155:134–143

16. Cascales L, Henriques ST, Kerr MC, Huang YH, Sweet MJ, Daly NL, Craik DJ (2011) Identification and characterization of a new family of cell-penetrating peptides: cyclic cell-penetrating peptides. J Biol Chem 286:36932–36943

17. Henriques ST, Craik DJ (2010) Cyclotides as templates in drug design. Drug Discov Today 15:57–64

18. Mylne JS, Chan LY, Chanson AH, Daly NL, Schaefer H, Bailey TL, Nguyencong P, Cascales L, Craik DJ (2012) Cyclic peptides arising by evolutionary parallelism via asparaginyl-endopeptidase-mediated biosynthesis. Plant Cell 24:2765–2778

19. Poth AG, Mylne JS, Grassl J, Lyons RE, Millar AH, Colgrave ML, Craik DJ (2012) Cyclotides associate with leaf vasculature and are the products of a novel precursor in petunia (Solanaceae). J Biol Chem 287:27033–27046

20. Poth AG, Colgrave ML, Lyons RE, Daly NL, Craik DJ (2011) Discovery of an unusual biosynthetic origin for circular proteins in legumes. Proc Natl Acad Sci U S A 108:1027–1032

21. Jennings C, West J, Waine C, Craik D, Anderson M (2001) Biosynthesis and insecticidal properties of plant cyclotides: the cyclic knotted proteins from Oldenlandia affinis. Proc Natl Acad Sci U S A 98:10614–10619

22. Gillon AD, Saska I, Jennings CV, Guarino RF, Craik DJ, Anderson MA (2008) Biosynthesis of circular proteins in plants. Plant J 53:505–515

23. Saska I, Gillon AD, Hatsugai N, Dietzgen RG, Hara-Nishimura I, Anderson MA, Craik DJ (2007) An asparaginyl endopeptidase mediates in vivo protein backbone cyclization. J Biol Chem 282:29721–29728

24. Nguyen GK, Wang S, Qiu Y, Hemu X, Lian Y, Tam JP (2014) Butelase 1 is an Asx-specific ligase enabling peptide macrocyclization and synthesis. Nat Chem Biol 10:732–738

25. Jagadish K, Borra R, Lacey V, Majumder S, Shekhtman A, Wang L, Camarero JA (2013) Expression of fluorescent cyclotides using protein trans-splicing for easy monitoring of cyclotide-protein interactions. Angew Chem Int Ed Engl 52:3126–3131

26. Austin J, Wang W, Puttamadappa S, Shekhtman A, Camarero JA (2009) Biosynthesis and biological screening of a genetically encoded library based on the cyclotide MCoTI-I. Chembiochem 10:2663–2670

27. Camarero JA, Kimura RH, Woo YH, Shekhtman A, Cantor J (2007) Biosynthesis of a fully functional cyclotide inside living bacterial cells. Chembiochem 8:1363–1366

28. Kimura RH, Tran AT, Camarero JA (2006) Biosynthesis of the cyclotide kalata B1 by using protein splicing. Angew Chem Int Ed 45:973–976

29. Hernandez JF, Gagnon J, Chiche L, Nguyen TM, Andrieu JP, Heitz A, Trinh Hong T, Pham TT, Le Nguyen D (2000) Squash trypsin inhibitors from Momordica cochinchinensis exhibit an atypical macrocyclic structure. Biochemistry 39:5722–5730

30. Sancheti H, Camarero JA (2009) "Splicing up" drug discovery. Cell-based expression and screening of genetically-encoded libraries of backbone-cyclized polypeptides. Adv Drug Deliv Rev 61:908–917

31. Zettler J, Schutz V, Mootz HD (2009) The naturally split Npu DnaE intein exhibits an extraordinarily high rate in the protein trans-splicing reaction. FEBS Lett 583:909–914

32. Iwai H, Zuger S, Jin J, Tam PH (2006) Highly efficient protein trans-splicing by a naturally split DnaE intein from Nostoc punctiforme. FEBS Lett 580:1853–1858

33. Jagadish K, Gould A, Borra R, Majumder S, Mushtaq Z, Shekhtman A, Camarero JA (2015) Recombinant expression and phenotypic screening of a bioactive cyclotide against alpha-synuclein-induced cytotoxicity in baker's yeast. Angew Chem Int Ed Engl 54:8390–8394

Chapter 5

Ribosomal Synthesis of Thioether-Bridged Bicyclic Peptides

Nina Bionda and Rudi Fasan

Abstract

Many biologically active peptides found in nature exhibit a bicyclic structure wherein a head-to-tail cyclic backbone is further constrained by an intramolecular linkage connecting two side chains of the peptide. Accordingly, methods to access macrocyclic peptides sharing this overall topology could be of significant value toward the discovery of new functional entities and bioactive compounds. With this goal in mind, we recently developed a strategy for enabling the biosynthesis of thioether-bridged bicyclic peptides in living bacterial cells. This method involves a split intein-catalyzed head-to-tail cyclization of a ribosomally produced precursor peptide, combined with inter-sidechain cross-linking through a genetically encoded cysteine-reactive amino acid. This approach can be applied to direct the formation of structurally diverse bicyclic peptides with high efficiency and selectivity in living *Escherichia coli* cells and provides a platform for the generation of combinatorial libraries of genetically encoded bicyclic peptides for screening purposes.

Key words Split intein, Bicyclic peptides, Unnatural amino acid, Amber stop codon suppression, Thioether-linked peptide macrocycles

1 Introduction

The structural diversity and breadth of biological activities exhibited by naturally occurring cyclic peptides have stimulated significant interest in this structural class for the development of chemical probes and therapeutics, in particular directed at challenging targets such as protein–protein and protein–nucleic acid interactions [1–3]. The growing interest in this area has translated into the development of various synthetic, semisynthetic, and biosynthetic strategies to obtain macrocyclic peptides and libraries thereof, including methods for obtaining head-to-tail, side-chain-to-tail, or side-chain-to-side-chain cyclic peptides that also comprise non-peptidic moieties and linkers [4–11]. Among the bioactive peptides found in nature, several exhibit a bicyclic structure wherein a head-to-tail cyclic structure is further constrained by an intramolecular linkage connecting two side chains of the peptide. Representative examples

Henning D. Mootz (ed.), *Split Inteins: Methods and Protocols*, Methods in Molecular Biology, vol. 1495,
DOI 10.1007/978-1-4939-6451-2_5, © Springer Science+Business Media New York 2017

58 Nina Bionda and Rudi Fasan

of these bicyclic peptides include α-amanitin, a potent inhibitor of eukaryotic RNA polymerases, the antimicrobial β-defensins, and cytotoxic plant-derived peptides belonging to the bouvardin and celogentin families. In these classes of molecules, the conformational rigidification imposed by the bicyclic backbone is often critical for their biological activity and beneficial toward increasing their cell permeability and proteolytic stability.

As part of our efforts toward developing versatile methodologies to access structurally diverse peptide-based macrocycles [12–16], we have recently devised a strategy to enable the ribosomal synthesis of 'natural product-like' bicyclic peptides, which feature a head-to-tail cyclic backbone along with an inter-side-chain thioether linkage [17, 18]. As illustrated in Fig. 1, this methodology couples split intein-mediated peptide circularization [19] with a spontaneous, intramolecular cross-linking reaction between a genetically encoded, cysteine-reactive unnatural amino acid (O-(2-bromoethyl)-tyrosine or O2beY) and a neighboring cysteine

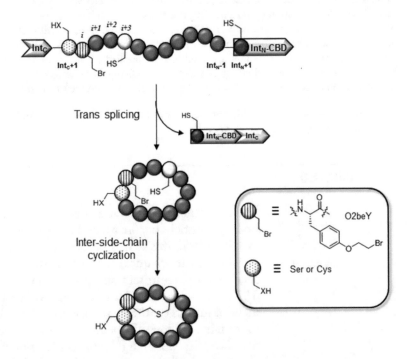

Fig. 1 Overall strategy for the ribosomal synthesis of thioether-bridged bicyclic peptides in *E. coli*. From the N- to C-terminus, the precursor protein comprises: the C-terminal domain of DnaE split intein (Int_C); a variable target peptide sequence; and the N-terminal domain of DnaE split intein fused to a chitin binding domain (Int_N-CBD). The target peptide sequence contains an initial Ser or Cys residue (*dotted circle*) at the 'Int_C + 1' site for mediating split intein-catalyzed head-to-tail cyclization and comprises the unnatural amino acid O-(2-bromoethyl)-tyrosine (O2beY, *circle with vertical lines*) and the reactive cysteine (*white circle*) for inter-side-chain cross-linking via thioether bond formation

residue present in the peptide sequence [17, 18]. Briefly, peptide head-to-tail circularization is achieving by framing a target peptide sequence between the C-terminal (Int_C) and the N-terminal domain (Int_N) of the naturally occurring split intein DnaE from *Synechocystis sp.* PCC6803 [19]. The overall mechanism of this reaction is described in Fig. 2 and it involves the association of the two intein fragments to promote a $N \rightarrow S$ acyl shift at the level of the cysteine residue at the Int_{N+1} position. This thioester intermediate is intercepted by a nucleophilic residue at the Int_{C+1} position (either serine or cysteine) to undergo a trans(thio)esterification followed by the asparagine-mediated release of the side-chain-to-head linked cyclic peptide which upon $O/S \rightarrow N$ acyl shift gives the desired head-to-tail cyclic peptide. While this method can provide a means to generate genetically encoded libraries of head-to-tail cyclic peptides (e.g., via genetic randomization of residues within the peptide target sequence), its scope has remained restricted to the formation of monocyclic peptides [9]. To enable the biosynthesis of bicyclic peptides, we envisioned the possibility of incorporating, within the peptide target sequence, an electrophilic unnatural amino acid that could react with a nearby cysteine residue to form a hydrolytically and redox stable inter-side-chain thioether bond (Fig. 1).

The design and realization of this strategy involved a series of important considerations. First, the reactivity of the unnatural amino acid has to be properly tuned in order to support the desired intramolecular thioether bond formation, while avoiding undesirable reactions with other nucleophiles present in the intracellular milieu (e.g., glutathione). Another important requirement is the availability of an efficient system for the ribosomal incorporation of such unnatural amino acid into the precursor

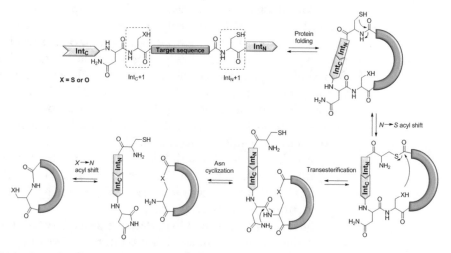

Fig. 2 Mechanism of split intein-catalyzed peptide circularization. Int_N and Int_C correspond to the N-domain and C-domain respectively, of DnaE split intein from *Synechocystis sp.* PCC6803. The $\text{Int}_N + 1$ cysteine and $\text{Int}_C + 1$ cysteine (or serine) residues are indicated

polypeptide expressed in *E. coli*. To this end, we established that *O*-(2-bromoethyl)-tyrosine (O2beY) has the desired reactivity profile and we engineered an orthogonal aminoacyl-tRNA synthetase (AARS)/tRNA pair for incorporation of O2beY in response to an amber stop codon (TAG) [17], respectively. The O2beY-specific AARS was derived from the *Methanocaldococcus jannaschii* tyrosyl-tRNA synthetase, which had been previously evolved for encoding a variety of tyrosine derivatives via the amber stop codon suppression technology [20], respectively. The O2beY-specific AARS was found to offer an excellent amber stop codon suppression efficiency (85%), enabling the expression of O2beY-containing proteins in high yields (e.g., ~40 mg/L for O2beY-containing yellow fluorescent protein).

Using a series of model polypeptide constructs produced in *E. coli*, it was established that O2beY is able to form a thioether linkage with a neighboring cysteine residue that is located from two to eight residues apart from O2beY with high efficiency (80–100% cyclization yield) [17]. Furthermore, the efficiency of the cross-linking reaction was found to be independent of the composition of the target peptide sequence and relative orientation of the O2beY/Cys pair (i.e., O2beY/Cys vs. Cys/O2beY) across the various constructs tested.

By combining split intein-mediated peptide circularization with the O2beY/Cys cross-linking strategy (Fig. 1), it has been possible to direct the spontaneous biosynthesis of natural product-like bicyclic peptides inside living *E. coli* cells [17, 18]. To illustrate such capability, a streptavidin binding motif (His-Pro-Gln) was introduced within the target peptide sequence of the precursor polypeptide, thus enabling the isolation and identification of the desired bicyclic products, along with potential acyclic or monocyclic byproducts, via streptavidin affinity capture followed by LC-ESI-MS and MS/MS analyses. Using this approach, bicyclic peptides comprising between 12 and 18 amino acid residues and featuring different O2beY/Cys connectivities (e.g., $i/i+3$ and $i/i+8$ linkages) and thus varying ring sizes could be produced in *E. coli* in good to excellent yields (70–97%) [17, 18]. Mechanistic studies showed that these bicyclic peptides are generated at the post-translational level primarily through head-to-tail cyclization followed by inter-side-chain cross-linking. In the presence of target sequences less prone to split intein-mediated circularization, however, the thioether bridging reaction was found to increase the overall yield of the bicyclization reaction, presumably by facilitating the *trans* splicing process [18].

The protocols provided here describe procedures for the cloning, production in *E. coli*, and isolation of two representative bicyclic peptides featuring a $i/i+3$ and a $i/i+8$ thioether linkage and comprising a streptavidin-binding motif for isolation from *E. coli* cells via affinity capturing. This general methodology can be readily adapted to generate libraries of genetically encoded bicyclic peptides which can be screened using in vitro assays [18] or, alternatively, using high throughput methods based on intracellular reporter [21] or selection systems [22], in order to identify bicyclic peptides with novel or improved functional properties.

2   Materials

2.1   Reagents

1. Chemically competent BL21(DE3) cells.

2. Ampicillin stock solution (1000×): Dissolve ampicillin sodium salt in water at 100 mg/mL. Filter-sterilize and store in aliquots at −80 °C for up to 6 months.

3. Chloramphenicol stock solution (1000×): Dissolve chloramphenicol in ethanol at 30 mg/mL. Store in aliquots at −80 °C for up to 6 months.

4. Sodium hydrogen phosphate (Na_2HPO_4).

5. Potassium dihydrogen phosphate (KH_2PO_4).

6. Sodium chloride (NaCl).

7. Ammonium chloride (NH_4Cl).

8. Lysogeny Broth (LB) medium: Dissolve 10 g tryptone, 5 g yeast extract, and 10 g NaCl in 1 L of deionized water. Autoclave the solution and store at room temperature.

9. M9 salts (5×) solution: Dissolve 64 g Na_2HPO_4, 15 g KH_2PO_4, 2.5 g NaCl, and 5 g NH_4Cl in 1 L of deionized water. Autoclave the solution and store at room temperature.

10. M9 medium: To a 1 L sterile flask, add 370 mL of distilled water, 100 mL of 5× M9 salt solution, 500 μL of 2 M $MgSO_4$, 50 μL 1 M $CaCl_2$, 5 mL glycerol, and 50 mL of 10% yeast extract (see Note 1).

11. Magnesium sulfate ($MgSO_4$) solution: Dissolve 24 g $MgSO_4$ in 100 mL of deionized water to make 2 M solution, autoclave and store at room temperature.

12. Calcium chloride ($CaCl_2$) solution. Dissolve 1.1 g $CaCl_2$ in 10 mL of deionized water to make 1 M solution, filter-sterilize and store at room temperature.

13. Sterile glycerol: Autoclave 100 mL glycerol and store at room temperature.

14. Arabinose stock solution (10%): Dissolve 1 g of arabinose in 10 mL of deionized water, filter-sterilize and store at −20 °C.

15. Chitin beads (New England Biolabs).

16. Streptavidin-coated agarose beads.

17. (L)-N-tert-butoxycarbonyl-tyrosine.

18. Dibromoethane.

19. Potassium carbonate (K_2CO_3).

20. pSFBAD09 and pJJDuet30 plasmids (Addgene #11963 and #11962, respectively).

21. Agarose.

22. Deoxynucleotide (dNTPs) mix solution.

23. High fidelity DNA Polymerase (e.g., Phusion HF DNA polymerase from NEB) and appropriate buffer solution (e.g., Phusion buffer HF 5× from NEB).

24. PCR Purification Kit (e.g., QIAquick from Qiagen).

25. Gel Extraction Kit (e.g., QIAquick from Qiagen).

26. Restriction enzymes *DpnI*, *NdeI*, and *KpnI* HF and appropriate buffer (e.g., CutSmart buffer 10× from NEB).

27. Calf intestinal alkaline phosphatase (CIP).

28. T4 DNA ligase and T4 DNA ligase reaction buffer.

29. Stock solution of 200 mM *O*-2-bromoethyl-tyrosine (O2beY) in water. The synthesis of O2beY is described below.

30. Phosphate buffer: 50 mM potassium phosphate solution, pH 8.0.

2.2 Solvents

1. Acetic acid (HOAc).

2. Trifluoroacetic acid (TFA).

3. Formic acid (FA).

4. Acetonitrile.

5. Hexanes.

6. Dimethylsulfoxide (DMSO).

7. Deionized water.

8. *N*-*N*-Dimethylformamide (DMF).

9. Dichloromethane (DCM).

10. Ethyl Acetate (EtOAc).

3 Methods

3.1 Synthesis of O-(2-Bromoethyl) Tyrosine (O2beY)

As described in the scheme reported in Fig. 3, the unnatural amino acid *O*-(2-bromoethyl)tyrosine (O2beY) can be prepared in two steps starting from commercially available *N*-Boc-tyrosine.

1. Transfer 20 mL of dry DMF to a 125-mL round-bottom flask equipped with a stirring bar.

2. To the DMF solution, add *N-tert*-butoxycarbonyl-tyrosine **1** (2 g, 7.1 mmol) and potassium carbonate (2.94 g, 21.3 mmol) under argon.

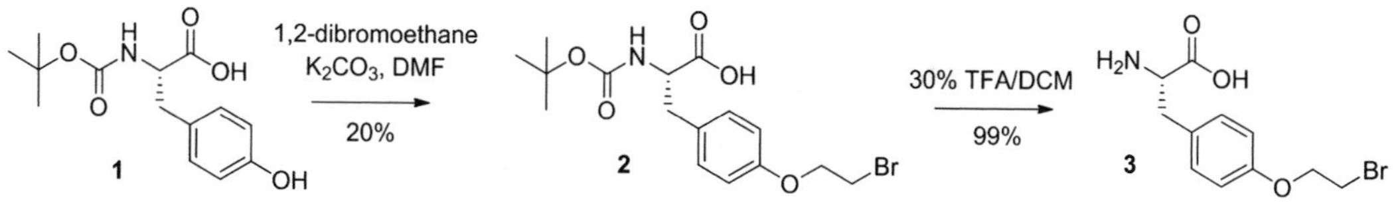

Fig. 3 Synthetic route for preparation of *O*-(2-bromoethyl)tyrosine (O2beY)

3. To the DMF solution of **step 2**, add dibromoethane (1.83 mL, 21.3 mmol) dropwise over 20 min. Stir the reaction mixture at room temperature.

4. Monitor the progress of the alkylation reaction by thin-layer chromatography (TLC) using hexane–ethyl acetate (7:3) as the mobile phase (R_f for **1**: 0.05; R_f for **2**: 0.5).

5. Upon completion of the reaction (approximately 18 h), filter the reaction mixture using a Buchner funnel.

6. Dilute the filtrate with 60 mL of water (*see* **Note 2**).

7. Acidify the water–DMF solution with HCl 0.1 M to pH 4.

8. Extract the acidified water–DMF solution with 2×100 mL of ethyl acetate. Combine the organic layers and dry them over sodium sulfate. Remove the organic solvent under reduced pressure in rotary evaporator to obtain **2** as a crude product (yellow oil).

9. Purify *O*-2-bromoethyl-*N*-*tert*-butoxycarbonyl-tyrosine **2** by flash column chromatography using a 10:9:1 mixture of hexanes–ethyl acetate–acetic acid as the solvent system.

10. Identify the fractions containing *O*-2-bromoethyl-*N*-*tert*-butoxycarbonyl-tyrosine **2** by TLC, combine them, and remove the solvent under reduced pressure in a rotary evaporator to yield **2** as an off-white powder (0.54 g, 20% yield).

11. Dissolve 10–15 mg of isolated **2** in deuterated chloroform ($CDCl_3$) and confirm its identity by ^1H NMR and mass spectrometry. ^1H NMR (400 MHz, $CDCl_3$) δ 1.38 (9H, s), 2.99 – 3.04 (2H, m), 3.43 (2H, t, J = 6 Hz), 3.58 (1H, t, J = 6 Hz), 4.22 (2H, t, J= 6.4 Hz), 6.80 (2H, d, J = 8.4 Hz), 7.05 (2H, d, J = 8.4) MS (ESI) calculated for $C_{14}H_{19}NO_5$ [M]$^+$: m/z 387.07, found 387.17.

12. For deprotection of the Boc protecting group (**2 → 3** in Fig. 3), add 2 g of purified **2** to 20 mL of 30% (v/v) trifluoroacetic acid in dichloromethane in a round-bottom flask equipped with a stirring bar. Stir the solution at 0 °C.

13. Monitor the reaction by TLC using hexane–ethyl acetate (7:3) as the mobile phase (R_f for **2**: 0.5; R_f for **3**: 0).

14. Upon completion of the reaction (approximately 2 h), remove the solvent under reduced pressure in a rotary evaporator.

15. Dissolve the crude residue in 10 mL of acetic acid and remove the solvent under reduced pressure in a rotary evaporator (*see* **Note 3**). Repeat this step one more time to yield the final product, *O*-2-bromoethyl-tyrosine **3**, as an off-white solid (1.7 g, 99 % yield).

16. Dissolve 10–15 mg of isolated **3** in deuterated methanol (CD$_3$OD) and confirm its identity by ^{1}H and ^{13}C NMR and mass spectrometry. ^{1}H NMR (400 MHz, CD$_3$OD) δ 3.02 – 3.23 (2H, m), 3.65 (2H, t, J = 5.6 Hz), 4.10 (1H, t, J= 5.8 Hz), 4.26 (2H, t, J = 5.6 Hz), 6.90 (2H, d, J = 8.4), 7.18 (2H, d, J = 8); ^{13}C NMR (125 MHz, CD$_3$OD) δ 28.9, 35.3, 54.1, 67.8, 114.8, 126.7, 130.2, 157.9, 167.6. MS (ESI) calculated for C$_{11}$H$_{14}$BrNO$_3$ [M + H]$^+$: m/z 288.02, found 288.51.

3.2 Cloning of the Expression Vectors for the Precursor Protein

The methodology described in this protocol involves the use of two plasmid vectors; a first one for directing the expression of the split intein-containing precursor protein and a second one encoding for the orthogonal AARS/tRNA pair for amber stop codon suppression with the unnatural amino acid O2beY. Various combinations of commercially available plasmid vectors for recombinant protein expression can be used for this purpose but the choice of the vectors must satisfy certain requirements. In general, the two plasmids need to carry two different and compatible origins of replication to allow for their stable co-existence in the bacterial cell. In addition, they need to carry two different antibiotic resistance markers to enable the selection of transformant *E. coli* cells containing both plasmids. *E. coli* cells that contain only one of the two vectors would either produce only the AARS/tRNA system and no precursor protein, or produce a truncated form of the precursor protein due to failed suppression of the amber stop codon introduced within the peptide target sequence.

In the present protocol, the plasmid encoding for the precursor protein is based on a pBAD vector system (Thermo Fischer Scientific), which contains a ColE1 origin of replication and an ampicillin resistance marker. The gene encoding for the precursor protein is under an arabinose-inducible promoter (AraBAD promoter) and is composed of: (a) Int$_C$-domain of DnaE split intein from *Synechocystis* sp., (b) a desired peptide target sequence containing a codon for encoding of the cysteine residue (TGT or TGC) and an amber stop codon (TAG) for encoding of the cysteine-reactive O2beY; (c) Int$_N$-domain of DnaE split intein from *Synechocystis* sp.; and (d) the chitin-binding domain (CBD) of chitinase A1 from *Bacillus circulans*.

The plasmid encoding for the amber stop codon suppression system is based on a pEVOL vector [23], which carries a p15A origin of replication and a chloramphenicol ampicillin resistance marker. This vector contains a first copy of the AARS gene under an arabinose-inducible promoter (AraBAD promoter), a second copy of the AARS gene under a constitutive *glnS* promoter, and a single copy of the cognate amber stop codon suppressor tRNA under a constitutive *proK* promoter. The Obe2Y-specific AARS gene is an engineered variant of *M. jannaschii* tyrosyl-tRNA synthetase carrying the following mutations: Y32G, E107P, D158A, L162A. The cognate tRNA is an engineered variant of *M. jannaschii* tyrosyl tRNA, in which the anticodon has been replaced with the triplet CUA for recognition of an amber stop codon.

While the plasmid/promoter configuration described above has proven effective toward producing bicyclic peptides in *E. coli* in good yields (>0.5 mg/L culture) according to the methods described herein, alternative configurations are likely to be equally viable provided the key elements are included and the plasmid compatibility issues are properly addressed. For example, O2beY-containing proteins have been successfully expressed using a different vector (e.g., pET vector) and from a gene under a different inducible promoter (e.g., IPTG-inducible T7 promoter) [17].

The following procedure was utilized to prepare pBAD-based vectors for the expression of the representative precursor proteins Z3C_O2beY and Z8C_O2beY described in Table 1. The oligonucleotide primers used in this protocol are described in Table 2 and were obtained by custom synthesis from IDT Technologies.

1. The genes encoding the N-terminal and C-terminal DnaE inteins from *Synechocystis* sp. were extracted from pSFBAD09 and pJJDuet30 plasmids [24] (Addgene #11963 and #11962, respectively). The pSFBAD09 vector also served as the host vector for insertion of the precursor protein gene.

2. For preparation of the Z3C_O2beY construct, PCR amplify the DnaE-Int$_C$-containing segment of the precursor protein gene using pSFBAD09 vector as the template, primer SICLOPPS_for as the forward primer, and primer Z3C_1/2_rev as the reverse primer. PCR solution: 10 µL Phusion buffer HF 5×, 36 µL deionized water, 0.5 µL template plasmid (=50 ng/µL), 1 µL SICLOPPS_for primer (10 µM), 1 µL Z3C_1/2_rev primer (10 µM), 1 µL dNTP mix, 0.5 µL Phusion polymerase (*see* **Note 4**). PCR method: 98 °C for

Table 1
Name and target sequences of the precursor protein constructs described in the protocols

Construct name	Target sequence[a]	Cys position[b]
Z3C_O2beY	S(**O2beY**)TN<u>C</u>HPQFANA	$i+3$
Z8C_O2beY	S(**O2beY**)TNVHPQF<u>C</u>NA	$i+8$

[a]Full-length precursor protein corresponds to Int$_C$-(target sequence)-Int$_N$-(Chitin Binding Domain)

[b]Position of reactive cysteine residue (underlined) in relation to O2beY (= i position)

Table 2
Oligonucleotide primers

Primer	Sequence
SICLOPPS_for	5′-CAGGTCATATGGTTAAAGTTATCGGTCGTCGATCC-3′
SICLOPPS_rev	5′-CAACAGGTACCTTTAATTGTACCTGCGTCAAGTAATGGAAAG-3′
Z3C_for	5′-CGCAGTTCGCGAACGCGTGCTTAAGTTTTGGCACCGAAATT-3′
Z3C_1/2_rev	5′-GGATGGCAGTTGGTCTAGCTATTGTGGGCGATAGCACCATTAGC-3′
Z3C_2/2_rev	5′-CGCGTTCGCGAACTGCGGATGGCAGTTGGTCTAGCTATTGTG-3′
Z8C_1/2_rev	5′-GATGCACGTTGGTCTAAGAATTGTGGGCGATAGCACCATTAGC-3′
Z8C_2/2_rev	5′-CGCGTTGCAGAACTGCGGATGCACGTTGGTCTAAGAATTG-3′
pBAD_seq_for	5′-GGATCCTACCTGACGCTTTTTATCG-3′

2 min, then 30 cycles: 98 °C for 15 s, 61 °C for 15 s, 72 °C for 30 s; then 72 °C for 2 min, then 4 °C. Analyze the in vitro synthesized dsDNA by resolving on 1.5 % agarose gel.

3. Add 0.5 μL of *Dpn* I to the reaction mixture of **step 2** (to digest the plasmidic DNA) and incubate at 37 °C for 2 h. Purify the product of the PCR reaction (0.12 kbp) using QIAquick PCR Purification Kit (Qiagen) according to the manufacturer's instructions.

4. Using the purified PCR product from **step 3** as the template, perform second PCR reaction using primer SICLOPPS_for as the forward primer and primer Z3C_2/2_rev as the reverse primer. PCR solution: 10 μL Phusion buffer HF 5×, 36 μL deionized water, 1 μL PCR product from **step 3** (50–100 ng/ μL), 1 μL SICLOPPS_for primer (10 μM), 1 μL Z3C_2/2_ rev primer (10 μM), 1 μL dNTP mix, 0.5 μL Phusion polymerase. PCR method: 98 °C for 2 min, then 30 cycles: 98 °C for 15 s, 58 °C for 15 s, 72 °C for 30 s; then 72 °C for 2 min, then 4 °C. Analyze the in vitro synthesized dsDNA by resolving on 1.5 % agarose gel.

5. Purify the product of the PCR reaction (0.13 kbp) in **step 4** using a PCR purification kit according to the manufacturer's instructions.

6. PCR amplify the DnaE-Int$_N$-containing segment of the precursor protein gene using pJJDuet30 vector as the template, primer Z3C_for as the forward primer and SICLOPPS_rev as the reverse primer. PCR solution: same as in **step 2** with the use of appropriate primers and template as described here. PCR method: 98 °C for 2 min, then 30 cycles: 98 °C for 15 s, 61 °C for 15 s, 72 °C

for 30 s; then 72 °C for 2 min, then 4 °C. Analyze the in vitro synthesized dsDNA by resolving on 1.5 % agarose gel.

7. Add 0.5 μL of *Dpn* I to the PCR reaction of **step 6** (to digest the plasmidic DNA) and incubate at 37 °C for 2 h. Purify the product of the PCR reaction (0.38 kbp) using a PCR purification kit according to the manufacturer's instructions.

8. Merge the genes isolated in **steps 5** and **6** by PCR Overlap Extension using SICLOPPS_for and SICLOPPS_rev as superprimers. PCR solution: 10 μL Phusion buffer HF 5×, 28 μL deionized water, 5 μL each of the gene segments (150–200 ng each in 1:1 ratio), 1 μL dNTP mix, 0.5 μL Phusion polymerase. PCR method: 98 °C for 2 min, then 6 cycles: 98 °C for 15 s, 55 °C for 15 s, 72 °C for 30 s; then add 1 μL of 1:1 mixture of SICLOPPS_for and SICLOPPS_rev primers; then 30 cycles: 98 °C for 15 s, 63 °C for 15 s, 72 °C for 30 s; then 72 °C for 2 min, then 4 °C. Analyze the in vitro synthesized dsDNA by resolving on 1.5 % agarose gel (*see* **Note 5**).

9. Purify the product of the PCR Overlap Extension reaction (0.51 kbp) in **step 8** using a PCR purification kit according to the manufacturer's instructions.

10. Digest the purified product of the PCR Overlap Extension reaction from **step 9** using *NdeI* and *KpnI* HF restriction enzymes. Double digest mixture: 100–300 ng DNA, 1 μL *NdeI*, 1 μL *KpnI*-HF, 5 μL 10× CutSmart buffer, deionized water to final volume of 50 μL. Leave the reaction in a water bath for 45 min.

11. Digest the pSFBAD09 vector using *NdeI* and *KpnI* restriction enzymes. Double digest mixture: 300–500 ng vector DNA, 1 μL *NdeI*, 1 μL *KpnI*-HF, 5 μL 10× CutSmart buffer, deionized water to final volume of 50 μL. Leave the reaction in a water bath for 45 min.

12. To the double digest reaction mixture from **step 11** add of 1 μL calf intestinal alkaline phosphatase (CIP) in order to dephosphorylate the 5′ and 3′ end of the digested vector and prevent circularization of the vector without the insert. Incubate the reaction mixture for additional 30 min in a water bath at 37 °C.

13. Load the digested insert (**step 10**) and digested vector (**step 12**) into separate wells of a 1.2 % agarose gel. Resolve the dsDNA products by electrophoresis and excise the bands corresponding to the insert and linearized vector. Combine the excised bands and extract the dsDNA by a gel extraction kit according to the manufacturer's instructions. Elute the DNA with 30 μL of the elution buffer provided with the kit.

14. Ligate the insert and vector together using T4 ligase. Ligation solution mixture: 12 μL deionized water, 1 μL T4 ligase, 2 μL

10× T4 ligase buffer, 5 μL DNA mixture from **step 13**. Leave the ligation reaction for 2 h at room temperature. Cloning of the insert into the *NdeI/KpnI* cassette in the pSFBAD09 vector result in the fusion of the CBD protein to the C-terminus of the DnaE-Int$_C$-(target sequence)-DnaE-Int$_N$ gene (Fig. 4).

15. Transform the purified ligation mixture (**step 14**) into chemically competent DH5α cells or equivalent *E. coli* cells for general cloning purposes. Typical procedure: add 5 μL of the ligation mixture to a 50 μL aliquot of chemically competent DH5α cells. Gently tap the tube and incubate the cells on ice. After 30 min, place the tube in a water bath pre-warmed at 42 °C for 30 s followed by transfer to ice for additional 5 min.

16. Add 500 μL of LB medium and incubate the cells with shaking for 45 min at 37 °C. Plate 100 μL of the cell culture onto LB agar plates containing ampicillin at 100 mg/L and incubate the plate at 37 °C for 14–18 h.

17. Pick a colony from the plate with a sterile toothpick and inoculate a culture tube containing 5 mL of LB medium supplemented with ampicillin at 100 mg/L. After overnight incubation at 37 °C, pellet the cells by centrifugation at 5000×*g*, and extract the plasmidic DNA from the cells using a plasmid miniprep kit according to the manufacturer's instruction.

18. Confirm the identity of the clone construct by DNA sequencing using pBAD_seq_for as sequencing primer. The DNA and protein sequence of the gene for expression of the precursor protein Z3C_O2beY is provided in Fig. 4.

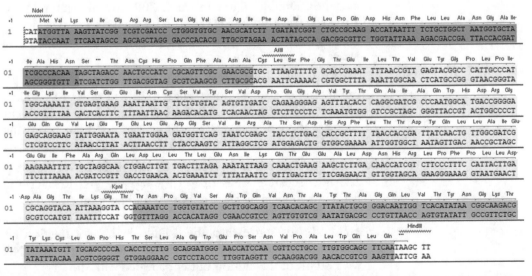

Fig. 4 DNA and protein sequence for the precursor protein Z3C_O2beY (Table 1). The target peptide sequence is highlighted in *pink* and is preceded by the Int$_C$ domain of DnaE split intein (*green*) and C-terminally fused to DnaE Int$_N$ domain (*yellow*) and the chitin binding domain affinity tag (*grey*)

19. The Z8C_O2beY construct can be cloned using an identical procedure (**steps 2** through **18**) with the exception that primer Z8C_rev_1/2 and Z8C_rev_2/2 are to be used instead in **step 6**.

3.3 Expression of the O2beY-Containing Precursor Proteins

The following section describes a typical procedure for the expression of the precursor protein constructs Z3C_O2beY and Z8C_O2beY described in Table 1. An important consideration concerning the choice of the *E. coli* strain for recombinant expression of these constructs is the use of a non-suppressor strain such as BL21 or BL21(DE3). *E. coli* strains characterized by genotypic signatures such as *supD, supH, supE, supF, supP*, or *supZ* contain mutated tRNAs that cause the incorporation of a natural amino acid (Ser, Gln, Tyr, etc.) in response to an amber stop codon, thus interfering with the incorporation of O2beY into the target sequence during the expression of the precursor protein. As illustrated by the model constructs described in Table 1, the precursor proteins should consist of, from the N- to the C-terminus, the DnaE Int_C-domain, followed by an arbitrary peptide target sequence containing the O2beY and cysteine residues spaced by one to seven interjecting residues, followed by the DnaE Int_N-domain fused to a chitin-binding domain (CBD). While the CBD domain is not essential for production of the bicyclic peptide, it is useful for isolation of the precursor protein and *trans* splicing product via chitin-affinity chromatography and subsequent analysis of the extent of split intein-mediated splicing of the precursor protein.

In terms of design principles, target peptide sequences comprising between 12 and 18 amino acid residues have been successfully cyclized in living *E. coli* cells yielding the desired bicyclic peptides in high yields (70–99%) [18]. It is reasonable to assume that shorter and longer sequences would be also amenable to bicyclization. As mentioned earlier, the inter-side-chain cross-linking reaction proceeds most efficiently (80–100%) when the O2beY and cysteine residues are separated by 2 up to 7–8 residues. Constructs with the O2beY/Cys pair in $i/i+10$ to $i/i+12$ relationships are still able to form the thioether bridge but with reduced efficiency (20–50%) [17]. The first residue of the peptide target sequence (i.e., 'Int_C+1 position', Fig. 1) is involved in the mechanism of split intein-catalyzed *trans* splicing (Fig. 2) and must be a serine or cysteine residue. In a side-by-side comparison experiment, precursor proteins containing a serine or cysteine residue at this site were found to lead to the desired bicyclic peptide with equal efficiency [18], suggesting that the bicyclization process is not affected by the inherently different nucleophilicity of these amino acids. If a cysteine residue occupies the Int_C+1 site, however, O2beY should be ideally positioned in the Int_C+2 position so to prevent alkylation of the mechanistically important cysteine by means of O2beY-borne bromoethyl group. In this $i/i+1$ arrangement, the O2beY/Cys cross-linking reaction was indeed found to be very inefficient (<5%), possibly due to the formation of a strained 14-membered macrocycle [17]. This structural requirement is eliminated when a serine residue is used as $Int_C + 1$ residue.

1. Add 50 ng of pBAD_Z3C (or pBAD_Z8C) plasmid and 50 ng µL of pEVOL_O2beY plasmid to a 50 µL aliquot of chemically competent BL21(DE3) *E. coli* cells. Gently tap the tube to mix the DNA and incubate the cells on ice for 30 min.

2. Place the cells in a heat block at 42 °C for 30 s followed by a quick transfer to ice for additional 5 min.

3. Add 500 µL of LB medium and incubate the cells with shaking for 45 min at 37 °C.

4. Plate 100 µL of the LB culture onto an LB agar plate containing 50 mg/L ampicillin and 30 mg/L chloramphenicol. Place the plate in an incubator at 37 °C for 14–18 h until visible colony growth is observed.

5. Pick a single colony from the plate with a sterile toothpick and inoculate a culture tube containing 5 mL of LB medium supplemented with 50 mg/L ampicillin and 30 mg/L chloramphenicol. Incubate the culture tube overnight with shaking at 37 °C.

6. Prepare 500 mL of M9 medium using presterilized solutions. Add ampicillin to 50 mg/L and chloramphenicol to 30 mg/L concentration. Mix well (*see* **Note 6**).

7. Inoculate 500 mL of M9 medium with the overnight cell culture from **step 5**. Incubate the flask at 37 °C with shaking (150–200 rpm) until OD_{600} reaches 0.6 (~3–4 h).

8. Once the desired cell density is reached, transfer the culture flask to room temperature. After 30 min, add l-arabinose to a final concentration of 0.1% (w/v) and O2beY to a final concentration of 2 mM from a stock solution of 0.2 M in water (*see* **Note 7**).

9. Grow the culture in a shaking incubator at 27 °C for 12 h followed by incubation at 37 °C for 3 h (*see* **Note 8**).

10. Harvest the cells by centrifugation at $5000 \times g$ for 10 min. Resuspend the cells in 50 mL of phosphate buffered saline solution (100 mM sodium phosphate, 150 mM NaCl, pH 7.5).

11. Lyse the cells by sonication. Method: amplitude 50, process time 30 s, pulse on 1 s, pulse off 10 s. Repeat this method two times.

12. Centrifuge the cell lysate at $18,000 \times g$ for 30 min at 4 °C. Use the clarified lysate immediately or flash freeze it in dry ice and store it at –80 °C. If frozen, use within 1 week.

3.4 Isolation and Characterization of the DnaE Trans Splicing Products

A key step in the formation of the bicyclic peptide products according to the general methodology of Fig. 1 is the *trans* splicing event occurring upon association of the N- and C-domain of DnaE split intein, which results in the head-to-tail cyclization of the target peptide sequence (Fig. 2). The presence of the CBD domain fused to the C-terminal end of the DnaE Int_N domain allows for the isolation of both the full-length precursor protein (Int_C-(target sequence)-Int_N-CBD) and spliced protein (Int_N-CBD/Int_C complex) via chitin-affinity capturing using chitin-coated agarose beads. Upon elution of these proteins from the chitin beads, the extent of *trans* splicing can be estimated by measuring the relative amount of full-length protein and spliced intein by SDS-PAGE analysis followed by gel band densitometry (Fig. 5a). The occurrence and extent of DnaE-catalyzed *trans* splicing can be assessed and further confirmed by LC-ESI-MS analysis of the chitin-affinity purified proteins (Fig. 5b).

For the constructs described in Table 1, the molecular weights for Int_N-CBD and the Int_N-CBD/Int_C complex are 19,699 Da and 23,533 Da, respectively, whereas the molecular weights for the full-length precursor proteins Z3C_O2beY and Z8C_O2beY are 24,892 Da and 24,920 Da, respectively. It should be noted that under standard LC-ESI-MS conditions using C_{18} reverse-phase chromatography in water/acetonitrile systems with 0.1 % formic acid and electrospray ionization (ESI), the non-covalent Int_N-CBD/ Int_C complex is sufficiently stable to be observed in the associated form (Fig. 5b). In contrast, under denaturing conditions such as in the presence of sodium dodecyl sulfate (Fig. 5a) or guanidinium chloride (Fig. 5c), the Int_N-CBD/Int_C complex gives rise to a large (19.7 kDa) and a small (3.8 kDa) protein fragment corresponding to the dissociated Int_N-CBD and Int_C polypeptides.

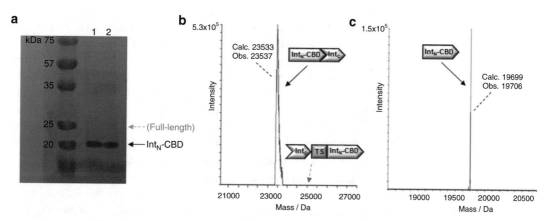

Fig. 5 Analysis of DnaE *trans* splicing products. (**a**) SDS-PAGE analysis of the protein construct Z3C_O2beY (*lane 1*) and Z8C_O2beY (*lane 2*) isolated using chitin beads. The bands corresponding to Int_N-CBD and full-length precursor protein are labeled. The latter is not visible due to complete splicing of the precursor protein. (**b**) Deconvoluted MS spectrum of chitin-affinity purified Int_N-CBD/Int_C complex from Z3C_O2beY sample. The position of the full-length precursor protein (if present) is also indicated. (**c**) Deconvoluted MS spectrum corresponding to the dissociated Int_N-CBD fragment upon treatment of the chitin-affinity purified Int_N-CBD/Int_C complex (from Z3C_O2beY sample) with 6 M guanidine hydrochloride

1. Transfer 0.5 mL of chitin beads to a 15 mL tube and spin them down at $1400 \times g$ for 1 min. Carefully remove the solvent, resuspend the beads in 2 mL of phosphate buffer and spin them down using the same conditions, repeat this process two more times.

2. To the chitin beads add 3 mL aliquot of lysed cells and incubate for 1 h on ice. Wash the beads two times with 2 mL of phosphate buffer, as described in **step 1**, each time remove as much of the buffer as possible without disturbing the beads. Avoid prolonged incubation of beads in the phosphate buffer to circumvent potential release of the proteins from the chitin beads.

3. Add 0.2 mL of acetonitrile–H_2O mixture (70:30 v/v) to the chitin beads and incubate at room temperature for 1 min to release any chitin-bound protein (*see* **Note 9**).

4. Carefully separate the eluate from the beads and lyophilize the collected eluate.

5. The identity of the proteins can be determined by LC-ESI-MS and/or SDS-PAGE analysis, Fig. 5. For LC-ESI-MS analysis dissolve the lyophilized eluates in a minimal amount (i.e., 100–200 μL) of water or desired buffer (i.e., 6 M guanidinium hydrochloride) with gentle mixing. For SDS-PAGE analysis dissolve the lyophilized eluates in a minimal amount of water.

3.5 Isolation and Characterization of the Macrocyclic Peptide Products

The precursor proteins described in Table 1 were designed to contain a His-Pro-Gln (HPQ) motif which is known to bind to streptavidin [25, 26] and which can be exploited to facilitate the isolation of the macrocyclic peptide products via capturing with streptavidin-coated agarose beads. In addition to the desired bicyclic peptide product (Fig. 1), a possible product of the reaction of Fig. 1 is the head-to-tail monocyclic peptide, which occurs whenever the O2beY/Cys cross-linking reaction does not proceed in a quantitative manner. Both the bicyclic and monocyclic peptide products (along with any undesired adduct, *see* **Note 10**) incorporate the HPQ motif and are captured during the streptavidin-binding procedure, while remaining distinguishable based on their different masses and behavior to MS/MS fragmentation. Accordingly, after elution from the streptavidin beads, these peptides products can be resolved and quantified by LC-ESI-MS analysis under standard analytical conditions (Fig. 6). In the following, a protocol for the isolation, identification, and quantification of the bicyclic and monocyclic peptide products generated upon in vivo cyclization of the protein precursor Z8C_O2beY is described.

1. Transfer 0.1 mL of the streptavidin-coated agarose slurry to a 15 mL tube and spin them down at $1400 \times g$ for 1 min. Carefully remove the solvent, resuspend the beads in 2 mL of phosphate buffer and spin them down using the same conditions, repeat this process four more times.

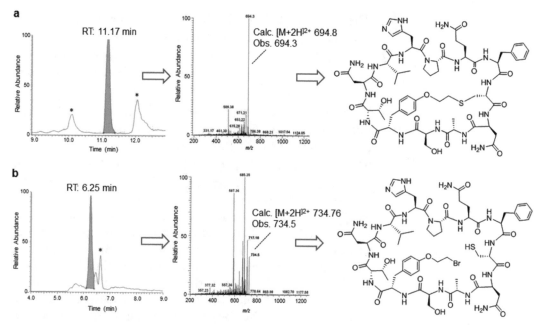

Fig. 6 LC – MS analysis of macrocyclic products obtained from in vivo cyclization of Z8C_O2beY precursor protein. (**a**) LC-MS extracted-ion chromatogram (*left panel*), MS/MS fragmentation spectrum (*middle panel*), and chemical structure (*right panel*) of bicyclic peptide isolated via streptavidin affinity from cells expressing the construct Z8C_O2beY. (**b**) Same as (**a**) for the head-to-tail cyclic peptide byproduct. Peaks labeled with *asterisk* correspond to unrelated multicharged ions from adventitious proteins

2. Transfer 3 mL aliquot of lysed cells to the tube containing streptavidin-coated agarose beads and incubate for 3 h under gentle shaking on ice (*see* **Note 11**). Wash the beads two times with 2 mL of the phosphate buffer as described in **step 1**. Carefully remove the solvent above the beads without disturbing them.

3. Incubate the agarose beads with 0.2 mL of acetonitrile–H_2O mixture (70:30 v/v) for 1 min to release any streptavidin-bound peptides and lyophilize the eluates (*see* **Note 11**).

4. Dissolve the lyophilized eluates in minimal amount of DMSO (i.e., 10–20 µL) and dilute with water to final concentration of DMSO of 10%. The identity of the eluted peptides can be determined by LC-ESI-MS analysis, Fig. 6. For measuring the relative amount of the monocyclic and bicyclic peptide products all of the potential peptide products need to be taken into consideration (*see* **Note 10**).

4   Notes

1. M9 medium needs to be freshly prepared each time using pre-sterilized components. If the M9 medium is made from non-sterile components and then autoclaved, the salts will precipitate.

2. The DMF reaction should be diluted with a minimum of 3× volumes of water to ensure successful product extraction.

3. Substitution of the TFA with acetic acid lowers the toxicity of the unnatural amino acid to bacterial cells during expression of the precursor proteins.

4. Keep all of the reagents for the PCR reactions at −20 °C and only take out just prior to use. Always add polymerase last, just prior to starting the PCR reaction.

5. The success of the PCR Overlap Extension reaction is highly dependent upon the quality of the PCR products that are to be joined together. If the gene segments are of low quality (e.g., as apparent from smearing of the corresponding band on the agarose gel), the PCR Overlap Extension reaction may not work or may lead to mixture of products. If necessary, higher quality gene segments can be prepared by optimizing the PCR conditions, e.g., by varying the annealing time and temperature or by redesigning the primers. Alternatively, the PCR product can be purified by gel extraction using commercial kits such as QIAquick Gel Extraction Kit (Qiagen).

6. Upon addition of $CaCl_2$, some precipitate may be observed. Mix well until it is completely dissolved.

7. The O2beY stock solution is a suspension, mix thoroughly prior to addition to bacterial cells. Store the O2beY stock solution at −20 °C.

8. The additional 3-h incubation step was found to promote DnaE-mediated *trans* splicing and thus head-to-tail cyclization of the peptide.

9. Chitin beads can be reused up to five times without significant loss of CBD binding capacity. To preserve their binding properties, the beads should be washed 3× with 5 mL of phosphate buffer immediately after treatment with the acetonitrile–H_2O mixture and can be stored in the same buffer at 4 °C for several weeks.

10. Other possible byproducts may consist of modified adducts of the head-to-tail monocyclic peptide resulting from nucleophilic displacement of O2beY side-chain 2-bromoethyl group with abundant cellular nucleophiles (i.e., cysteine, glutathione) or water. While the possible occurrence of these side-reactions should not be ignored, these adducts were never observed across several different target peptide sequences [17, 18], supporting the high chemoselectivity and regioselectivity of the O2beY-mediated cross-linking reaction.

11. Streptavidin-coated agarose beads can be reused up to three times without significant loss of binding capacity. To preserve their binding properties, the beads should be washed 5× with 5 mL of phosphate buffer immediately after treatment with the acetonitrile–H_2O mixture and can be stored in the same buffer at 4 °C for several weeks.

Acknowledgments

This work was supported by the US National Institutes of Health (grant R21 CA187502). MS instrumentation was supported by the US National Science Foundation (grants CHE-0840410 and CHE-0946653).

References

1. Driggers EM, Hale SP, Lee J, Terrett NK (2008) The exploration of macrocycles for drug discovery—an underexploited structural class. Nat Rev Drug Discov 7(7):608–624

2. London N, Raveh B, Schueler-Furman O (2013) Druggable protein-protein interactions—from hot spots to hot segments. Curr Opin Chem Biol 17(6):952–959

3. Bionda N, Fasan R (2015) Peptidomimetics of α-helical and β-strand protein binding epitopes. In: Czechtizky W, Hamley P (eds) Small molecule medicinal chemistry. Strategies and technologies. Wiley, Hoboken, NJ, pp 431–464

4. White CJ, Yudin AK (2011) Contemporary strategies for peptide macrocyclization. Nat Chem 3(7):509–524

5. Hipolito CJ, Suga H (2012) Ribosomal production and in vitro selection of natural product-like peptidomimetics: the FIT and RaPID systems. Curr Opin Chem Biol 16(1–2):196–203

6. Smith JM, Frost JR, Fasan R (2013) Emerging strategies to access peptide macrocycles from genetically encoded polypeptides. J Org Chem 78(8):3525–3531

7. Frost JR, Smith JM, Fasan R (2013) Design, synthesis, and diversification of ribosomally derived peptide macrocycles. Curr Opin Struct Biol 23(4):571–580

8. Baeriswyl V, Heinis C (2013) Polycyclic peptide therapeutics. ChemMedChem 8(3):377–384

9. Lennard KR, Tavassoli A (2014) Peptides come round: using SICLOPPS libraries for early stage drug discovery. Chemistry 20(34):10608–10614

10. Bowers AA (2012) Biochemical and biosynthetic preparation of natural product-like cyclic peptide libraries. Med Chem Commun 3(8):905–915

11. Josephson K, Ricardo A, Szostak JW (2014) mRNA display: from basic principles to macrocycle drug discovery. Drug Discov Today 19(4):388–399

12. Smith JM, Vitali F, Archer SA, Fasan R (2011) Modular assembly of macrocyclic organo-peptide hybrids using synthetic and genetically encoded precursors. Angew Chem Int Ed Engl 50(22):5075–5080

13. Satyanarayana M, Vitali F, Frost JR, Fasan R (2012) Diverse organo-peptide macrocycles via a fast and catalyst-free oxime/intein-mediated dual ligation. Chem Commun (Camb) 48(10):1461–1463

14. Frost JR, Vitali F, Jacob NT, Brown MD, Fasan R (2013) Macrocyclization of organo-peptide hybrids through a dual bio-orthogonal ligation: insights from structure-reactivity studies. Chembiochem 14(1):147–160

15. Smith JM, Hill NC, Krasniak PJ, Fasan R (2014) Synthesis of bicyclic organo-peptide hybrids via oxime/intein-mediated macrocyclization followed by disulfide bond formation. Org Biomol Chem 12(7):1135–1142

16. Frost JR, Jacob NT, Papa LJ, Owens AE, Fasan R (2015) Ribosomal synthesis of macrocyclic peptides in vitro and in vivo mediated by genetically encoded aminothiol unnatural amino acids. ACS Chem Biol 10(8):1805–1816

17. Bionda N, Cryan AL, Fasan R (2014) Bioinspired strategy for the ribosomal synthesis of thioether-bridged macrocyclic peptides in bacteria. ACS Chem Biol 9(9):2008–2013

18. Bionda N, Fasan R (2015) Ribosomal synthesis of natural-product-like bicyclic peptides in Escherichia coli. Chembiochem 16(14):2011–2016

19. Scott CP, Abel-Santos E, Wall M, Wahnon DC, Benkovic SJ (1999) Production of cyclic peptides and proteins in vivo. Proc Natl Acad Sci U S A 96(24):13638–13643

20. Liu CC, Schultz PG (2010) Adding new chemistries to the genetic code. Annu Rev Biochem 79:413–444

21. Nordgren IK, Tavassoli A (2014) A bidirectional fluorescent two-hybrid system for monitoring protein-protein interactions. Mol Biosyst 10(3):485–490

22. Horswill AR, Savinov SN, Benkovic SJ (2004) A systematic method for identifying small-molecule modulators of protein-protein interactions. Proc Natl Acad Sci U S A 101(44):15591–15596

23. Young TS, Ahmad I, Yin JA, Schultz PG (2010) An enhanced system for unnatural amino acid mutagenesis in *E. coli*. J Mol Biol 395(2):361–374

24. Zuger S, Iwai H (2005) Intein-based biosynthetic incorporation of unlabeled protein tags into isotopically labeled proteins for NMR studies. Nat Biotechnol 23(6):736–740

25. Katz BA (1995) Binding to protein targets of peptidic leads discovered by phage display: crystal structures of streptavidin-bound linear and cyclic peptide ligands containing the HPQ sequence. Biochemistry 34(47):15421–15429

26. Naumann TA, Savinov SN, Benkovic SJ (2005) Engineering an affinity tag for genetically encoded cyclic peptides. Biotechnol Bioeng 92(7):820–830

Chapter 6

Preparation of Semisynthetic Peptide Macrocycles Using Split Inteins

Shubhendu Palei and Henning D. Mootz

Abstract

Cyclic peptide are highly desired molecules not only for basic research but also for many biomedical and pharmacological applications. Due to their potentially superior physicochemical properties as compared to their linear counterparts, they are considered as ideal candidates for studying protein–protein interactions, among others. Most of the methods developed in recent years to prepare cyclic peptide focus either on a synthetic or a recombinant route. While the former provides access to diversified, noncanonical peptide, including unnatural and D-amino acid, for example, the latter can harness the power of genetic randomization to generate and select from large peptide libraries. Only few approaches have been reported to prepare semisynthetic macrocycles that would benefit from both the advantages associated with synthetic and genetically encoded parts. We describe in this chapter a chemo-enzymatic method to make semisynthetic cyclic peptide in vitro from two fragments using protein *trans*-splicing and bioorthogonal oxime ligation.

Key words Antibiotics, Cyclic peptide, Intein, Oxime ligation, Protein *trans*-splicing, *Ssp* DnaB intein

1 Introduction

Cyclic peptide are an important class of molecules that are drawing attention as novel therapeutic target molecules in recent years. One can distinguish between truly cyclic peptide exhibiting head-to-tail-cyclization of the backbone and head-to-side-chain or side-chain-to-side-chain cyclized peptide. Importantly, all these forms are covalently constrained and thereby conformationally restricted. This topological feature can confer them, as compared to their linear counterparts, improved binding affinity to target molecules as well as increased stability and cell permeability to biomolecular interfaces [1, 2]. Representing natural products of the cyclic peptide class include cyclosporine, daptomycin, the lantibiotics family, cyclotides and mammalian Θ-defensins [3]. These are synthesized naturally either in a non-ribosomal fashion with a complex enzymatic machinery or involve a set of dedicated enzymes for

Henning D. Mootz (ed.), *Split Inteins: Methods and Protocols*, Methods in Molecular Biology, vol. 1495,
DOI 10.1007/978-1-4939-6451-2_6, © Springer Science+Business Media New York 2017

posttranslational processing of a ribosomally synthesized precursor peptide. In both cases, the enzymes involved display a high degree of substrate specificity. Hence, reprogramming these biosynthetic pathways by genetic manipulation is typically very difficult [4–7].

New bioactive compounds can be selected by screening of large peptide libraries. Typically these libraries can either be achieved by synthetic or genetic means [2]. The non-proteinogenic residues such as D-amino acids or other pharmacophore groups can be integrated into the macrocycle using the synthetic peptide library, which can improve the physicochemical and binding properties of the lead compound. On the other hand, cyclic peptide from genetically encoded templates can be generated by various ways, however, only few approaches allow for the straight-forward coupling between genotype and phenotype necessary for subsequent selection procedures [8]. The probably most widely used way is phage display with peptide being cyclized in the side-chain-to-side-chain format by exploring two cysteine residues in the sequence [9]. Truly head-to-tail cyclized peptide can be obtained using ligases like sortase A or butelase [10, 11] or an split intein arranged with the peptide sequence in intramolecular fashion known as the SICLOPPS approach [12].

Recently several techniques have combined the advantage of genetic encoding with various ways to introduce non-proteinogenic building blocks either co-translationally or posttranslationally to make semisynthetic macrocycles. For example, using in vitro translation various unnatural and modified amino acids can be incorporated into the ribosomally produced peptide sequence, and crosslinkers have been used to effect cyclization [13–15]. Charging of the tRNA with the non-cognate amino acids is achieved either by chemical synthesis, by an engineered aminoacyl-tRNA synthetase or by flexizymes [8]. This technique has been successfully used to identify high-affinity cyclic peptide inhibitors. However, the technical complexity of reprogramming the ribosomal translation for each new building block poses restrictions to wider applications. Bifunctional or trifunctional cyclic peptide have been achieved by bioorthogonal modifications of cysteines with short organic linkers in combination with phage-displayed peptide [16, 17]. The latter can also be structurally expanded using chemical modification or unnatural amino acid incorporation by amber stop codon suppression to contain non-proteinogenic building blocks [18, 19]. Unnatural amino acid mutagenesis has also been used to incorporate chemical linker moieties that react with an intein-generated thioester to affect biorthogonal cyclization of a ribosomally synthesized peptide backbone [20, 21]. However, the synthetic parts used in these techniques served mostly only as a linker for cyclization and have not been explored so far to incorporate important structural elements like D-amino acid, N-methyl amino acids, or active pharmacophores.

We have developed a novel fragment-based approach to generate semisynthetic cyclic peptide by linking synthetic and genetically encoded parts [22]. The former being chemical in nature can be produced by solid phase peptide synthesis and the latter generated by recombinant bacterial expression of DNA. The two fragments are ligated by native peptide bond formation using protein *trans*-splicing followed by an intramolecular bioorthogonal oxime ligation (Fig. 1). Various and multiple non-proteinogenic building blocks can be incorporated into the synthetic part. The chemoenzymatic nature of protein *trans*-splicing reaction with the high affinity of the intein N- and C-terminal fragments (Int^N and Int^C) allows to perform the reaction in the micromolar range [23]. Furthermore, the oxime bond is stable under reducing conditions; a potential advantage over disulfide bonds used typically for side-chain-to-side-chain cyclized peptide.

Protein *trans*-splicing is an autocatalytic process where two complementary intein fragments (Int^N and Int^C) reconstitute into an active form and link their flanking sequences (the exteins Ex^N and Ex^C) by excising themselves during the reaction [24, 25]. Split inteins have been used to prepare semisynthetic peptide or proteins where one intein fragment is short enough to be synthesized by solid phase and the other fragment can be expressed recombinantly in *E. coli* [26]. In the described protocol we use the M86 mutant of the *Ssp* DnaB split intein with an Int^N of just 11 amino acids in length that can be easily synthesized by solid phase synthesis [23,

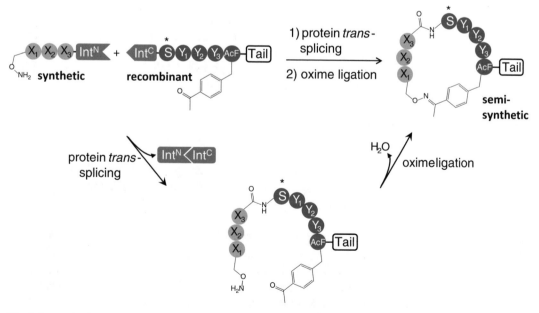

Fig. 1 General scheme for preparing semisynthetic cyclic peptide. The *asterisk* indicates the conserved serine residue required for splicing with the M86 mutant of the *Ssp* DnaB intein. X = building blocks in the synthetic fragment; Y = genetically encoded, proteinogenic amino acids

27, 28]. Furthermore, this intein system is optimized for rapid and efficient spicing, high sequence tolerance, and has high affinity between Int^N and Int^C fragments ($K_d = 0.1$ µM) [23]. The catalytic serine at +1 position of the intein is the only invariable residue. The trans-splicing reaction alone is very similar to the approach for preparing semisynthetic proteins with this intein system, i.e., by ligating synthetic moieties introduced in the Ex^N part of the Ex^N-Int^N peptide to the N terminus of the protein of interest. For the oxime ligation to facilitate cyclization, aminooxy and ketone functionalities are introduced at the terminal residues (*see* **Note 1**). Figure 1 illustrates the reaction scheme of the approach reported herein [22]. The resulting peptide is best described as cyclic peptide of the head-to-side-chain type. The aminoxy group is incorporated into the synthetic part by solid phase synthesis, whereas p-acetylphenylalanine (AcF) bearing a ketone group is incorporated via the amber stop codon suppression technique developed by Peter G. Schultz and coworkers [29]. The plasmid pEVOL-AcF encodes for an orthogonal aminoacyl-tRNA synthetase and tRNA pair which recognizes and incorporates AcF in response to the amber STOP codon (TAG) of the encoding gene. Both the *trans*-splicing reaction and oxime ligation are spontaneous and occurs without a catalyst once the fragments are mixed. However, given the rates of the *trans*-splicing reaction [23] and the oxime ligation [30], the splice reaction occurs first at the neutral pH ligating the two exteins (Ex^N and Ex^C) with a native peptide bond. Thereafter the linear splice product is shortly exposed to pH~4 for rapid cyclization by the oxime reaction. In this chapter, we describe the preparation of the synthetic Int^N and recombinant Int^C pieces including the sequence parts to be incorporated into one particular sample cyclic peptide [22]. We provide the protocol for performing the *trans*-splicing reaction and the oxime ligation and detail the analytical procedures to monitor the reactions by SDS-PAGE and HPLC.

2 Material

All solutions were prepared using double distilled water at 25 °C. All reagents and solutions were stored at room temperature unless otherwise mentioned.

2.1 Plasmid Construction

1. Plasmid pSP06 (encoding SBP-M86^C-SIEGSGG-AcF-H$_6$, based on vector backbone pET16b), plasmid pEVOL-AcF (encoding aminoacyl tRNA synthetase/tRNA pair for incorporation of unnatural amino acid-AcF [29]; a kind gift from Peter G. Schultz, Scripps Research Institute, California, USA).

2. Restriction enzymes: *Bsr*GI, *Nhe*I and their corresponding buffers and primers.

3. Gel-purification kit to purify DNA fragments from agarose gels.

4. 1.25× TE buffer: 12.5 mM Tris–HCl, 1.25 mM EDTA, 62.5 mM NaCl, pH 8 (for annealing the synthetic primer inserts).

5. Forward and reverse primers encoding the peptide sequence.

6. T4-DNA ligase.

7. *E. coli* DH5α cells (heat-competent).

8. LB-agar plates: 5 g/L NaCl, 10 g/L tryptone, 5 g/L yeast extract, pH 7.5 with 2% agar.

2.2 Protein Expression and Purification

1. Electro-competent *E. coli* BL21 Gold (DE3) cells.

2. LB medium: 5 g/L NaCl, 10 g/L tryptone, 5 g/L yeast extract, pH 7.5.

3. Ampicillin stock solution: 100 mg/L ampicillin, sterile filtered.

4. Chloramphenicol stock solution: 34 mg/L chloramphenicol, sterile filtered.

5. Isopropyl-β-thiogalactoside (IPTG) stock solution: 400 mM IPTG, sterile filtered.

6. L-arabinose (solid).

7. p-Acetylphenylalanine (AcF) (*see* **Note 2**).

8. Ni^{2+}-NTA agarose beads.

9. Ni^{2+}-NTA buffer A: 50 mM Tris–HCl, 300 mM NaCl, pH 8.0.

10. Imidazole stock solution: 5 M imidazole, pH 8.0.

11. Strep-Tactin Sepharose beads (e.g., IBA Life Science) for affinity purification of proteins tagged with streptavidin-binding peptide.

12. Buffer W: 100 mM Tris–HCl, 150 mM NaCl, 1 mM EDTA, pH 8.0.

13. Elution buffer: 2.5 mM D-Desthiobiotin in buffer W, pH 8.0.

14. 4× SDS sample buffer: 500 mM Tris–HCl, 8% (w/v) SDS, 40% (v/v) glycerol, 20% (v/v) β-mercaptoethanol, 4 g/L bromophenol blue, pH 6.8.

15. DTT.

16. Dialysis tube (MWCO = 14 kDa).

2.3 Solid-Phase Peptide Synthesis

1. The following building blocks were used: Fmoc-Ala-OH.H_2O, Boc-aminooxyacetic acid (Boc-Ao), Fmoc-Asp(OtBu)-OH, Fmoc-Cys(Trt)-OH, Fmoc-Gly-OH, Fmoc-Ile-OH, Fmoc-Leu-OH, Fmoc-Lys-OH, Fmoc-Ser(OtBu)-OH, Fmoc-Trp(Boc)-OH.

2. Fmoc-L-Ala-Wang-TG resin with 0.25 mmol/g substitution.

3. Coupling reagents: 0.5 M HBTU and Oxymapure in DMF.

4. Activating base: 2 M DIPEA in NMP.

5. Fmoc deprotecting solution: 20% piperidine in DMF.

6. Syringe reactor with filter.

7. Reagent K: 81.5% (v/v) TFA, 5% (v/v) thioanisole, 5% (w/v) phenol, 5% (v/v) ddH$_2$O, 2.5% (v/v) EDT, 1% TIS.

8. DMF, DCM (peptide synthesis grade).

9. Diethyl ether (keep at –20 °C).

2.4 Protein Trans-Splicing and Peptide Cyclization

1. Splice buffer: 50 mM Tris–HCl, 300 mM NaCl, 1 mM EDTA, pH 7.0.

2. TCEP stock solution: 50 mM TCEP. Dissolve 14.3 mg TCEP in 1 mL splice buffer. Then adjust the pH to pH 7 using NaOH.

3. 4× SDS sample buffer: *see* Subheading 2.2, **step 14**.

4. 10% TFA (v/v) in splice buffer.

2.5 Analysis and Purification of Cyclic Peptide

1. Analytical reversed phase column (e.g., Zorbax 300SB C3; 3.5 μm, 3.0 × 150 mm).

2. Mobile phase A: ddH$_2$O + 0.1% (v/v) TFA.

3. Modile phase B: acetonitrile + 0.1% (v/v) TFA.

4. TA30 solution: 30% (v/v) acetonitrile in water + 0.1% TFA.

5. CHCA matrix: saturated solution of α-cyano-4-hydroxycinnamic acid in TA30 solution.

3   Method

3.1 Plasmid Construction

Use expression vector pSP06 (*see* Fig. 2) to express the Int^C fusion proteins (SBP-Int^C-SIEGSGG-AcF-H$_6$). Figure 2 illustrates the procedure to insert two annealed oligonucleotides coding for the desired peptide sequence, in this example SIEGSGG-AcF, While designing these variable constructs, the catalytic serine residue immediately following the Int^C fragment must be kept for protein *trans*-splicing (*see* **Notes 1** and **3**).

1. Digest plasmid pSP06 by using restriction enzymes *Bsr*GI and *Nhe*I.

2. Gel-purify the fragment of the vector backbone (5997 bp) using standard methods.

3. Anneal forward and reverse oligonucleotides encoding the peptide insert (Table 1). Mix 20 μL of 100 μM stock of oligonucleotides with 80 μL of 1.25× TE buffer to make a final 20 μM solution in 1× TE buffer. Mix 10 μL of 20 μM concentration of each primer in a PCR cup (final concentration = 10 μM) and anneal the complementary primers in a

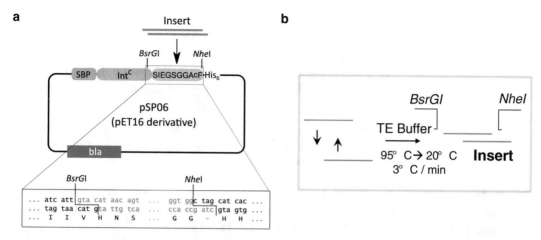

Fig. 2 Cloning strategy of the expression vector. (**a**) To ligate the desired variable insert encoding recombinant part of the macrocycle into the expression vector, two unique restriction sites, *BsrGI* and *NheI* were devised in the Int^c and C-extein regions, respectively. The *BsrGI* site was included in the codons for amino acids V152 and H153 of the intein-coding region (N154 is the last amino acid of the DnaB intein; the following serine is the catalytic +1 residue). The *NheI* site was devised in the Gly-AcF sequence of extein (*see* also **Note 3**). (**b**) The desired inserts are ordered as synthetic oligonucleotides, annealed into double stranded DNA with the correct sticky overhangs for the two restriction sites. This cloning strategy gives a simple and rapid access to variable peptide inserts of the recombinant part. Note that in the example described here plasmid pSP06 both serves as the vector backbone for insert insertion and is the target plasmid at the same time

Table 1
Synthetic oligonucleotides[a]

Forward primer	5′-GTACATAACAGTATTGAAGGATCCGGTGG-3′
Reverse primer	5′-CTAGCCACCGGATCCTTCAATACTGTTAT-3′

[a]The sequence encoding the variable peptide insert is *underlined*

thermocycler as per the following protocol to rule out any unspecific pairing. Heat the PCR cup at 95 °C for 4 min, then lower down the temperature each 3 °C for every 40 s until it reaches 20 °C. Then set a storage temperature at 4 °C. Dilute the annealed insert with ddH$_2$O in 1:50 ratio and photometrically determine the concentration.

4. Ligate the pSP06 backbone fragment and the annealed insert in 1:5 M ratio using T4-DNA ligase overnight at 25 °C.

5. Add 10 μL of the ligation mixture to heat competent *E. coli* DH5α cells (50 μL) thawed on ice and incubate 30 min on ice.

6. For heat transformation, incubate the cells for 30 s at 42 °C in a water bath and latter on ice for 2 min, followed by 1 mL addition of LB or SOC medium without antibiotic (*see* **Note 4**). Let the cells grow for 1 h at 37 °C.

7. Spin down the cells at $5000 \times g$, discard the supernatant and spread the cells on a preincubated (37 °C, 1 h) LB-agar plate containing 100 μg/L ampicillin.

8. Incubate the plate overnight at 37 °C until visible colonies are seen. Pick five colonies and grow overnight in LB medium containing 100 μg/L ampicillin.

9. Prepare the plasmid of the respective clones by using any standard method and verify the sequence of the plasmid by restriction digest and sequencing.

3.2 Protein Expression and Purification

E. coli BL21 Gold (DE3) cells are co-transformed with plasmids pSP06 and pEVOL-AcF for the incorporation of unnatural amino acid AcF into the protein product SBP-Int^C-SIEGSGG-AcF-H$_6$ (*see* Table 2 for complete amino acid sequence) by the amber stop codon suppression technique. The suppression of the amber stop codon is typically not quantitative and will also result in the formation of truncation product (SBP-Int^C-SIEGSGG) (*see* **Note 5**). Hence, the C-terminal His$_6$ tag serves to purify the full length protein (*see* **Note 6**). Moreover, by using *NheI* and *HindIII* sites His$_6$ tag can be exchanged with a larger POI (e.g., Trx-His$_6$, *see* **Note 3**) for making cyclic peptide–protein conjugate [22].

1. Co-transform electro-competent *E. coli* BL21 (DE3) cells with the generated plasmid pSP06 and pEVOL-AcF. Grow the bacteria in LB or SOC medium without antibiotic for 1 h (*see* **Note 4**). Spread the transformed cells onto an LB-agar plate containing 100 μg/L ampicillin and 34 μg/L chloramphenicol. Incubate the plate overnight at 37 °C to grow visible colonies.

2. Pick a single colony and transfer it to 12 mL liquid LB medium containing 100 μg/L ampicillin and 34 μg/L chloramphenicol and incubate overnight in a shaker or roller incubator.

3. Inoculate 600 mL LB medium containing 100 μg/L ampicillin and 34 μg/L chloramphenicol with the overnight culture

Table 2
Protein sequence of the Int^c construct

Protein	Sequence[a]
SBP-Int^C-SIEGSGG-AcF-H6	*MDEKTTGWRGGHVVEGLAGELEQLRARLEHHPQGQREP*GASGGG GSSSNNNNNNNNNNLGIEGRISEF<u>STGKRVPIKDLLGEKDFEIWA INEQTMKLESAKVSRVFSTGKKLVYTLKTRLGRTIKATANHRFLTI DGWKRLDELSLKEHIALPRKLESSSLQLAPEIEKLPQSDIYWDPIV SITETGVEEVFDLTVPGLRNFVANDIIVHNS</u>IEGSGG(AcF) HHHHHH

[a]SBP sequence is in *italics*; Int^C(12-154) is *underlined*

in a ratio of 1:50 (v/v). Incubate the culture in a rotary shaker at 37 °C until it reaches an $OD_{600} = 0.6$–0.7.

4. Remove 500 μL of sample (pre-induced) for SDS-PAGE analysis. Centrifuge the cells ($20,000 \times g$, 1 min), discard the supernatant and resuspend the cells in 60 μL of 1× SDS sample buffer. Store sample at –20 °C.

5. Reduce the temperature of the incubator to 27 °C. Add AcF as a solid to a final concentration of 1 mM to the culture and let it dissolve for 15 min in the shaker.

6. Induce the culture with IPTG to a final concentration of 0.4 mM (600 μL of 400 mM IPTG stock solution) and 0.2% L-(+)-arabinose (solid, w/v). Incubate the culture overnight (18 h) at 27 °C to allow high expression of protein. These parameters can be optimized for different proteins.

7. After 18 h, take another 500 μL of sample (post-induced) and treat as described for the pre-induced sample (**step 4**).

8. For harvesting the cells, centrifuge the culture ($4000 \times g$, 30 min) in a precooled centrifuge rotor at 4 °C. Discard the supernatant. Resuspend the pellet in 10 mL of Ni^{2+}-NTA buffer A with 20 mM imidazole (40 μL from a 5 M imidazole stock solution). At this point the cells can either be stored at –80 °C for a prolonged period of time or directly proceed to the next step for Ni^{2+}-NTA protein purification.

9. Lyse the cells using any preferred method (mechanical or enzymatic). Centrifuge the cell lysate (30 min, $22,000 \times g$ and 4 °C) to separate the soluble fraction from insoluble material and cell debris.

10. Pass the supernatant through a pre-equilibrated Ni^{2+}-NTA column (bed volume 2 mL, pre-equilibrated with Ni^{2+}-NTA buffer A with 20 mM imidazole) (*see* **Note 7**). Take a sample of 9 μL of the supernatant and flow-through fractions for latter analysis and mix each with 3 μL of 4× SDS sample buffer.

11. Wash the column using ten column volumes (CV; 20 mL) of Ni^{2+}-NTA buffer A with 20 mM imidazole, followed by 5 CV (10 mL) of Ni^{2+}-NTA buffer A with 40 mM imidazole.

12. Elute the protein with 4 CV (8 mL) of Ni^{2+}-NTA buffer A with 250 mM imidazole. Collect eight fractions each of 1 mL and store them at 4 °C. Take a sample of 9 μL from each fraction for further analysis of purity by SDS-PAGE and mix with 3 μL of 4× SDS sample buffer.

13. Boil all the samples at 95 °C for 10 min and analyze them by SDS-PAGE. For better resolution, apply 1 μL of pre- and post-induction, pellet samples, 2 μL of supernatant and flow-through samples, and 4 μL of wash and elution fractions. Analyze the gel by Coomassie staining.

14. Pool all elution fractions containing the SBP-Int^C-SIEGSGG-AcF-H$_6$ fusion protein and pass through a Strep-Tactin Sepharose column (bed volume of 2 mL) pre-equilibrated with buffer W. Take 9 μL sample of pooled elution fraction and flow-through followed by mixing of 3 μL 4× SDS sample buffer.

15. Wash the Strep-Tactin Sepharose column with 10 CV (20 mL) of buffer W. Elute the protein with 4 CV of elution buffer (2.5 mM d-Desthiobiotin in buffer W). Collect eight fractions of 1 mL volume and store at 4 °C. Take 9 μL sample of wash and elution fractions followed by addition of 3 μL 4× SDS sample buffer.

16. Boil all the samples at 95 °C for 10 min and analyze them by SDS-PAGE. Load 4 μL of each boiled sample on gel. Analyze the gel by Coomassie staining.

17. Pool all elution fractions containing purified and concentrated protein. Dialyze the protein three times in the cold room (4 °C); first with splice buffer with 2 mM DTT followed by splice buffer without DTT and finally with splice buffer with 10% glycerol (v/v). Glycerol is an antifreeze agent which protects the proteins from precipitation during freezing.

18. Determine the concentration of protein photometrically using A$_{280}$ and aliquot it in samples of 50–100 μL. Flash-freeze the aliquots in liquid nitrogen and store at –80 °C. For latter use, thaw them on ice.

3.3 Solid-Phase Peptide Synthesis

Synthesize the aminoxy-Int^N peptide (Ao-SG-CISGDSLISLA-SWKA, Int^N sequence underlined) using Fmoc-based solid phase synthesis. The amino acids SG represent the native flanking residues upstream of the intein. The lysine is used to improve the solubility and the tryptophan for measuring concentration of the purified peptide by A$_{280}$. The peptide could be synthesized using local peptide synthesis facilities or ordered from commercial suppliers. Here, we describe the standard protocol for Fmoc-based solid phase peptide synthesis, which we have opted for synthesis of the aminoxy-Int^N (hereafter named as Ao-Int^N) peptide in a 0.1 mmol scale.

1. Synthesize the peptide using Fmoc-L-Ala-Wang-TG resin (substitution 0.25 mmol/g, e.g., Iris Biotech). Perform the coupling of Fmoc protected amino acids in 5:1 molar excess to the resin. Activate the amino acids using five equivalent HBTU + Oxyma Pure in DMF (0.5 M), and ten equivalent DIPEA/NMP (2 M). Use 20% piperidine/DMF for deprotection of Fmoc group.

2. Remove the resin from the peptide synthesizer and transfer into a syringe reactor with filter (10 mL). Wash the resin with 3–4 reactor volume (RV) of DMF followed by 3–4 RV of DCM. Dry the resin at least for 2 h using vacuum. At this point the resin can be stored at 4 °C for 3–4 weeks before the cleavage of the peptide from the resin.

3. Cleave the peptide from the dried resin using Reagent K. For cleavage, treat the resin with 10 mL of Reagent K for 1 h. Precipitate the peptide by addition of four volumes (40 mL) of −20 °C cold diethyl ether. Incubate the mixture at −20 °C for 30 min and pellet down the peptide by centrifugation at 4 °C for 30 min. Repeat this washing step at least two times. Afterwards, dissolve the pellet in 10 mL ddH$_2$O and dry using a lyophilizer.

4. To purify the crude peptide, dissolve it in 5% acetonitrile–ddH$_2$O with 0.1% TFA and purify by preparative HPLC on a C18 column using standard ddH$_2$O–acetonitrile (0.1% TFA) gradient.

5. Check the purified fractions by LC-MS. Pool them together and dry using a lyophilizer. Store the dried peptide at −20 °C for future use.

3.4 Protein Trans-Splicing and Peptide Cyclization

The M86 mutant split intein described here typically gives 80–95% of yield of protein *trans*-splicing reaction within 2–4 h (*see* **Notes 1** and **8**) [23].

1. Prepare 100 μM stock solution of Ao-Int^N peptide by dissolving it in splice buffer (pH 7.0).

2. Mix the Ao-Int^N peptide and the SBP-Int^C-SIEGSGG-AcF-H$_6$ protein in a 3:1 ratio (45 μM and 15 μM, respectively) in splice buffer with 2 mM TCEP (final concentration, *see* **Note 9**). Table 3 shows a typical scheme for pipetting. Add the peptide as the last component to start the reaction (*see* **Note 10**). After mixing all the components, immediately take a 0 h sample of 9 μL and mix it with 3 μL of 4× SDS sample buffer. Incubate the remaining reaction mixture at 25 °C. Take out samples after 1 and 4 h, and treat them as described above (*see* **Note 11**).

3. To accelerate the cyclization by oxime ligation, add TFA to 100 μL of the *trans*-splicing reaction to a final concentration of 0.1% (v/v). Incubate the reaction mixture at 25 °C for 30 min (*see* **Note 12**).

Table 3
Pipetting scheme for protein *trans*-splicing reaction

Components	Stock conc	Volume (μL)	Final conc
Int^C protein	40 μM	37.5	15 μM
Ao-Int^N peptide	100 μM	45	45 μM
TCEP	50 mM	4	2 mM
Splice buffer		13.5	
Total		100	

3.5 Analysis of Trans-Splicing Reaction and Purification of Cyclic Peptide

1. Boil all the samples at 95 °C for 10 min and analyze them by SDS-PAGE. Load 4 µL of each boiled sample on gel. Analyze the gel by Coomassie staining to estimate the conversion of precursor due to splicing (*see* Fig. 3 and **Note 8**). Note that the splice product (cyclic peptide) is too small to be detected on Coomassie gel, hence subjected to HPLC analysis.

2. For HPLC analysis and purification of the cyclic peptide, add 5 % acetonitrile to the cyclization reaction mixture (as starting analytical HPLC condition). Use water–acetonitrile mobile phase and Zorbax 300SB C3 column at flow rate 0.6 mL/min or any semipreparative HPLC column (*see* Fig. 3c). Collect the corresponding peak for cyclic peptide and measure the mass of the cyclic peptide by MALDI-TOF-MS. Alternatively, use a coupled LC-MS instrument to analyze the mass of cyclic peptide.

3. For MALDI-TOF-MS spot 0.5 µL of saturated CHCA matrix on the stainless steel MALDI plate and let it dry for few minutes at room temperature. Mix 0.5 µL of the HPLC cyclic peptide fraction with 1.5 µL of saturated CHCA matrix. Add 1 µL of this mixture to the dried matrix on MALDI plate. Measure the MALDI-TOF by using a linear reflex mode (Fig. 3d).

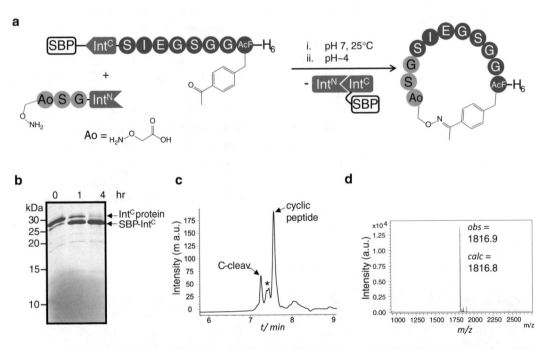

Fig. 3 Analysis of protein *trans*-splicing and peptide cyclization. (**a**) Scheme of the specific reaction described here. (**b**) Coomassie-stained gel of the *trans*-splicing reaction showing the consumption of precursor protein. (**c**) HPLC analysis of the reaction mixture. (**d**) MALDI-TOF-MS analysis of the cyclic peptide peak. Masses of [M + H]$^+$ are given in Da. *Asterisk* = product with same mass as the cyclic peptide corresponding to the other *E/Z* isoform of the oxime linkage

4   Notes

1. In this protocol we describe one particular Int^N peptide and one particular Int^C protein partner to generate one specific cyclic peptide. In this example, the flanking residues before the Int^N (SG) and after the Int^C (SIEG) are adapted to the native sequence context of the intein as to potentially observe the best splicing activity. However, only the serine following the Int^C fragment should be strictly conserved, as this is the catalytic +1 residue in the protein splicing pathway. The other residues flanking the intein (according to the intein nomenclature the –1, –2, –3, etc., and +2, +3, +4, etc. positions) can be used quite flexible because this intein has been optimized by directed evolution to be very general [23, 31]. However, a systematic and quantitative analysis on different amino acids in these positions has not yet been reported. A proline at the –1 position might not be accepted [23]. Less preferred residues might lead to somewhat lower yields, slower splicing rates and more N- or C-terminal cleavage side reactions. In other studies we have reported various splice products originating from different amino acids at the N- and C-extein positions [31, 32]. In our initial study on the method described in this paper we also generated several cyclic peptide with varying amino acids at the flanking positions [22]. We have also reported a different building block to incorporate the aminooxy moiety. N-ß-(amino-oxyacetyl)-L-diaminopropionic acid (Dpr) was used to generate side-chain-to-side-chain cyclized lipopeptides [22] (*see* also **Note 8**).

2. Unnatural amino acid (AcF) can be purchased from several commercial suppliers. A chemical synthesis protocol was published in the original report [29].

3. The C-terminal His-tag can be exchanged with a larger protein, e.g., thioredoxin (Trx) for preparation of semisynthetic cyclic peptide that is C-terminally conjugated to a protein. For this, the unique restriction sites *Nhe*I and *Hind*III can be used to ligate similarly digested Trx-His$_6$ PCR amplified insert. The cloning strategy depicted in Fig. 2 could also be used to introduce randomized DNA to generate a cyclic peptide library. Other designs are possible, for example by using a *Bam*HI site downstream of the TAG stop codon to enable fully randomized sequences between the codon for the catalytic serine and the TAG stop codon [22].

4. SOC medium may be superior to LB medium when high transformation efficiency is desired.

5. We have typically observed about 30% truncation and 70% suppression product, but this number is likely to depend on the sequence context and other expression parameters like temperature, cells, medium and concentration of AcF.

6. To further purify the protein one can optionally perform a second affinity chromatography using the N-terminal SBP tag (as described in Subheading 3.2, **steps 14** and **15**).

7. Perform the protein purification at 4 °C to reduce the degradation of protein by cellular proteases.

8. The *trans*-splicing reaction is always associated with certain levels of N- and C-terminal cleavage as side reactions [23]. C-terminal cleavage is caused by premature succinimide formation due to cyclization of Asn at the last position of the intein. N-terminal cleavage is a result of a nucleophilic attack onto the initially formed linear thioester by water or other nucleophiles, i.e., thiols. The N-terminal cleavage can also induce C-terminal cleavage [33]. These side reactions can be reduced drastically by using a lower temperature 8 °C for the *trans*-splicing reaction, however compromising with the slower reaction rate [22, 34] (*see* also **Note 9**).

9. We prefer TCEP over DTT because thiol based reducing agents can act as good nucleophiles and cleave the thioester intermediate during splicing.

10. Int^C-protein is used at a concentration of 10–50 μM (typically 15 μM) with an access of aminooxy-Int^N peptide (20–30 μM higher than Int^C-protein, typically 45 μM). These concentrations are used for better analysis of splice reaction by SDS-PAGE or HPLC. However, note that the splice reaction can also be performed with lower concentrations. The K_d of the two intein halves (Int^N and Int^C) is 0.1 μM [23].

11. Even after adding the 4× SDS loading buffer N- and C-cleavage of the intein as side reactions can still take place take place due to the high concentration of the nucleophile β-mercaptoethanol in the loading buffer.

12. For the examples we studied in detail [22], we always observed quantitative cyclization of the *trans*-splicing products indicating that intramolecular oxime ligation under these conditions was very efficient, yet the rate was dependent on the size of the cyclic peptide.

Acknowledgements

We thank Peter G. Schultz (Scripps Research Institute, La Jolla) for providing the plasmid for AcF incorporation. We acknowledge financial support from the DFG (MO1073/3-2, SPP1623 and Cells in Motion cluster EXC1003) and the International Graduate School of Chemistry in Münster (GSC-MS; Ph.D. stipend to S. P.).

References

1. Clardy J, Walsh C (2004) Lessons from natural molecules. Nature 432(7019):829–837
2. Driggers EM et al (2008) The exploration of macrocycles for drug discovery--an underexploited structural class. Nat Rev Drug Discov 7(7):608–624
3. Arnison PG et al (2013) Ribosomally synthesized and post-translationally modified peptide natural products: overview and recommendations for a universal nomenclature. Nat Prod Rep 30(1):108–160
4. Trauger JW et al (2000) Peptide cyclization catalysed by the thioesterase domain of tyrocidine synthetase. Nature 407(6801):215–218
5. Tseng CC et al (2002) Characterization of the surfactin synthetase C-terminal thioesterase domain as a cyclic depsipeptide synthase. Biochemistry 41(45):13350–13359
6. Grunewald J, Sieber SA, Marahiel MA (2004) Chemo- and regioselective peptide cyclization triggered by the N-terminal fatty acid chain length: the recombinant cyclase of the calcium-dependent antibiotic from Streptomyces coelicolor. Biochemistry 43(10):2915–2925
7. Mootz HD, Schwarzer D, Marahiel MA (2000) Construction of hybrid peptide synthetases by module and domain fusions. Proc Natl Acad Sci U S A 97(11):5848–5853
8. Passioura T et al (2014) Selection-based discovery of druglike macrocyclic peptides. Annu Rev Biochem 83:727–752
9. Smith GP, Petrenko VA (1997) Phage display. Chem Rev 97(2):391–410
10. Antos JM et al (2009) A straight path to circular proteins. J Biol Chem 284(23):16028–16036
11. Nguyen GK et al (2014) Butelase 1 is an Asx-specific ligase enabling peptide macrocyclization and synthesis. Nat Chem Biol 10(9):732–738
12. Scott CP et al (1999) Production of cyclic peptides and proteins in vivo. Proc Natl Acad Sci U S A 96(24):13638–13643
13. Morimoto J, Hayashi Y, Suga H (2012) Discovery of macrocyclic peptides armed with a mechanism-based warhead: isoform-selective inhibition of human deacetylase SIRT2. Angew Chem Int Ed Engl 51(14):3423–3427
14. Kawakami T et al (2009) Diverse backbone-cyclized peptides via codon reprogramming. Nat Chem Biol 5(12):888–890
15. Schlippe YVG et al (2012) In vitro selection of highly modified cyclic peptides that act as tight binding inhibitors. J Am Chem Soc 134(25):10469–10477
16. Timmerman P et al (2005) Rapid and quantitative cyclization of multiple peptide loops onto synthetic scaffolds for structural mimicry of protein surfaces. Chembiochem 6(5):821–824
17. Heinis C et al (2009) Phage-encoded combinatorial chemical libraries based on bicyclic peptides. Nat Chem Biol 5(7):502–507
18. Day JW et al (2013) Identification of metal ion binding peptides containing unnatural amino acids by phage display. Bioorg Med Chem Lett 23(9):2598–2600
19. Ng S, Jafari MR, Derda R (2012) Bacteriophages and viruses as a support for organic synthesis and combinatorial chemistry. ACS Chem Biol 7(1):123–138
20. Smith JM et al (2011) Modular assembly of macrocyclic organo-peptide hybrids using synthetic and genetically encoded precursors. Angew Chem Int Ed Engl 50(22):5075–5080
21. Satyanarayana M et al (2012) Diverse organo-peptide macrocycles via a fast and catalyst-free oxime/intein-mediated dual ligation. Chem Commun (Camb) 48(10):1461–1463
22. Palei S, Mootz HD (2016) Cyclic peptides made by linking synthetic and genetically encoded fragments. Chembiochem 17(5):378–382
23. Appleby-Tagoe JH et al (2011) Highly efficient and more general cis- and trans-splicing inteins through sequential directed evolution. J Biol Chem 286(39):34440–34447
24. Volkmann G, Mootz HD (2013) Recent progress in intein research: from mechanism to directed evolution and applications. Cell Mol Life Sci 70(7):1185–1206
25. Shah NH, Muir TW (2014) Inteins: nature's gift to protein chemists. Chem Sci 5:446–461
26. Mootz HD (2009) Split inteins as versatile tools for protein semisynthesis. Chembiochem 10(16):2579–2589
27. Ludwig C et al (2006) Ligation of a synthetic peptide to the N terminus of a recombinant protein using semisynthetic protein trans-splicing. Angew Chem Int Ed Engl 45(31):5218–5221
28. Ludwig C, Schwarzer D, Mootz HD (2008) Interaction studies and alanine scanning analysis of a semi-synthetic split intein reveal thiazoline ring formation from an intermediate of the protein splicing reaction. J Biol Chem 283(37):25264–25272
29. Wang L et al (2003) Addition of the keto functional group to the genetic code of Escherichia coli. Proc Natl Acad Sci U S A 100(1):56–61
30. Dirksen A, Hackeng TM, Dawson PE (2006) Nucleophilic catalysis of oxime ligation. Angew Chem Int Ed Engl 45(45):7581–7584
31. Wasmuth A, Ludwig C, Mootz HD (2013) Structure-activity studies on the upstream splice junction of a semisynthetic intein. Bioorg Med Chem 21(12):3495–3503

32. Böcker JK et al (2015) Generation of a genetically encoded, photoactivatable intein for the controlled production of cyclic peptides. Angew Chem Int Ed Engl 54(7):2116–2120

33. Binschik J, Zettler J, Mootz HD (2011) Photocontrol of protein activity mediated by the cleavage reaction of a split intein. Angew Chem Int Ed Engl 50(14):3249–3252

34. Thiel IV et al (2014) An atypical naturally split intein engineered for highly efficient protein labeling. Angew Chem Int Ed Engl 53(5):1306–1310

Chapter 7

Semisynthesis of Membrane-Attached Proteins Using Split Inteins

Stefanie Hackl, Alanca Schmid, and Christian F.W. Becker

Abstract

The site-selective installation of lipid modifications on proteins is critically important in our understanding of how membrane association influences the biophysical properties of proteins as well as to study certain proteins in their native environment. Here, we describe the use of split inteins for the C-terminal attachment of lipid-modified peptides to virtually any protein of interest (POI) via protein *trans*-splicing (PTS). To achieve this, the protein of interest is expressed in fusion with the N-terminal split intein segment and the C-terminal split intein segment is prepared by solid phase peptide synthesis. A synthetic peptide carrying two lipid chains is also made chemically to serve as a membrane anchor and subsequently linked to the C-terminal split intein by native chemical ligation. Proteins of interest for our work are the prion protein as well as small GTPases; however, extensions to other POIs are possible. Detailed information for the C-terminal introduction of a lipidated membrane anchor (MA) peptide using split intein systems from *Synechocystis* spp. and *Nostoc punctiforme* for the Prion protein (PrP, as a challenging protein of interest) and the enhanced green-fluorescent protein (eGFP, as an easily trackable target protein) are provided here.

Key words Membrane-associated proteins, Synthetic membrane anchor, Protein semisynthesis, Split inteins, Prion protein, Protein *trans*-splicing, Lipid-coated particles, Liposomes, Green fluorescent protein

1 Introduction

The association of proteins to cellular membranes is often achieved by covalent attachment of lipids to amino acid side chains or the protein N-terminus. This type of posttranslational modification with a variety of simple lipids such as myristoyl, palmitoyl, farnesyl, geranylgeranyl or more complex molecules such as cholesterol, phosphatidylethanolamine (PE), and glycosylphosphatidylinositol (GPI) anchors provides an effective strategy to control protein localization within the (eukaryotic) cell, in cellular compartments as well as extracellularly. These modifications severely influence the biophysical properties of the target proteins and can be critically important for physiological as well as for pathological functions.

Henning D. Mootz (ed.), *Split Inteins: Methods and Protocols*, Methods in Molecular Biology, vol. 1495,
DOI 10.1007/978-1-4939-6451-2_7, © Springer Science+Business Media New York 2017

An example for the latter is the conversion of cellular, predominantly α-helical prion protein (PrP) into its toxic, β-sheet enriched isoform, termed scrapie PrP. In the absence of a GPI anchor this conversion process is either completely blocked or at least significantly reduced [1–5]. The proximity of PrP to negatively charged head groups of the phospholipids constituting the lipid bilayer is also thought to facilitate the conversion process [6, 7].

A prominent example for lipid attachment to maintain important physiological function of proteins can be found with small GTPases such as Ras isoforms that relocate between intracellular membranes dependent on their lipidation status [8, 9] or the Rab proteins for which temporary shielding of lipid modifications allows cellular relocalization [10, 11].

Other complex lipid modifications, mainly found in bacterial lipoproteins in which phosphatidic acid reacts with cysteine side chains in combination with N-terminal acylation or in which trans-amidation leads to attachment of proteins to the peptidoglycan could also be mimicked by protein *trans*-splicing using a membrane anchor peptide but no examples are discussed here.

The number of lipids attached to a single protein can vary between one and three, whereas two lipid moieties are typically sufficient to tightly link proteins to membranes as demonstrated for Ras variants and for other membrane anchoring peptides [12–17].

Many innovative approaches for the site-selective introduction of lipid groups into proteins have been described over the past decade and all of them face similar challenges, namely the hydrophobic character of all lipid modifications and the related difficulties when handling these groups and the resulting modified proteins. In addition some of the chemical linkages formed are not very stable, especially the thioester bonds between cysteine side chains and palmitoyl groups. Therefore direct attachment of lipids using chemical and enzymatic approaches is often avoided and lipidated peptides are prepared that are ligated to the proteins of interest [18–22]. Lipidated peptides are in many cases much easier to handle and can be equipped with temporary solubilizing tags such as poly-lysine, poly-arginine, or polyethylene glycol (PEG) [23–27]. Here, we describe the use of split inteins to introduce a lipidated membrane anchoring peptide into proteins of interest via protein *trans*-splicing. The lipidated peptide can be viewed as a general tool inducing tight membrane-association of proteins of interest upon successful *trans*-splicing.

Such systems have the advantage of working under non-denaturing conditions allowing the use of folded POIs at low (μM) concentrations [28–32] and that they can lead to direct attachment of the POI to any target membrane, such as vesicles, membrane-coated particles and cells [33–35]. However, *trans*-splicing yields depend on the solubility and folding status of the

POI-split intein fusion, the flanking sequences of the exteins as well as on the choice of the split intein [36–38].

Scheme 1 summarizes the strategy described here, in which the POI is genetically fused to an N-terminal split intein segment and expressed in *E. coli*. Affinity tags such as a hexahistidine allow straightforward purification. The C-terminal split intein segment is generated by solid phase peptide synthesis and linked to the membrane anchor peptide via native chemical ligation [39]. PTS reactions are carried out with the C-terminal split intein-membrane anchor fusion dissolved in detergent-containing aqueous buffers or tightly associated to liposomes or lipid-coated particles.

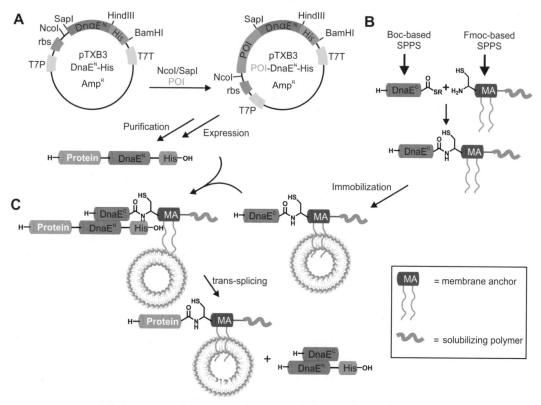

Scheme 1 General strategy for the attachment of protein of interest (POIs in *green*) to membranes using protein *trans*-splicing. (**a**) Cloning strategy based on the pTXB3 vector is outlined. A similar strategy can be pursued with the commercially available pTXB1 vector (New England Biolabs). The POI is inserted into this vector N-terminally of the N-terminal split intein segment (DnaE[N] in *red*) and a suitable affinity tag. Expression is carried out in *E. coli* and purification is achieved via the C-terminal His-tag. (**b**) Solid phase peptide synthesis is used to generate the C-terminal split intein segment (DnaE[C] in *red*) and the membrane anchor peptide (MA in *grey*). (**c**) The membrane anchor is added to liposomes or other membrane systems and protein *trans*-splicing occurs over a time frame of 2–24 h to give the membrane attached POI carrying the membrane anchor and the non-covalently associated split intein

2   Materials

All chemicals used here are purchased at the highest quality available unless otherwise stated. Lyophilized, synthetic peptides are stored at –20 °C. Expressed proteins in solution are kept at –80 °C after being shock frozen in liquid nitrogen. All aqueous buffers are prepared with double deionized water and sterilized by filtration through filter units with a pore size of 0.2 μm. Please consider all necessary precautions when handling chemicals such as neat trifluoroacetic acid (TFA) or hydrogen fluoride (HF) during peptide synthesis [40]. Proteins of interest may require additional safety measure, as the prion protein described here.

2.1   Peptide Synthesis

1. *N,N*-dimethylformamide (DMF) (peptide synthesis grade).

2. Fmoc-Ala-Wang-resin.

3. Deprotection solution: 20 % (v/v) piperidine in DMF.

4. Succinic anhydride.

5. 0.5 M HOBt solution: Weigh 2.30 g 1-hydroxybenzotriazole (HOBt) (*see* **Note 1**) and add DMF to a volume of 30 mL. Mix. Store at 4 °C.

6. *N,N*-diisopropylethylamine (DIEA).

7. 0.5 M CDI solution: Weigh 2.43 g 1,1'-carbonyldiimidazole (CDI) and add DMF to a volume of 30 mL. Mix. Store light protected at 4 °C.

8. 4,7,10-trioxatridecane-1,13-diamine.

9. Amino acids for Fmoc-based SPPS: Fmoc-Lys(Boc)-OH, Fmoc-Lys(Mtt)-OH, Fmoc-Ala-OH, Fmoc-Ser(tBu)-OH, Fmoc-Gln(Trt)-OH, Fmoc-Phe-OH, Fmoc-Tyr(tBu)-OH, Fmoc-Leu-OH, Fmoc-Glu(OtBu)-OH, Fmoc-Asn(Trt)-OH, Fmoc-Gly-OH, Boc-Cys(Trt)-OH.

10. 0.5 M HBTU solution: Weigh 2.84 g *N,N,N',N'*-tetramethyl-*O*-(1*H*-benzotriazol-1-yl)uronium hexafluorophosphate (HBTU) and add DMF to a volume of 15 mL. Mix. Store at 4 °C.

11. Dichloromethane (DCM) (HPLC grade).

12. Methyltrityl (Mtt) cleavage solution: 1 % (v/v) trifluoroacetic acid (TFA) and 1 % (v/v) triisopropylsilane (TIS) in DCM (*see* **Note 2**).

13. Palmitoyl chloride.

14. Triethylamine.

15. Cleavage cocktail for fluorenylmethoxycarbonyl (Fmoc) and *tert*-butyloxycarbonyl (Boc) solid phase peptide synthesis (SPPS), respectively: 92.5 % (v/v) TFA, 5 % TIS, 2.5 % dH$_2$O, and neat HF with 5 % (v/v) *p*-cresol, respectively. For the latter, a teflon apparatus with calcium oxide (CaO) trap to handle neat HF (highly toxic) is required.

16. Cooled diethylether.

17. Buffer A: ddH$_2$O with 0.1% (v/v) TFA.

18. Buffer B: acetonitrile (ACN) with 0.08% (v/v) TFA.

19. 6 M Gdn-HCl buffer (pH 4.7): Weigh 573.24 g guanidine hydrochloride in a 1 L graduated beaker, add water to a volume of 1 L and mix.

20. Boc-Leu-PAM resin.

21. S-trityl mercaptopropionic acid.

22. Trityl (Trt) cleavage solution: 2.5% (v/v) water and 2.5% (v/v) triisopropylsilane (TIS) in TFA.

23. Amino acids for Boc-based SPPS: Boc-Asn(Xan)-OH, Boc-Ala-OH, Boc-Ile-OH, Boc-Gly-OH, Boc-Leu-OH, Boc-Phe-OH, Boc-His(Tos)-OH, Boc-Asp(OcHx)-OH, Boc-Gln(Xan)-OH, Boc-Pro-OH, Boc-Arg(Tos)-OH, Boc-Val-OH, Boc-Ser(Bzl)-OH, Boc-Lys(2ClZ)-OH, Boc-Met-OH.

24. Native chemical ligation (NCL) buffer: 6 M guanidine hydrochloride, 300 mM NaPi, pH 7.5. Prepare a stock solution of 1 M NaPi, pH 7.5 through dissolving 2.60 g NaH$_2$PO$_4$·H$_2$O and 21.76 g Na$_2$HPO$_4$·7H$_2$O in 100 mL water. Weigh 5.73 g guanidine hydrochloride, add 3 mL of 1 M NaPi, pH 7.5 and water to a volume of 10 mL. Mix the buffer and degas before immediate use. Addition of 17 mg/mL dodecylphosphocholine (DPC) is optional.

25. Thiophenol.

26. ß-mercaptoethanol.

2.2 Expression of Intein Fusion Constructs

1. Modified pTXB3 vector (New England Biolabs) containing the DnaE N-terminal split intein sequence (DnaE^N) from *Synechocystis* sp. (*Ssp*) or *Nostoc punctiforme* (*Npu*).

2. *E. coli* BL21 (DE3) RIL cells.

3. LB medium: 10 g/L tryptone, 5 g/L yeast extract, 10 g/L NaCl. Autoclave at 121 °C for 20 min. Add ampicillin or chloramphenicol as indicated.

4. Ampicillin stock solution: 100 mg/mL ampicillin in water, sterile filtered.

5. Chloramphenicol stock solution: 30 mg/mL chloramphenicol in ethanol, sterile filtered.

6. Isopropyl-β-D-thiogalactopyranoside (IPTG) stock solution: 1 M IPTG in water, sterile filtered.

7. 10× PBS: 1.4 M NaCl, 27 mM KCl, 100 mM Na$_2$HPO$_4$, 18 mM KH$_2$PO$_4$, pH 7.3. Dissolve 81.8 g NaCl, 2 g KCl, 17.8 g Na$_2$HPO$_4$, 2.5 g KH$_2$PO$_4$ in 1 L water. Mix and adjust pH to 7.3.

8. 1 M EDTA, pH 8: Weigh 146.12 g EDTA in a graduated beaker and add water for 500 mL. Mix and adjust pH with 2 M NaOH to pH 8.

9. lysis buffer: 1× PBS buffer, 0.1 mM EDTA, pH 7.4. Use 100 mL of 10× PBS buffer, pH 7.3, 100 mL of 1 M EDTA, pH 8 and add water to a volume of 1 L. Mix.

10. 1 M Tris–HCl, pH 8: Weigh 121.14 g Tris–HCl (hydroxymethyl) aminomethane (Tris–HCl) in a graduated beaker and add water to 1 L. Mix and adjust pH 8 with 6 M HCl (*see* **Note 3**).

11. Triton X-100.

12. Resolubilizing buffer: 8 M guanidine hydrochloride, 50 mM Tris–HCl, pH 8.0. Weigh 764.32 g guanidine hydrochloride in a graduated beaker, add 50 mL of 1 M Tris–HCl, pH 8 and water to 1 L. Mix.

13. Nickel nitrilotriacetic acid (Ni-NTA) resin: ion metal affinity chromatography (IMAC) sepharose 6 fast flow (GE) loaded with Ni.

14. 6 M Gdn-HCl buffer (pH 8): 6 M guanidine hydrochloride, 50 mM Tris–HCl, pH 8.0. Weigh 573.24 g guanidine hydrochloride in a 1 L graduated beaker, add 50 mL of 1 M Tris–HCl, pH 8 and water to a volume of 1 L, and mix.

15. Elution buffer: 6 M guanidine hydrochloride, 50 mM Tris–HCl, 250 mM imidazole, pH 8.0. Weigh 573.24 g guanidine hydrochloride, 17.02 g imidazole in a 1 L graduated beaker, add 50 mL of 1 M Tris–HCl, pH 8 and water to a volume of 1 L, and mix.

16. Slide-A-Lyzer cassettes with a molecular weight cutoff (MWCO) of 3.5 kDa (Thermo Fisher).

17. refolding buffer: 0.6 M L-arginine, 50 mM Tris–HCl, 5 mM glutathione reduced (GSH)/0.5 mM glutathione oxidized (GSSG), pH 8.6. Dissolve 1.05 g Arg, 15 mg GSH, 3 mg GSSG in 0.5 mL 1 M Tris–HCl, pH 8 and add water to a volume of 10 mL. Mix and adjust pH.

18. Storage buffer: 50 mM Tris–HCl, pH 7.5. Dilute 1 M Tris–HCl, pH 8 and adjust pH.

2.3 Trans-Splicing

1. Tris(2-carboxyethyl)phosphine (TCEP).

2. Sodium mercaptoethanesulfonate.

3. 3, 2-dioleoyl-sn-glycero-3-phosphocholine (DOPC).

4. Chloroform.

5. Gases: argon and helium.

6. Splicing buffer: 50 mM Tris–HCl, pH 7.5. Dilute 1 M Tris–HCl, pH 8 and adjust pH with 1 M HCl.

7. Silica particles with an average particle diameter of 800 nm (Evident Technologies).

8. Methanol.

9. 1 M KOH: Weigh 5.6 g KOH in a graduated beaker, add water to a volume of 100 mL, and mix.

10. Detergent containing splicing buffer: 50 mM Tris–HCl, 20 mM octyl-β-D-glucoside (OG), pH 7.5. Dissolve 58.47 mg OG in 0.5 mL 1 M Tris–HCl, pH 8 and add water to a volume of 10 mL. Mix and adjust pH.

11. SDS sample buffer: 500 mM Tris–HCl, 6% (w/v) sodium dodecyl sulfate (SDS), 35% (v/v) glycerin, 3.55% (v/v) β-mercaptoethanol, 0.05% (w/v) bromophenol blue, pH 6.8.

12. 6 M Gdn-HCl buffer (pH 4.7): Weigh 573.24 g guanidine hydrochloride in a 1 L graduated beaker, add water to a volume of 1 L and mix.

13. Materials for Ni-NTA purification: *see* Subheading 2.2. Expression of intein fusion constructs, 13–15.

14. Superdex 75 (GE).

15. 6 M Gdn-HCl buffer (pH 8): 6 M guanidine hydrochloride, 50 mM Tris–HCl, pH 8.0. Weigh 573.24 g guanidine hydrochloride in a 1 L graduated beaker, add 50 mL of 1 M Tris–HCl, pH 8 and water to a volume of 1 L, and mix.

16. Sucrose.

17. Folding buffer: 20 mM NaOAc, 20 mM OG, 3 mM GSH/0.3 mM GSSG, pH 5. Dissolve 9.2 mg GSH, 1.8 mg GSSG, 16.4 mg NaOAc, 58.47 mg OG in 10 mL water. Mix and adjust pH.

18. Storage buffer: 20 mM NaOAc, 20 mM OG, pH 5. Dissolve 16.4 mg NaOAc, 58.47 mg OG in 10 mL water. Mix and adjust pH.

19. Slide-A-Lyzer cassettes with a molecular weight cutoff (MWCO) of 3.5 kDa (Thermo Fisher).

20. Vivaspin 500, PES (GE) filters with MWCO of 10 kDa.

3 Methods

Carry out all procedures at room temperature and use only L-amino acids in peptide synthesis, unless otherwise stated.

3.1 Peptide Synthesis and Ligation

3.1.1 Synthesis of Membrane Anchor Peptide H-CKGENLYFQSK$_{Palm}$ AAK$_{Palm}$K-$_{(PPO)3}$-A-OH

1. Swell 0.2 mmol of Fmoc-Ala-Wang-resin in DMF for 2 h (*see* **Note 4**).

2. For Fmoc deprotection treat the resin with deprotection solution for 3 min and then 7 min. Then, perform flow wash (*see* **Note 5**) for 1 min with DMF to completely remove piperidine.

3. For coupling of polyethyleneglycol polyamide oligomer (PPO) dissolve 10 eq. of succinic anhydride in 5 eq. of 0.5 M HOBt in DMF and 6 eq. N,N-diisopropylethylamine (DIEA) and

incubate the resin for 30 min. After flow wash with DMF, equilibrate the resin with 2 mL of 0.5 M CDI (*see* **Note 6**) in DMF and incubate with additional 8 mL for 30 min. Next to subsequent flow wash with DMF, incubate the activated carboxylic group with a solution of 5 mL 0.5 M HOBt and 5 mL 4,7,10-trioxatridecane-1,13-diamine for 30 min (*see* **Note 7**). After final flow wash with DMF, repeat these three steps twice more to obtain the trimer of PPO.

4. For coupling of the Fmoc protected amino acids Lys(Boc), Lys(Mtt), Ala, Ser(tBu), Gln(Trt), Phe, Tyr(tBu), Leu, Glu(OtBu), Asn(Trt), Gly and Boc protected Cys(Trt) (*see* **Note 8**) activate 2.5 eq. with 2.38 eq. of 0.5 M HBTU (*see* **Note 9**) in DMF and 5 eq. DIEA, and incubate the resin for 30 min. After each coupling step, flow wash the resin with DMF for 1 min and deprotect Fmoc with deprotection solution for 3 min and then 7 min. Then, perform flow wash for 1 min with DMF and go on with coupling the following amino acid in the sequence.

5. Perform palmitoylation of the ε-amino groups at Lys(11) and Lys(14) after coupling of Fmoc-Ser(tBu)-OH. Hence, first change solvent from DMF to DCM and incubate the resin for 30 min. Remove selectively the methyltrityl (Mtt) protecting groups through repetitive addition of Mtt cleavage solution for 10 min and flow wash of the resin with DCM for 1 min. Stop after the disappearance of the yellow color of the solution, indicating Mtt cleavage. To enhance nucleophilicity of the protonated amino groups at acidic pH, flow wash the resin with 20 mL of 1% DIEA in DCM and DCM for 1 min. Next, prepare a suspension of 20 eq. palmitoyl chloride, 20 eq. HOBt and 22 eq. triethylamine in 24 mL DCM:DMF (3:1) solution (*see* **Note 10**) and add it to the resin. Leave the reaction o/n. Then, flow wash the resin with DCM, change solvent back to DMF and incubate the resin for 30 min. After that, go on with coupling of the remaining amino acids of the peptide as described above.

6. When the synthesis is finished, wash the resin vigorously with DCM, dry it in a vacuum desiccator o/n and finally cleave the peptide from the resin (*see* **Note 11**) through incubation with the cleavage cocktail for 3 h. Then, obtain the crude peptide through precipitation with cooled diethyl ether, subsequent dissolving in buffer A:B (7:3) and lyophilizing.

7. For purification dissolve the lyophilized crude peptide in 6 M Gdn-HCl buffer (pH 4.7) and purify with high performance liquid chromatography (HPLC). Identify pure fractions via electrospray ionization mass spectrometry (ESI-MS) and check with analytical HPLC.

3.1.2 Synthesis of C-Terminal Intein Segment

H-MVKVIGRRSLGVQRIFDIGLPQDHNFLLANGAIAAN-SR

1. Swell 0.2 mmol of Boc-Leu-PAM resin for preparing thioester generating resin [41] in DMF for 30 min.

2. First remove Boc with 2× 1 min of 4 mL neat TFA. Then, perform flow wash for 1 min with DMF to remove all TFA.

3. Couple 12 eq. *S*-trityl mercaptopropionic acid activated with 11 eq. 0.5 M HBTU in DMF and 24 eq. DIEA for 15 min. Flow wash the resin with DMF for 1 min. Deprotect Trt with Trt cleavage solution for 2× 1 min and flow wash the resin with DMF for 1 min.

4. For coupling of the Boc protected amino acids Asn(Xan), Ala, Ile, Gly, Leu, Phe, His(Tos), Asp(OcHx), Gln(Xan), Pro, Arg(Tos), Val, Ser(Bzl), Lys(2ClZ), Met activate 10 eq. with 9 eq. of 0.5 M HBTU (*see* **Note 12**), and incubate the resin for 15 min. Start with Boc protected Asn. After each coupling step, flow wash the resin with DMF for 1 min and deprotect Boc with 2× 1 min of 4 mL neat TFA. Then, perform flow wash for 1 min with DMF to remove all TFA and go on with coupling the following amino acid in the sequence.

5. When the desired sequence is assembled, treat the resin with 10 mL of the cleavage cocktail (*see* **Note 11**) at 0 °C in a teflon apparatus for 1 h to cleave the peptide from the resin (*see* **Note 13**). Next, remove HF under vacuum and trap HF with excess CaO. Then, obtain the crude peptide through washing with 3× 20 mL of cooled diethyl ether, filtering and subsequent dissolution in buffer A:B (1:1). Lyophilize.

6. For purification dissolve the lyophilized crude peptide in 6 M Gdn-HCl buffer (pH 4.7) and purify with HPLC. Identify pure fractions via ESI-MS and check with analytical HPLC.

3.1.3 Native Chemical Ligation of C-Terminal Intein and Membrane Anchor Peptide

1. Carry out the native chemical ligation reaction of the synthetic peptides in NCL buffer with 1 % (v/v) thiophenol as ligation mediator and optional addition of 17 mg/mL dodecylphosphocholine (DPC) at pH 7.4–7.7 using concentrations of 5–6 mM of C- and N-terminal peptides, respectively [39].

2. Quench the reaction by addition of three volume equivalents of ligation buffer and 20 % (v/v) of β-mercaptoethanol.

3. For purification use HPLC and identify pure fractions via ESI-MS.

3.1.4 HPLC Analysis

Perform HPLC analysis of all peptides on an analytical reversed phase (RP) C4 column at a flow rate of 1 mL/min over 30 min with a gradient from 5 to 80 % (v/v) buffer B in buffer A. Determine masses by ESI-MS operating in positive ion mode.

3.2 Protein Expression and Purification

3.2.1 Cloning

Clone DNA encoding for the POI into a modified pTXB3 vector containing the DnaE N-terminal split intein sequence (DnaE^N) from *Synechocystis* sp. (*Ssp*) or *Nostoc punctiforme* (*Npu*) and an additional C-terminal 6× His tag (available from the authors) in the following steps:

1. Design primers for the POI containing recognition sites for the restriction enzymes *NcoI* and *SapI*, respectively (*see* **Note 14**).

2. Digest both, the vector and the template with these enzymes.

3. Ligate the resulting fragments into the described vector (rPrP-DnaE^N-His).

4. Transfer this vector for efficient expression into *E. coli* BL21(DE3) RIL cells.

3.2.2 Expression

Work under sterile conditions until the cells are harvested.

1. Inoculate BL21(DE3) RIL cells into 10 mL of LB medium containing 0.1 mg/mL of ampicillin and 0.03 mg/mL chloramphenicol overnight at 37 °C and 170 rpm.

2. Transfer the entire culture into 1 L of LB medium containing 0.1 mg/mL ampicillin and 0.03 mg/mL chloramphenicol. Allow growing at 37 °C and 170 rpm until an optical density at 600 nm (OD_{600nm}) of 0.6 is reached.

3. Induce expression by addition of isopropyl-β-D-thiogalactopyranoside (IPTG) to a final concentration of 1 mM. From here on, allow the cultures to grow for another 4 h.

4. Harvest the cells by centrifugation at $8980 \times g$ and 4 °C for 30 min.

5. Resuspend in lysis buffer (*see* **Note 15**) and disrupt by a microfluidizer (*see* **Note 16**).

6. After centrifugation at $48,300 \times g$, 4 °C for 30 min, sodium dodecylsulfate polyacrylamide gel electrophoresis (SDS-PAGE) analysis of pellet and supernatant reveals where the overexpressed fusion protein can be found (*see* **Note 17**). If found in inclusion bodies, *see* **step 7** and if found in supernatant perform non-denaturing affinity purification (*see* **step 9**).

7. To remove membrane debris wash the pellet with 1× PBS buffer containing 0.1% Triton X-100, and after that 2× with 1× PBS buffer to remove the detergent. Centrifuge after each washing step at $48,300 \times g$ 4 °C for 30 min.

8. Solubilize the pellet in solubilizing buffer for 4 h at 4 °C.

9. Load the solubilized inclusion bodies, with 6× His tag after DnaE^N for purification, onto a Ni-NTA column, which has been equilibrated with 6 M Gdn buffer (pH 8) (*see* **Note 18**). Wash the column with five column volumes (CVs) of washing buffer and elute the fusion protein with elution buffer. Check the fractions with SDS-PAGE and pool accordingly.

10. Dialyze the obtained fractions using Slide-A-Lyzer cassettes with a molecular weight cutoff (MWCO) of 3.5 kDa against 6 M Gdn-HCl buffer (pH 8) to remove the high amount of imidazole for further refolding or direct storage of the fusion protein (*see* **Note 19**).

3.2.3 Folding

1. Dilute unfolded POI-DnaE^N-His fusion proteins obtained in 6 M Gdn-HCl buffer (pH 8) tenfold to a final concentration of 0.1 mg/mL into the refolding buffer (*see* **Note 20**).

2. Incubate the resulting solution for 12 h at 4 °C.

3. Remove L-arginine and the GSH/GSSG redox agents by dialysis against the storage buffer, using Slide-A-Lyzer cassettes with a molecular weight cutoff (MWCO) of 3.5 kDa.

3.3 Protein Trans-Splicing

3.3.1 Liposomes and Lipid-Coated Particles (See Note 21)

1. Dissolve 2 mg of 1,2-dioleoyl-sn-glycero-3-phosphocholine (DOPC) in 1 mL chloroform and transfer under argon into a 25 mL round bottom flask. Evaporate the solvent under rotation in a stream of helium.

2. Dry the resulting lipid layer in high vacuum for at least 2 h.

3. Resuspend the DOPC layer in 1 mL of splicing buffer.

4. Transfer the suspension into a 15 mL tube and keep it under argon.

5. Sonicate the lipid suspension until the solution becomes opalescent.

6. Centrifuge at 100,000×g and use the clear supernatant for splicing experiments on liposomes.

7. For lipidation of silica particles with an average particle diameter of 800 nm use a 16-fold excess of lipids to nanoparticles.

8. Centrifuge the necessary volume of nanoparticles stock solution at 775×g in a microcentrifuge for 3 min.

9. Resuspend the particles in 1 mL of methanol, sonicate for 5 min and centrifuge at 775×g in a microcentrifuge for 3 min.

10. Resuspend the particles in 1 mL of 1 M KOH and centrifuge at 775×g in a microcentrifuge for 3 min.

11. Wash at least five times extensively with 1 mL of splicing buffer with repetitive centrifugation under conditions as above.

12. Mix the liposomes and the particles and incubate under gentle agitation for 2 h.

3.3.2 Trans-Splicing Reactions

1. Carry out protein *trans*-splicing of POI-DnaE^N and DnaE^C-membrane anchors (MA) in solution at equimolar concentrations of 5 μM in detergent containing splicing buffer in the presence of 50 mM sodium mercaptoethanesulfonate (MESNA).

2. Stop the reaction by addition of SDS sample buffer for SDS-PAGE analysis or by transfer into 6 M Gdn-HCl buffer (pH 4.7).

3. First, perform purification by affinity chromatography using Ni-NTA columns with 6 M Gdn-HCl buffer (pH 8). Unreacted POI-DnaE^N-His and DnaE^N-His are retained on the Ni-NTA material.

4. Perform size exclusion chromatography using Superdex 75 to separate lipidated POI from unreacted DnaE^C-MA not in complex with DnaE^N-His in 6 M Gdn-HCl buffer (pH 8).

(a) Alternatively to **steps 1–4**, attach DnaE^C-MA to DOPC liposomes or lipid-coated silica particles. Mix 6.7 μmol liposomes, 0.3 μmol DnaE^C-MA with a fivefold excess of POI-DnaE^N for 12 h (*see* **Note 22**).

(b) Separate POI-modified liposomes from starting materials by ultracentrifugation in a sucrose gradient from 5 to 20%.

(c) Analyze by SDS-PAGE (*see* **Note 23**).

3.3.3 Folding of PrP-MA Constructs

1. Dissolve PrP with membrane anchor (PrP-MA) in 6 M Gdn-HCl buffer (pH 8) to a concentration of 0.5 mg/mL (*see* **Note 24**).

2. Perform a dilution with folding buffer to a final concentration of 2.5 M guanidine hydrochloride in steps of 5, 4, 3.5 M guanidine hydrochloride with 1 h incubation time between each at 4 °C under continuous gentle agitation.

3. Incubate for 3 days at 4 °C under continuous gentle agitation.

4. Remove guanidine hydrochloride and the GSH/GSSG redox agents by dialysis against the storage buffer at 4 °C, using Slide-A-Lyzer cassettes with a molecular weight cutoff (MWCO) of 3.5 kDa (*see* **Note 25**).

5. Centrifuge at 4 °C, $14,000 \times g$ for 15 min. In case of precipitation, take off supernatant.

6. For higher concentration of the sample use Vivaspin 500, PES (GE) filters with MWCO of 10 kDa (*see* **Note 26**) and centrifuge at 4 °C, $13,500 \times g$.

4   Notes

1. In our experience HOBt works best, but as it can cause allergic reactions and since its availability is limited, Oxyma pure might be used instead.

2. As DCM is volatile, we find that it is best to prepare this fresh.

3. pH is strongly dependent on temperature.

4. We find that it is more efficient to use preloaded Wang resin than coupling the first amino acid to the linker.

5. Wash under a continuous stream of DMF with an appropriate level of vacuum keeping the resin covered with solvent.

6. We find that it is best to prepare this fresh each day.

7. After coupling of 4,7,10-trioxatridecane-1,13-diamine the resin becomes sticky and requires vigorous washing with DMF.

8. Through usage of a Boc protected Cys as last amino acid in the sequence, separate Fmoc removal can be saved and Boc can be removed simultaneously in the cleavage of the peptide from the resin.

9. We recommend not to store and use it longer than 2–3 days.

10. First dissolve 20 eq. HOBt in 24 mL DCM:DMF (3:1) solution, then 20 eq. palmitoyl chloride, and finally add 22 eq. triethylamine, which leads to white precipitate. Vortex this mixture vigorously.

11. Use 1.25 mL of cleavage cocktail for 100 mg of dry peptide resin.

12. In case of Gln coupling flow wash the resin with DCM before and after Boc deprotection by using TFA, instead of DMF, to avoid possible high-temperature induced pyroglutamate formation.

13. For neat HF, please follow safety rules!

14. Example primer design [44]: POI = eGFP.
    eGFP-NcoI-for: 5′-CTA GCT AGC **CAT GG**G GGT GAG CAA GGG CGA GG -3′.
    eGFP-SapI-rev: 5′-GCC AAA ACT GA**G CTC TTC** GGC AAT AAT CCG -3′.

    Bold characters show the recognition site of the respective restriction enzymes. Keep in mind that *Sap*I is cutting outside the recognition site in a sequence independent region. Hence, not all created overhangs are complementary.

15. Use a homogenizer.

16. Work at 10 °C with 1.81 kbar and equilibrate the microfluidizer before usage. We use a benchtop microfluidizer (Constant Systems Ltd.).

17. Inclusion bodies require solubilization in chaotrope containing buffers.

18. Before loading the Ni-NTA column we recommend one more centrifugation step at 48,300×g, 4 °C for 30 min and using the supernatant to avoid blocking of the column.

19. Hydrate the Slide-A-Lyzer cassettes before usage and use 500× sample volume of 6 M Gdn-HCl buffer (pH 8) during dialysis.

20. No general folding protocol can be provided since it heavily depends on the properties of the POI. The protocol provided here has worked for several POI-intein fusion constructs in our hands.

21. In order to achieve high *trans*-splicing yields add low concentrations (5–10 µM) of reducing agents such as MESNA or tris(2-carboxyethyl)phosphine (TCEP) to the reaction mixtures. In the absence of these reducing agents PTS yields are decreased by 20–50%, most likely due to oxidative modification of side chain thiol groups required for successful *trans*-splicing. PTS reactions with the DnaE^C-MA construct attached to DOPC vesicles and subsequent addition of POI-DnaE^N led to similar yields as described for the detergent-containing splicing buffer. We also studied the influence of short amino acid sequences of the native extein sequences from the DnaE^C-split intein, which have been

reported to be required for efficient trans-splicing yields [42, 43]. A direct comparison of the 36-amino acid DnaE^C intein with a DnaE^C intein containing three original extein residues (amino acids CFN) provided no evidence for more efficient PTS when these extra residues were included. This might be based on the fact that the interaction of the DnaE^C-MA fusion construct exhibits a much weaker interaction with POI-DnaE^N (3.5 μM) than measured for other, less hydrophobic/membrane bound C-terminal split intein segments (42 nM) [42, 44]. The use of other split inteins can also lead to faster reaction rates and/or higher yields as demonstrated in Fig. 1b for eGFP as the POI and the *Npu* DnaE split intein. Therefore testing different intein systems is always recommended.

22. This leads to immobilization of 0.1–0.2 μmol of POI on liposomes or lipid-coated particles.

23. Successful PTS is demonstrated by stripping the lipids and proteins from the particles using SDS sample buffer. Figure 2b depicts a PTS reaction with PrP as POI on an SDS-PAGE.

We would like to point out here that expressed protein ligation (EPL) [45] for which POI thioesters are isolated, often proves very useful for protein semisynthesis. Lately, artificially fused DnaE inteins with specific mutations have been described as very fast and efficient inteins for EPL [28, 37]. Figure 2b shows a comparison of the widely used *Mxe* GyrA intein and fused *Npu* DnaE intein. The latter one provides improved reaction yields even for a challenging (refolded) PrP-DnaE fusion construct. Therefore EPL often is an alternative route to PTS if larger amounts are required.

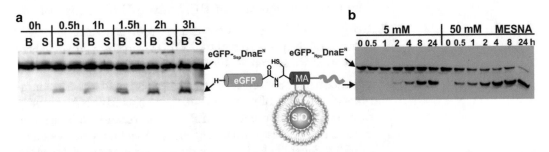

Fig. 1 *Trans*-splicing reactions on lipid-coated particles and liposomes. (**a**) Time course of the immobilization of eGFP on lipid-coated particles (silica particles with a diameter of 800 nm) using eGFP fused to the *Ssp* DnaE^N split intein (1.5 μM) and the respective DnaE^C-peptide linked to the membrane anchor MA (0.3 μM). The reaction reaches its maximum yield after 3–5 h at room temperature. eGFP in supernatant (S) and on beads (B) was detected by western blotting. (**b**) Time course of the reaction of eGFP fused to the *Npu* DnaEN split intein segment with the respective *Npu* DnaE^C-peptide linked to the membrane anchor MA both at 5 μM. The membrane peptide is linked to DOPC liposomes. Addition of 50 mM MESNA leads to a significant increase in reaction rates

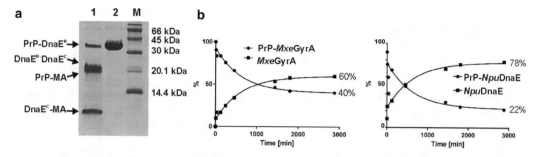

Fig. 2 (**a**) SDS-PAGE analysis of a *trans*-splicing reaction of PrP(aa 90-231) fused to *Ssp* DnaE^N intein with *Ssp* DnaE^C linked to the membrane anchor MA. The DnaE^N-DnaE^C complex appears at a similar molecular weight as the desired PrP-MA *trans*-splicing product. (**b**) Comparison of cleavage rates of the *Mxe* GyrA intein and the artificially assembled *Npu* DnaE intein [28] for generating PrP-α-thioester for expressed protein ligation

24. We find that significant higher concentrations lead to more pronounced precipitation of PrP-MA constructs during folding process, whereas too low concentrations make it hard to proof proper folding.

25. Hydrate the Slide-A-Lyzer cassettes before usage and use 500× sample volume of storage buffer during dialysis. Exchange the storage buffer 3×, after 2× 3 h and o/n incubation.

26. In our experience best yields are achieved with this material, whereas different filters can lead to loss of protein.

References

1. Priola SA, McNally KL (2009) The role of the prion protein membrane anchor in prion infection. Prion 3(3):134–138

2. Caughey B, Raymond GJ (1991) The scrapie-associated form of PrP is made from a cell surface precursor that is both protease- and phospholipase-sensitive. J Biol Chem 266(27):18217–18223

3. Puig B, Altmeppen H, Glatzel M (2014) The GPI-anchoring of PrP: implications in sorting and pathogenesis. Prion 8(1):11–18

4. Bate C, Tayebi M, Williams A (2010) The glycosylphosphatidylinositol anchor is a major determinant of prion binding and replication. Biochem J 428(1):95–101. doi:10.1042/BJ20091469

5. Chu NK, Shabbir W, Bove-Fenderson E, Araman C, Lemmens-Gruber R, Harris DA, Becker CF (2014) A C-terminal membrane anchor affects the interactions of prion proteins with lipid membranes. J Biol Chem 289(43):30144–30160. doi:10.1074/jbc.M114.587345

6. Wang F, Wang X, Yuan CG, Ma J (2010) Generating a prion with bacterially expressed recombinant prion protein. Science 327(5969):1132–1135. doi:10.1126/science.1183748

7. Deleault NR, Piro JR, Walsh DJ, Wang F, Ma J, Geoghegan JC, Supattapone S (2012) Isolation of phosphatidylethanolamine as a solitary cofactor for prion formation in the absence of nucleic acids. Proc Natl Acad Sci U S A 109(22):8546–8551. doi:10.1073/pnas.1204498109

8. Rocks O, Peyker A, Kahms M, Verveer PJ, Koerner C, Lumbierres M, Kuhlmann J, Waldmann H, Wittinghofer A, Bastiaens PI (2005) An acylation cycle regulates localization and activity of palmitoylated Ras isoforms. Science 307(5716):1746–1752

9. Rocks O, Gerauer M, Vartak N, Koch S, Huang ZP, Pechlivanis M, Kuhlmann J, Brunsveld L, Chandra A, Ellinger B, Waldmann H, Bastiaens PI (2010) The palmitoylation machinery is a spatially organizing system for peripheral membrane proteins. Cell 141(3):458–471. doi:10.1016/j.cell.2010.04.007

10. Rak A, Pylypenko O, Durek T, Watzke A, Kushnir S, Brunsveld L, Waldmann H, Goody RS, Alexandrov K (2003) Structure

of Rab GDP-dissociation inhibitor in complex with prenylated YPT1 GTPase. Science 302(5645):646–650. doi:10.1126/science.1087761

11. Wu YW, Oesterlin LK, Tan KT, Waldmann H, Alexandrov K, Goody RS (2010) Membrane targeting mechanism of Rab GTPases elucidated by semisynthetic protein probes. Nat Chem Biol 6(7):534–540

12. Schelhaas M, Nagele E, Kuder N, Bader B, Kuhlmann J, Wittinghofer A, Waldmann H (1999) Chemoenzymatic synthesis of biotinylated Ras peptides and their use in membrane binding studies of lipidated model proteins by surface plasmon resonance. Chem Eur J 5(4):1239–1252

13. Brunsveld L, Kuhlmann J, Alexandrov K, Wittinghofer A, Goody RS, Waldmann H (2006) Lipidated ras and rab peptides and proteins—synthesis, structure, and function. Angew Chem Int Ed Engl 45(40):6622–6646

14. Grogan MJ, Kaizuka Y, Conrad RM, Groves JT, Bertozzi CR (2005) Synthesis of lipidated green fluorescent protein and its incorporation in supported lipid bilayers. J Am Chem Soc 127(41):14383–14387

15. Hicks MR, Gill AC, Bath IK, Rullay AK, Sylvester ID, Crout DH, Pinheiro TJT (2006) Synthesis and structural characterization of a mimetic membrane-anchored prion protein. FEBS J 273(6):1285–1299

16. Olschewski D, Seidel R, Miesbauer M, Rambold AS, Oesterhelt D, Winklhofer KF, Tatzelt J, Engelhard M, Becker CFW (2007) Semisynthetic murine prion protein equipped with a GPI anchor mimic incorporates into cellular membranes. Chem Biol 14(9):994–1006

17. Filchtinski D, Bee C, Savopol T, Engelhard M, Becker CF, Herrmann C (2008) Probing ras effector interactions on nanoparticle supported lipid bilayers. Bioconjug Chem 19(9):1938–1944

18. Hang HC, Linder ME (2011) Exploring protein lipidation with chemical biology. Chem Rev 111(10):6341–6358. doi:10.1021/cr2001977

19. Bader B, Kuhn K, Owen DJ, Waldmann H, Wittinghofer A, Kuhlmann J (2000) Bioorganic synthesis of lipid-modified proteins for the study of signal transduction. Nature 403(6766):223–226

20. Pylypenko O, Rak A, Reents R, Niculae A, Sidorovitch V, Cioaca MD, Bessolitsyna E, Thoma NH, Waldmann H, Schlichting I, Goody RS, Alexandrov K (2003) Structure of Rab escort protein-1 in complex with Rab geranylgeranyltransferase. Mol Cell 11(2):483–494

21. Huang YC, Li YM, Chen Y, Pan M, Li YT, Yu L, Guo QX, Liu L (2013) Synthesis of autophagosomal marker protein LC3-II under

detergent-free conditions. Angew Chem Int Ed Engl 52(18):4858–4862. doi:10.1002/anie.201209523

22. Olschewski D, Becker CF (2008) Chemical synthesis and semisynthesis of membrane proteins. Mol Biosyst 4(7):733–740

23. Melnyk RA, Partridge AW, Yip J, Wu Y, Goto NK, Deber CM (2003) Polar residue tagging of transmembrane peptides. Biopolymers 71(6):675–685

24. Johnson ECB, Kent SBH (2007) Towards the total chemical synthesis of integral membrane proteins: a general method for the synthesis of hydrophobic peptide-(alpha)thioester building blocks. Tetrahedron Lett 48(10):1795–1799

25. Becker CF, Oblatt-Montal M, Kochendoerfer GG, Montal M (2004) Chemical synthesis and single channel properties of tetrameric and pentameric TASPs (template-assembled synthetic proteins) derived from the transmembrane domain of HIV virus protein u (Vpu). J Biol Chem 279(17):17483–17489

26. Marsac Y, Cramer J, Olschewski D, Alexandrov K, Becker CFW (2006) Site-specific attachment of polyethylene glycol-like oligomers to proteins and peptides. Bioconjug Chem 17(6):1492–1498

27. Shen F, Huang YC, Tang S, Chen YX, Liu L (2011) Chemical synthesis of integral membrane proteins: methods and applications. Isr J Chem 51(8–9):940–952. doi:10.1002/ijch.201100076

28. Shah NH, Dann GP, Vila-Perello M, Liu Z, Muir TW (2012) Ultrafast protein splicing is common among cyanobacterial split inteins: implications for protein engineering. J Am Chem Soc 134(28):11338–11341. doi:10.1021/ja303226x

29. Vila-Perello M, Liu Z, Shah NH, Willis JA, Idoyaga J, Muir TW (2013) Streamlined expressed protein ligation using split inteins. J Am Chem Soc 135(1):286–292. doi:10.1021/ja309126m

30. Shah NH, Muir TW (2014) Inteins: nature's gift to protein chemists. Chem Sci 5(1):446–461. doi:10.1039/C3SC52951G

31. Volkmann G, Mootz HD (2013) Recent progress in intein research: from mechanism to directed evolution and applications. Cell Mol Life Sci 70(7):1185–1206. doi:10.1007/s00018-012-1120-4

32. Wood DW, Camarero JA (2014) Intein applications: from protein purification and labeling to metabolic control methods. J Biol Chem 289(21):14512–14519. doi:10.1074/jbc.R114.552653

33. Ritchie TK, Grinkova YV, Bayburt TH, Denisov IG, Zolnerciks JK, Atkins WM, Sligar SG (2009) Chapter 11 - reconstitution of

membrane proteins in phospholipid bilayer nanodiscs. Methods Enzymol 464:211–231. doi:10.1016/S0076-6879(09)64011-8

34. Chu NK, Becker CF (2009) Semisynthesis of membrane-attached prion proteins. Methods Enzymol 462:177–193

35. Dhar T, Mootz HD (2011) Modification of transmembrane and GPI-anchored proteins on living cells by efficient protein trans-splicing using the Npu DnaE intein. Chem Commun (Camb) 47:3063–3065

36. Mootz HD (2009) Split inteins as versatile tools for protein semisynthesis. Chembiochem 10(16):2579–2589

37. Zettler J, Schütz V, Mootz HD (2009) The naturally split Npu DnaE intein exhibits an extraordinarily high rate in the protein trans-splicing reaction. FEBS Lett 583(5):909–914

38. Durek T, Becker CF (2005) Protein semi-synthesis: new proteins for functional and structural studies. Biomol Eng 22(5–6):153–72

39. Dawson PE, Muir TW, Clark-Lewis I, Kent SBH (1994) Synthesis of proteins by native chemical ligation. Science 266:776–779

40. Muttenthaler M, Albericio F, Dawson PE (2015) Methods, setup and safe handling for anhydrous hydrogen fluoride cleavage in Boc solid-phase peptide synthesis. Nat Protoc 10(7):1067–1083. doi:10.1038/nprot.2015.061

41. Hackeng TM, Griffin JH, Dawson PE (1999) Protein synthesis by native chemical ligation: expanded scope by using straightforward methodology. Proc Natl Acad Sci U S A 96(18):10068–10073

42. Martin DD, Xu MQ, Evans TC Jr (2001) Characterization of a naturally occurring trans-splicing intein from Synechocystis sp. PCC6803. Biochemistry 40(5):1393–1402

43. Lockless SW, Muir TW (2009) Traceless protein splicing utilizing evolved split inteins. Proc Natl Acad Sci 106(27):10999–11004

44. Chu NK, Olschewski D, Seidel R, Winklhofer KF, Tatzelt J, Engelhard M, Becker CF (2010) Protein immobilization on liposomes and lipid-coated nanoparticles by protein trans-splicing. J Pept Sci 16(10):582–588

45. Muir TW, Sondhi D, Cole PA (1998) Expressed protein ligation: a general method for protein engineering. Proc Natl Acad Sci U S A 95(12):6705–6710

Chapter 8

Protein Chemical Modification Inside Living Cells Using Split Inteins

Radhika Borra and Julio A. Camarero

Abstract

Methods to visualize, track, measure, and perturb or activate proteins in living cells are central to biomedical efforts to characterize and understand the spatial and temporal underpinnings of life inside cells. Although fluorescent proteins have proven to be extremely useful for in vivo studies of protein function, their utility is inherently limited because their spectral and structural characteristics are interdependent. These limitations have spurred the creation of alternative approaches for the chemical labeling of proteins. We describe in this protocol the use of fluorescence resonance emission transfer (FRET)-quenched DnaE split-inteins for the site-specific labeling and concomitant fluorescence activation of proteins in living cells. We have successfully employed this approach for the site-specific in-cell labeling of the DNA binding domain (DBD) of the transcription factor YY1 using several human cell lines. Moreover, we have shown that this approach can be also used for modifying proteins in order to control their cellular localization and potentially alter their biological activity.

Key words Split-intein, Protein *trans*-splicing, *Npu* intein, Protein labeling, Fluorescence

1 Introduction

Understanding the roles of specific proteins in cellular processes is a fundamental goal of molecular biology [1–3]. Methods to label and visualize proteins inside living cells are extremely useful in the study of localization, movement, interactions, and microenvironments of proteins in living cells. Although fluorescent proteins have revolutionized such studies, they have numerous shortcomings, which have spurred the creation of alternative approaches to chemically label proteins in living cells. These next generation approaches combine the genetic targeting capabilities of fluorescent proteins with the diversity and environmental sensitivity of fluorescent small molecules and/or other biophysical probes. Most of the available techniques, however, provide only limited temporal resolution for labeling of biomolecules in living cells.

Henning D. Mootz (ed.), *Split Inteins: Methods and Protocols*, Methods in Molecular Biology, vol. 1495,
DOI 10.1007/978-1-4939-6451-2_8, © Springer Science+Business Media New York 2017

Ideally these approaches should be modular, thus making possible the introduction of a wide variety of fluorophores or other type of biophysical probes. The kinetics of the labeling reaction should be fast enough to provide temporal resolution to satisfy the most time-sensitive biological assays. It should also allow spatial control during the in-cell labeling process. The labeling reaction should introduce minimal modifications on the target protein in order to preserve its original structure and biological function. Finally, it should make possible the simultaneous introduction of different probes onto multiple target proteins for simultaneous tracking purposes.

One of the most promising approaches for in-cell protein labeling involves the use of intein-mediated protein *trans*-splicing (Fig. 1) [5]. Protein *trans*-splicing is a naturally occurring post-translational

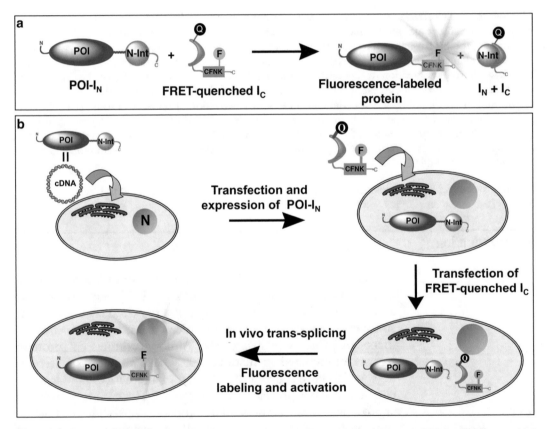

Fig. 1 (a) Site-specific labeling and fluorescence activation of a protein of interest (POI) by FRET-quenched protein *trans*-splicing. Key to this approach is the introduction of fluorescence quencher (**Q**) into the I_C polypeptide, which blocks the fluorescence signal of the fluorophore (**F**) located at the C-terminus of the I_C polypeptide before protein *trans*-splicing happens. When protein *trans*-splicing occurs the fluorophore is covalently attached to the C-terminus of the POI triggering its fluorescence. The use of this approach for in-cell modification and fluorescence tagging of proteins minimizes the fluorescence background from the unreacted I_C polypeptide thus facilitating the optical tracking of the labeled protein inside the cell. (**b**) Scheme showing the approach used for in-cell labeling of a POI with a fluorophore inside a live cell using protein *trans*-splicing (figure modified from [4])

modification similar to protein splicing with the difference being that the intein self-processing domain is split in two fragments, called N-intein (I_N) and C-intein (I_C), respectively [6, 7]. These two intein fragments are inactive individually, however, they can bind each other with high specificity under appropriate conditions to form a functional protein-splicing domain. Split mini-inteins have been widely used by our group and others for the site-specific modification of proteins in vitro [8–11] and in living cells [4, 12, 13]. In-cell labeling of proteins can be easily accomplished by expressing the protein of interest fused to the I_N fragment. The second half of the split intein can be chemically synthesized to contain any chemical probe at the C-extein moiety, and then introduced into the cells by using peptide transducing domains (PTD) [4, 12].

Intein-mediated labeling of proteins is highly modular allowing the covalent site-specific incorporation of a myriad of biophysical probes into proteins [9–11]. The kinetics of protein splicing is also relatively fast, with a number of split-inteins having reaction times in the order of several minutes [10, 14, 15]. Moreover, the recent development of conditional protein splicing, both through chemical and photochemical means, makes possible the chemical modification of proteins in living cells with temporal and spatial control [16–19].

One of the best-characterized naturally occurring split-inteins are α-subunit DNA polymerase III (DnaE) intein [20], with many known orthologs with high sequence homology in many cyanobacteria species (Fig. 2a) [14, 23]. The DnaE split-inteins are characterized for having I_N and I_C fragments with ≈120 and ≈30 residues, respectively. The relatively small size of the I_C fragment facilitates its chemical synthesis thus allowing the use of synthetic I_C fragments bearing different biophysical probes in the C-extein segment to be used for the chemical modification of proteins through protein *trans*-splicing [4, 10, 17, 18].

We will use in this protocol the *Nostoc punctiforme* PCC73102 (*Npu*) DnaE split-intein. This particular DnaE split-intein has one of the highest rate reported for protein trans-splicing ($\tau_{1/2} \approx 60$ s) [15] and a high splicing yield [15, 24]; and therefore is ideal for in-cell protein labeling purposes.

The use of protein *trans*-splicing for the site-specific labeling of proteins with fluorogenic dyes for in-cell tracking purposes requires that the labeling process must be linked to the simultaneous activation of fluorescence (Fig. 1). This can be accomplished by making use of fluorescence resonance emission transfer (FRET)-quenched DnaE split-inteins for the site-specific labeling and concomitant fluorescence activation of proteins in living cells. In this protocol we use fluorescein and dabcyl as fluorescence donor and FRET-quencher, respectively (Fig. 2), but any other combination of donor and quencher could be also used.

The fluorescein group is introduced at the C-terminus of the first four residues (Cys-Phe-Asn-Lys) of the C-extein, which are

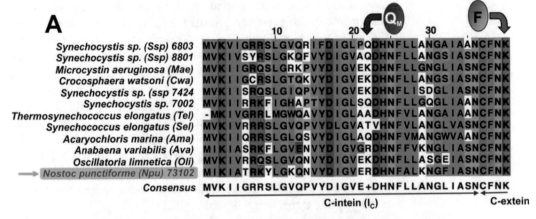

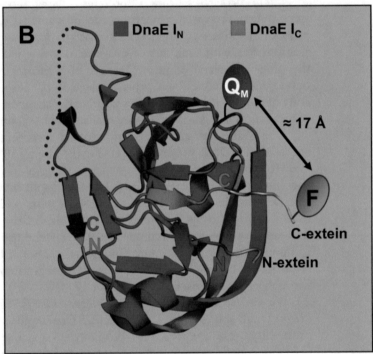

Fig. 2 Design of FRET-quenched DnaE split inteins. (**a**) Multiple sequence alignment of the DnaE I_C for different species indicating the positions used for the introduction of the quencher group in the I_C polypeptide. Multiple sequence alignment was performed using T-Coffee and visualized using Jalview [21]. Molecular representations of the DnaE inteins were generated using the PyMol software package. (**b**) Crystal structure of the *Npu* DnaE intein in the pre-spliced state (PDB code: 2KEQ) [22]. DnaE I_C and I_N are shown in *red* and *blue* respectively. The structural secondary elements are also shown. The position used to place the quencher and fluorophore groups at the I_C and C-extein, as well as the distances are indicated (figure modified from ref. [4])

required for efficient *trans*-splicing (Fig. 2) [4] The dabcyl group (Q_M, Fig. 2) is introduced on residue 22 of the *Npu* DnaE I_C polypeptide (peptide **IC**, Table 1). This position is in close proximity to the C-extein (≈ 17 Å, Fig. 2b) and provides an excellent FRET-quenching (>99% FRET-quenching) [4].

Table 1
Sequence of the DnaE C-intein (I$_C$) polypeptide used in this protocol
Standard single code letters are used for the peptides sequences. Single-letter codes B and X stand for norleucine and 6-amino hexanoic acid, respectively. 5-(Iodoacetamide)-fluoresceine (IAF) is used to introduce fluorescein into specific Lys or Cys residues, respectively. Dabcyl, Cam, Cam-Fl stand for 4-dimethylaminoazobenzene-4′-carboxyl, carboxamidomethyl, and fluorescein-carboxamidomethyl, respectively. Residues in *blue*, *magenta*, and *yellow* represent the *Npu* DnaE I$_C$ intein, *Npu* DnaE C-extein, and NLS peptide, respectively. The quencher and fluorophore groups are in *red* and *green* respectively

Peptide	Name	Compound	Sequence	Molecular Weight Found (Expected[a])/Da
Npu	Q$_M$-I$_C$-NLS-Fl	I$_C$	Ac-BIKIATRKYLGKQN VYDIGVEK(Dabcyl)DHNF ALKNGFIASNCFNKXC (Cam-Fl)XPKKKRKV-NH$_2$	6449.0±0.1 (6447.6)

[a]Average molecular weight

In this chapter we describe the protocol for in-cell C-terminal labeling of the DNA binding domain (DBD) of the transcription factor Yin Yang 1 (YY1) in live U2OS and HeLa cells. YY1 is a ubiquitously distributed multifunctional transcription factor belonging to the GLI-Kruppel class of zinc finger proteins [25]. The protein is involved in repressing and activating a diverse number of promoters including negative regulation of p53, thus making it of particular interest [26, 27]. In the example described in this chapter the DBD of YY1 was labeled at its C-terminal with a fluorophore (fluorescein) and a nuclear localization (NLS) signal peptide to demonstrate the potential of this technique to control the localization of and biological function of a protein/protein domain. The protocol described uses humans U2OS cells but it could be easily adapted to any other mammalian cell line.

It is important to note that this approach is highly modular and can be used for in-cell labeling of proteins with other biophysical probes or peptide sequences required for activity. In addition, the use of different orthogonal split inteins should also make possible the simultaneous labeling of multiple proteins with different probes.

Before performing the labeling experiment in-cell, it is advisable to test the labeling reaction in vitro first. This protocol also provides instructions on how to evaluate labeling by protein *trans*-splicing in vitro.

2 Materials

All solutions are prepared using ultrapure water with a resistivity of 18 MΩ×cm at 25 °C and analytical grade reagents. All reagents and solutions are stored at room temperature unless indicated otherwise.

2.1 Instruments	1. Water bath able to operate at 42 °C and 94 °C.

2. Table-top micro centrifuge capable of operating at $15,000 \times g$.

3. Microbiology incubator set at 37 °C.

4. Temperature controlled incubator Shaker.

5. Orbital shaker.

6. Polymerase chain reaction thermocycler.

7. Agarose gel electrophoresis unit.

8. Electrophoresis power pack able to operate up to 250 V.

9. UV–visible spectrophotometer.

10. Sonicator to lyse *E. coli* cells.

11. High speed centrifuge (e.g., Sorvall RC *5C* Plus, Thermo Fisher scientific).

12. SDS-PAGE electrophoresis apparatus.

13. Centrifuge tubes of 0.5, 1.5, 15, and 30 mL of capacity.

14. 5 mL polypropylene columns.

15. Class II, type A2 biosafety cabinets.

16. Gel and Blot Imaging System.

17. Fluorescence microscope.

18. CO_2 incubator.

2.2 Cloning of YY1-I$_N$ Construct

1. Synthetic DNA primers used to amplify genes encoding *Npu* DnaE I$_N$ and DBD-YY1 (20 nmol scale, HPLC purified) (Table 2).

2. Genomic DNA from *Nostoc punctiforme* strain ATCC 29133/ PCC 73102 (obtained from ATCC).

3. DNA encoding human YY1 proteins (cDNA clone IMAGE: 5261384) (can be obtained from many sources, e.g., Genscript, USA).

Table 2
DNA oligonucleotides used to generate the dsDNA encoding the YY1-I$_N$ and DnaE I$_N$

Primer name	Nucleotide sequence
p5-YY1	5′- AAA GAA GAT CAT ATG CCA AGA ACA ATA GCT TGC CCT C-3′
p3-YY1	5′- C TTC GGA TCC CTG GTT GTT TTT GGC CTTA GC-3′
p5-I$_N$	5′- CTA GTC GAC AAG CTT TTA AGT TTG CGG AAT ATT GTT TAA G CTA TG -3′
p3-I$_C$	5′- TTT GCG GCC GCT TAA TTC GGC AAA TTA TCA ACC CGC AT -3′
p5-YY1-IN	5′- AGG GGT ACC ACC ATG GGC AGC AGC -3′
p3-YY1-IN	5′- GT GGT GCT CGA GTG CGG CCG CA -3′

4. TE buffer: 10 mM Tris–HCl, 1 mM EDTA, pH 8.0.

5. Vent DNA polymerase, TaqDNA polymerase, dNTPs solution, 10× ThermoPol PCR buffer, 10× Taq DNA polymerase buffer.

6. Restriction enzymes *NotI*, *NdeI*, *BamHI*, *KpnI*, and *SalI*.

7. PCR Purification Kit (e.g.,, QIAquick from QIAGEN).

8. Miniprep Kit (e.g.,, QIAprep from QIAGEN).

9. Gel Extraction Kit (e.g.,, QIAquick from QIAGEN).

10. Chemical competent DH5α cells.

11. Expression plasmid pET28a (Novagen-EMD Millipore).

12. Mammalian Expression vector pcDNA4/TO/myc-His (Invitrogen).

13. T4 DNA ligase and T4 DNA ligase buffer.

14. Ampicillin stock solution: 100 mg ampicillin/mL in pure H_2O, sterilized by filtration. Store in 1 mL aliquots at –20 °C.

15. Kanamycin stock solution: 25 mg kanamycin/mL in pure H_2O, sterilized by filtration. Store in 1 mL aliquots at –20 °C.

16. Chloramphenicol stock solution: 34 mg chloramphenicol/mL in EtOH. Store in 1 mL aliquots at –20 °C.

17. LB medium: 25 g of LB broth was dissolved in 1 L of pure H_2O and sterilized by autoclaving at 120 °C for 30 min.

18. LB medium-agar: 3.3 g of LB agar was suspended in 100 mL of pure H_2O and sterilized by autoclaving at 120 °C for 30 min. To prepare plates, allow LB medium-agar to cool to ≈50 °C, then add 0.1 mL of antibiotic stock solution.

19. SOC Medium: 20 g of tryptone, 5 g yeast extract, 0.5 g NaCl, and 0.186 g KCl were suspended into 980 mL of pure water and sterilized by autoclaving at 120 °C for 30 min. Dissolve 4.8 g $MgSO_4$, 3.603 g dextrose in 20 mL of pure H_2O and filter-sterilize over a 45 μm filter and add to the autoclaved medium.

2.3 Bacterial Expression YY1-I_N Construct

1. Chemical competent BL21 (DE3) and Origami2 (DE3) cells (EMD Millipore).

2. Isopropyl-thio-β-D-galactopyranoside (IPTG): Prepare a stock solution of 1 M analytical grade IPTG in H_2O and sterilize by filtration over 45 μm filter. Store at –20 °C.

3. Lysis buffer: 0.1 mM EDTA, 25 mM sodium phosphate, 150 mM NaCl, 10 % (v/v) glycerol, pH 7.4.

4. Phosphate buffer saline (PBS): 25 mM sodium phosphate, 150 mM NaCl, pH 7.4.

5. 100 mM phenylmethylsulfonyl fluoride (PMSF) in EtOH (better to prepare fresh before use).

6. 4× SDS-PAGE sample buffer: 1.5 mL of 1 M Tris–HCl buffer at pH 6.8, 3 mL of 1 M DTT (dithiothreitol) in pure H_2O, 0.6 g of sodium dodecyl sulfate (SDS), 30 mg of bromophenol blue, 2.4 mL of glycerol, bring final volume to 7.5 mL.

7. SDS-PAGE sample buffer: dilute four times 4× SDS-PAGE sample buffer in pure H_2O and add 20% (v/v) 2-mercaptoethanol. Prepare fresh.

8. SDS-4–20% PAGE gels, 1× SDS running buffer.

9. Gel stain: Coomassie brilliant blue or Gelcode® Blue (Thermo scientific, USA) or silver stain kit.

2.4 In Vitro Trans-Splicing

1. Pure labeled DnaE I_C polypeptide shown in Table 1. The synthetic peptides used in this study were generated in-house but they could be ordered from any chemical supplier specialized in providing synthetic peptides.

2. Trans-splicing buffer: 0.5 mM EDTA, 1 mM tris-(2-carboxyethyl)phosphine (TCEP), 50 mM NaH_2PO_4, 250 mM NaCl, pH 7.0 (*see* **Note 1**).

2.5 In-Cell Trans-Splicing Reaction

1. U2OS cells, available from ATCC (ATCC® HTB-96™).

2. HeLa cells, available from ATCC (ATCC® CCL-2™).

3. Dulbecco's modified eagle medium (DMEM) containing 10% heat inactivated FBS (fetal bovine serum), 1% L-glutamine, and 1% penicillin–streptomycin solution.

4. RIPA BUFFER: 50 mM Tris–HCl, 150 mM NaCl buffer, pH 8.0, 1% NP-40, 0.5% sodium deoxycholate, 1% Triton X-100.

5. TBST buffer: 50 mM Tris–HCl, 150 mM NaCl, pH 7.6.

6. Nonpyrogenic sterile 100 × 20 mm polystyrene plates.

7. Nonpyrogenic sterile 35 mm glass bottom plates (e.g., MatTek).

8. Transfection reagent for mammalian cells (e.g., Fugene-6 from Promega or equivalent).

9. Complete protease cocktail (e.g., Thermo Scientific).

10. PVDF membrane for western blotting.

11. 5% skim milk in TBST buffer.

12. Anti-His Antibody (e.g., murine IgG). Store at –20 °C.

13. Secondary antibody (e.g., horseradish peroxidase-conjugated anti-murine IgG, Vector Lab).

14. ECL kit (e.g., Life Technologies).

15. Chariot protein delivery reagent (Active Motif).

3 Methods

3.1 Cloning of DNA Encoding Npu DnaE I_N into Expression Vector pET28a(+)

1. Amplify by PCR the gene containing the *Npu* DnaE I_N (residues 770–876, UniProtKB: B2J066) using a plasmid containing the DnaE gene from *Nostoc punctiforme* (Strain ATCC 2913/PCC 73102) as template using primers p5-I_N and p3-I_N (Table 2). The 5′-primer encodes a *Sal* I restriction site. The 3′-primer introduces a *NotI* restriction site and stop codon. Carry out the PCR reaction as follows: 40 µL sterile pure H₂O, 1 µL of DNA template ($\approx$10 ng/µL), 5 µL of 10× ThermoPol reaction buffer, 1.0 µL of dNTP solution (10 mM each), 1 µL of p5-I_N primer solution (0.2 µM), 1 µL of p3-I_N r primer solution (0.2 µM), and 1 µL Vent DNA polymerase (2 units). PCR cycle conditions used: initial denaturation at 94 °C for 5 min followed by 30 cycles (94 °C denaturation for 30 s, annealing at 52 °C for 45 s, and extension at 72 °C for 60 s) and final extension at 72 °C for 10 min.

2. Purify the PCR amplified fragment encoding *Npu* Dna I_N using a PCR purification kit following the manufacturer instructions and quantify it by UV absorption (for a 1-cm pathlength, an optical density at 260 nm (OD_{260}) of 1.0 equals to a concentration of 50 µg/mL solution of dsDNA).

3. Digest plasmid pET28a(+) (Novagen-EMD Millipore) and PCR-amplified gene encoding the DnaE I_N polypeptide with restriction enzymes *SalI* and *NotI*. Use a 0.5 mL centrifuge tube and add 5 µL of 10× restriction buffer (e.g., NEB buffer 2.1 from New England Biolabs), add enough pure sterile water to have a final volume reaction of 50 µL, add $\approx$10 µg of the corresponding dsDNA to be digested and finally add 1 µL (20 units) of restriction enzyme *NotI*. Incubate at 37 °C for 3 h. Then, add 1 µL (20 units) of restriction enzyme *SalI* to the same tube and incubate at 37 °C for 1 h.

4. Purify the double digested PCR-product and pET28a plasmid by agarose (0.8 % and 2 % agarose gels for pET28a(+) and PCR product should be used, respectively) gel electrophoresis. Cut out the bands corresponding to the double digested DNA and purify the DNA using a gel extraction kit. Elute DNA from spin columns with TE buffer and quantify using UV.

5. Ligate double digested pET28a(+) and PCR-product encoding DnaE I_N. Use a 0.5 mL centrifuge tube, add $\approx$100 ng of *SalI*, *NotI*-digested pET28a, $\approx$50 ng of *SalI*, *NotI*-digested PCR-amplified DNA encoding DnaE I_N, enough pure sterile H₂O to make a final reaction volume of 20 µL, 2 µL of 10× T4 DNA ligase buffer, 1 µL of 10 mM ATP, and 1 µL (400 units) T4 DNA ligase. Incubate at 16 °C overnight.

6. Transform the ligation mixture into DH5α competent cells. ≈100 µL of chemical competent cells are thawed on ice and mixed with the ligation mixture (20 µL) for 30 min. Heat-shock the cells are heat-shocked at 42 °C for 45 s and then keep on ice for an extra 10 min. Add 900 µL of SOC medium and incubate at 37 °C for 1 h in an orbital shaker. Plate 100 µL on LB agar plate containing kanamycin (25 µg/mL) and incubate the plate at 37 °C overnight.

7. Pick up several colonies (most of the times five colonies should be enough) and inoculate into 5 mL of LB medium containing kanamycin (25 µg/mL). Incubate tubes at 37 °C overnight in an orbital shaker.

8. Pellet down cells and extract DNA using a miniprep kit following the manufacturer protocol and quantify plasmid using UV spectroscopy.

9. Verify the presence of DNA encoding DnaE I_N in each colony using PCR and the same conditions described in **step 1** of this section.

10. Screen colonies containing DNA encoding DnaE I_N for protein expression (**steps 11** through **17**).

11. Transform chemical competent BL21 (DE3) cells with plasmids containing the DNA encoding DnaE I_N (*see* **Note 2**). Transformed cells are plated on LB plate containing kanamycin (25 µg/mL) and incubated at 37 °C overnight.

12. Resuspend the colonies from one plate in 1 mL of LB and inoculate 100 mL of LB containing kanamycin (25 µg/mL) in a 250 mL flask.

13. Grow cells in an orbital shaker incubator at 37 °C for 2–3 h to reach mid-log phase (OD at 600 nm ≈ 0.5). Add IPTG to reach a final concentration of 1 mM and incubate cells for 3 h at 37 °C.

14. Pellet 1 mL of cells by centrifugation at $6000 \times g$ for 15 min at 4 °C.

15. Discard the supernatant and resuspend pellets in fresh SDS-PAGE sample buffer.

16. Heat samples at 94 °C for 5 min and separate the soluble cell lysate fraction by centrifugation at $15,000 \times g$ for 20 min at 4 °C.

17. Analyze the soluble fraction by SDS-PAGE analysis to estimate the protein expression of the DnaE I_N intein in each clone analyzed. The DnaE I_N polypeptide should give a band around 15 kDa.

3.2 Cloning of DNA Encoding YY1-I_N Construct into Expression Vector pET28a

1. Amplify the gene containing the DNA binding domain of YY1 by PCR using the cDNA for human YY1 (cDNA clone IMAGE: 5261384) as template using primers p5-YY1 and p3-YY1 (Table 2). The 5′-primer and 3′-primer introduce *NdeI* and *BamHI* restriction sites, respectively. Carry out the

PCR reaction as follows: 40 µL sterile pure H_2O, 1 µL of DNA template (≈10 ng/µL), 5 µL of 10× ThermoPol reaction buffer, 1.0 µL of dNTP solution (10 mM each), 1 µL of p5-YY1 primer solution (0.2 µM), 1 µL of p3-YY1 primer solution (0.2 µM), and 1 µL Vent DNA polymerase (2 units). Use the following PCR cycle conditions: initial denaturation at 94 °C for 5 min followed by 30 cycles (94 °C denaturation for 30 s, annealing at 52 °C for 45 s, and extension at 72 °C for 60 s) and final extension at 72 °C for 10 min.

2. Purify the PCR amplified fragment encoding YY1 using a PCR purification kit following the manufacturer instructions and quantify by UV spectroscopy.

3. Digest plasmid pET28a encoding DnaE I_N (pET-I_N) (obtained in Subheading 3.1) and PCR-amplified gene encoding YY1 construct with restriction enzymes *Nde* I and *BamH* I. Use a 0.5 mL centrifuge tube and add 5 µL of 10× restriction buffer (e.g., NEB buffer 3.1 from New England Biolabs), add enough pure sterile water to have a final volume reaction of 50 µL, add ≈10 µg of the corresponding dsDNA to be digested and finally add 1 µL (20 units) of restriction enzyme *NdeI*. Incubate at 37 °C for 3 h. Then, add 1 µL (20 units) of restriction enzyme *BamH* I to the same tube and incubate at 37 °C for 1 h.

4. Purify the double digested PCR-product and pET-I_N plasmid by agarose (0.8% and 2% agarose gels for pET-I_N and PCR product should be used, respectively) gel electrophoresis. Cut out the bands corresponding to the double digested DNA and purify them using a gel extraction kit. Elute DNA from the spin columns with TE buffer and quantify using UV spectroscopy.

5. Ligate double digested pET-I_N and PCR-product encoding the DBD of YY1. Use a 0.5 mL centrifuge tube, add ≈100 ng of *NdeI, BamHI*-digested pET-I_N, ≈50 ng of *NdeI, BamHI*-digested PCR-amplified DNA encoding the DBD of YY1, enough pure sterile H_2O to make a final reaction volume of 20 µL, 2 µL of 10× T4 DNA ligase buffer, 1 µL of 10 mM ATP, and 1 µL (400 units) T4 DNA ligase. Incubate at 16 °C overnight.

6. Transform the ligation mixture into DH5α competent cells, plate them on a LB agar plate containing kanamycin (25 µg/mL), and incubate the plate at 37 °C overnight as described previously (Subheading 3.1, **steps 5** and **6**).

7. Pick up several colonies (most of the times five colonies should be enough) and screen for the presence of DNA inserts encoding the DBD of YY1 by PCR as described previously (Subheading 3.1, **steps 7** through **9**).

8. Screen colonies containing DNA encoding the YY1-I_N construct for protein expression (**steps 9** through **12**).

9. Transform chemical competent Rosetta (DE3) cells with plasmids containing the DNA encoding YY1-I_N (see **Note 3**). Transformed cells are plated on LB plate containing kanamycin (25 µg/mL) and chloramphenicol (34 µg/mL) incubated at 37 °C.

10. Resuspend the colonies from 1 plate in 1 mL of LB and inoculate 100 mL of LB containing kanamycin (25 µg/mL) and chloramphenicol (34 µg/mL) in a 250 mL flask.

11. Grow cells in an orbital shaker incubator at 37 °C for 2–3 h to reach mid-log phase (OD at 600 nm ≈ 0.5). Add IPTG to reach a final concentration of 1 mM and incubate cells for 3 h at 37 °C.

12. Pellet, lyse cells and analyze protein expression level of YY1-I_N construct by SDS-PAGE as described in Subheadings 3.1, **steps 14** and **17**. The YY1-I_N construct should give a band around 30 kDa.

3.3 Bacterial Expression of the YY1-I_N Construct

1. Transform chemical competent Rosetta (DE3) cells with plasmid containing the DNA encoding YY1-I_N (plasmid pET-YY1-I_N). Transformed cells are plated on LB plate containing kanamycin (25 µg/mL) and chloramphenicol (34 µg/mL) and incubated at 37 °C overnight (see **Note 4**).

2. Resuspend the colonies from two plates in 2 mL of LB and inoculate 1 L of LB containing kanamycin (25 µg/mL) and chloramphenicol (34 µg/mL) in a 2.5 L flask.

3. Grow cells in an orbital shaker incubator at 37 °C for 2–3 h to reach mid-log phase (OD at 600 nm ≈ 0.5). Add IPTG to reach a final concentration of 1 mM and incubate cells for 3 h at 37 °C.

4. Pellet cells by centrifugation at 6000 × g for 15 min at 4 °C. Discard the supernatant and process the pellet immediately (see **Note 5**).

3.4 Purification of YY1-I_N

1. Protein YY1-I_N was purified from inclusion bodies. Resuspend cell pellet with 30 mL of lysis buffer containing 1 mM PMSF. Lyse cells by sonication on ice using 25 s bursts spaced 30 s each (see **Note 6**). Repeat the cycle six times (see **Note 7**).

2. Separate the soluble cell lysate fraction by centrifugation at 15,000 × g for 20 min at 4 °C. Store the pellets at −80 °C in case they need to be reprocessed.

3. Resuspend the insoluble fraction with lysis buffer (30 mL) containing detergent (0.1% Triton X-100). Separate soluble fraction by centrifugation at 15,000 × g for 20 min at 4 °C. Repeat this process two more times.

4. Wash the insoluble fraction with lysis buffer without detergent as described in **step 3**.

5. Extract YY1-I_N with 10 mL of PBS containing 4 M urea (see **Note 8**).

6. Characterize protein purity by SDS-PAGE analysis and determine concentration by UV spectroscopy ($\varepsilon^{280} = 17,180$ M^{-1} cm^{-1}). Use immediately for in vitro *trans*-splicing experiments.

3.5 In Vitro Labeling of YY1 Using Protein Trans-Splicing

1. Pure polypeptide **IC** (Table 1) and YY1-I_N fusion protein are combined in freshly prepared and degassed *trans*-splicing buffer to a concentration of ≈1 μM and 0.1 μM, respectively. Dissolve I_C peptide in pure 30% acetonitrile in pure H_2O containing 0.1% trifluoroacetic acid (TFA) to obtain a stock solution of ≈1 mg/mL (≈160 μM). Add 3.1 μL of I_C stock solution and the appropriate amount of YY1-I_N dissolved in lysis buffer containing 4 M urea (Subheading 3.4, **step 6**) to 500 μL of fresh *trans*-splicing buffer. Keep the reaction at room temperature with occasional gently shaking.

2. Take aliquots (50 μL) at different times (1, 3, 7, 10, 15, 30, and 45 min) and rapidly quench them by adding 15 μL of 4× SDS-PAGE sample buffer and heating them at 94 °C for 2 min.

3. Monitor the progression of the *trans*-splicing reaction by using SDS-PAGE (Fig. 3). Load 25 μL of each time point onto an SDS-4–20% PAGE gel. Run the samples at 125 V for about 1 h and 30 min in 1× SDS running buffer. Remove SDS with pure water. Visualize trans-splice reaction by epifluorescence (e.g., Storm 860 Molecular Imager), and silver stain kit (e.g., Thermo scientific, USA) following manufacturer protocol (Fig. 3) (*see* **Note 9**).

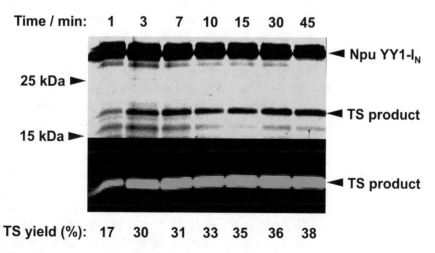

Fig. 3 SDS-PAGE analysis of the in vitro protein *trans*-splicing/labeling reaction between FRET-quenched DnaE I_C (Table 1) and YY1-I_N. Protein detection was performed by silver staining (*top*) and epifluorescence (*bottom*). TS = trans-splicing

3.6 Cloning of DNA Encoding YY1-I$_N$ Construct into Mammalian Expression Vector pcDNA4/TO/myc-His

1. Amplify the DNA encoding the DBD of YY1 fused to the N-terminus of the DnaE I$_N$ (YY1-I$_N$) by PCR using plasmid pET-YY1-I$_N$ (Subheading 3.2) as template using primers p5-YY1-I$_N$ and p3-YY1-I$_N$ (Table 2). The 5′-primer and 3′-primer introduce *KpnI* and *NotI* restriction sites, respectively. Carry out the PCR reaction as follows: 40 μL sterile pure H$_2$O, 1 μL of DNA template (≈10 ng/μL), 5 μL of 10× ThermoPol reaction buffer, 1.0 μL of dNTP solution (10 mM each), 1 μL of p5-YY1-I$_N$ primer solution (0.2 μM), 1 μL of p3-YY1-I$_N$ primer solution (0.2 μM), and 1 μL Vent DNA polymerase (2 units). Use the following PCR cycle conditions: initial denaturation at 94 °C for 5 min followed by 30 cycles (94 °C denaturation for 30 s, annealing at 52 °C for 45 s, and extension at 72 °C for 60 s) and final extension at 72 °C for 10 min.

2. Purify the PCR amplified fragment encoding the YY1-I$_N$ fusion protein using the a PCR purification kit following the manufacturer instructions and quantify it by UV–visible spectroscopy.

3. Digest plasmid pcDNA4/TO/myc-His (Invitrogen) and PCR-amplified gene encoding YY1-I$_N$ with restriction enzymes *NotI* and *KpnI*. Use a 0.5 mL centrifuge tube and add 5 μL of 10× restriction buffer (e.g., NEB buffer 2.1 from New England Biolabs), add enough pure sterile water to have a final volume reaction of 50 μL, add ≈10 μg of the corresponding dsDNA to be digested, add 1 μL (20 units) of restriction enzyme *NotI* and 1 μL (20 units) of restriction enzyme *KpnI*. Incubate at 37 °C for 16 h.

4. Purify the double digested PCR-product and pcDNA4 plasmid by agarose (0.8% and 2% agarose gels for pCDNA4 and PCR product should be used, respectively) gel electrophoresis. Cut out the bands corresponding to the double digested DNA and purify using a gel extraction kit. Elute DNA from the spin columns with TE buffer and quantify using UV–visible spectroscopy.

5. Ligate double digested pcDNA4 and PCR-product encoding YY1-I$_N$. Use a 0.5 mL centrifuge tube, add ≈100 ng of *KpnI*, *NotI*-digested pcDNA4, ≈50 ng of *KpnI*, *NotI*-digested PCR-amplified DNA encoding YY1-I$_N$, enough pure sterile H$_2$O to make a final reaction volume of 20 μL, 2 μL of 10× T4 DNA ligase buffer, 1 μL of 10 mM ATP and 1 μL (400 units) T4 DNA ligase. Incubate at 16 °C overnight.

6. Transform the ligation mixture into DH5α competent cells. Thaw ≈100 μL of chemical competent cells on ice and mix with the ligation mixture (20 μL) for 30 min. Heat-shock the cells at 42 °C for 45 s and then keep on ice for an extra 10 min. Add 900 μL of SOC medium and incubate at 37 °C for 1 h in an orbital shaker. Plate 100 μL on LB agar plate containing ampicillin (100 μg/mL) and incubate the plate at 37 °C overnight.

7. Pick up several colonies (most of the times five colonies should be enough) and inoculate into 5 mL of LB medium containing ampicillin (100 µg/mL). Incubate tubes at 37 °C overnight in an orbital shaker.

8. Pellet down cells and extract DNA using a miniprep kit following the manufacturer protocol and quantify plasmid using UV–visible spectroscopy.

9. Verify the presence of DNA encoding YY1-I_N construct in each colony using PCR. Carry out the PCR reaction out as follows: 40 µL sterile pure H_2O, 1 µL of plasmid DNA ($\approx$50 ng/µL), 5 µL of 10× TaqDNA polymerase buffer, 1.0 µL of dNTP solution (10 mM each), 1 µL primer p5-YY1-I_N solution (0.2 µM), 1 µL of primer p3-YY1-I_N (0.2 µM), and 1 µL Taq DNA polymerase (5 units).

10. Use the following PCR cycle conditions: initial denaturation at 94 °C for 5 min followed by 30 cycles (94 °C denaturation for 30 s, annealing at 56 °C for 45 s, and extension at 72 °C for 60 s) and final extension at 72 °C for 10 min.

3.7 Expression of YY1-I_N Fusion Protein in U2OS and HeLa Cells

1. Grow U2OS and HeLa cells in DMEM containing 10% heat-inactivated FBS, 1% L-glutamine, and 1% penicillin–streptomycin solution in a humidified incubator under 5% CO_2 atmosphere at 37 °C.

2. The transfection conditions and expression conditions should be optimized for each protein and cell line. Grow U2OS and HeLa cells in 100 mm nonpyrogenic sterile polystyrene plates to $\approx$60% confluency.

3. Transiently transfect the cells with plasmid pcDNA4-YY1-I_N (1 µg DNA) using transfection reagent (e.g., Fugene-6) following the manual instructions. The transfected cells are then incubated for 24 h at 37 °C.

4. Expression of YY1-I_N protein is estimated by western-blot analysis (**steps 5–14**).

5. Wash cells twice with PBS. Discard PBS wash. Resuspend cells in 0.2 mL RIPA buffer in the presence of 0.1% SDS, 1 mM PMSF and complete protease cocktail. Use a cell scrapper to dislodge cells.

6. Transfer the cell suspension to 1.5 mL centrifuge tubes and incubate on ice for 10 min.

7. Separate the insoluble and soluble fractions by centrifugation at 15,000 × g in a microcentrifuge at 4 °C for 30 min.

8. Incubate 100 µL of cell lysate with 20 µL of 4× SDS-PAGE sample buffer containing 20% mercaptoethanol. Heat the sample at 94 °C for 2 min.

9. Analyze the cell lysate samples using SDS-PAGE. Load samples (25 µL) onto an SDS-4–20 % PAGE gel. Run the samples at 125 V for about 1 h and 30 min in 1× SDS running buffer.

10. Transfer proteins from PAGE gel to a PVDF membrane using standard protocol.

11. Block membrane with 5 % skim milk in TBST buffer for 1 h at room temperature. Add primary anti-His antibody (e.g., murine anti-His IgG, Invitrogen) (*see* **Note 10**) in TBST buffer containing 5 % skim milk and incubate overnight at 4 °C on an orbital shaker.

12. Wash the membrane three times with TBST buffer for 5 min. Add secondary antibody (e.g., horseradish peroxidase-conjugated anti-murine IgG, Vector Lab) in TBST buffer containing 5 % skim, and incubate at room temperature for 1 h.

13. Wash membrane three times with TBST buffer for 5 min. Develop membrane with ECL mix (Life Technologies) following manufacturer instructions.

14. Visualize bands using a molecular imager system (e.g., Storm 860, Amersham Biosciences) or using an X-ray film (Fig. 4).

3.8 In-Cell Labeling Using Protein Trans-Splicing

1. Grow U2OS and HeLa cells in either 35 mm glass bottom plates (when fluorescence microscopy is required) or 100 mm nonpyrogenic sterile polystyrene plates to 40–50 % confluency (*see* Subheading 3.7, **steps 1–3**) and then transiently transfected with plasmid pcDNA4-YY1-I_N (1 µg DNA) using a transfection reagent (e.g., Fugene-6) following the manual instructions. Incubate transfected cells for 24 h at 37 °C.

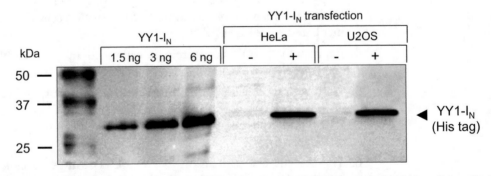

Fig. 4 Expression of protein YY1-I_N in U2OS (and HeLa) cells. Cells were collected after 24 h post-transfection, lysed and the soluble fraction analyzed and quantified by western blot using an anti-His antibody

Fig. 5 (continued) of the labeled-YY1 to the nuclear compartment. (c) Quantification of labeling yield for in-cell *trans*-splicing reaction. Identification of labeled YY1 DBD protein and quantification of in-cell *trans*-splicing yield was performed by western blot (*right panel*) and epifluorescence (*left panel*), respectively. *Bar* represents 25 µm in panels A and B. TS = trans-splicing

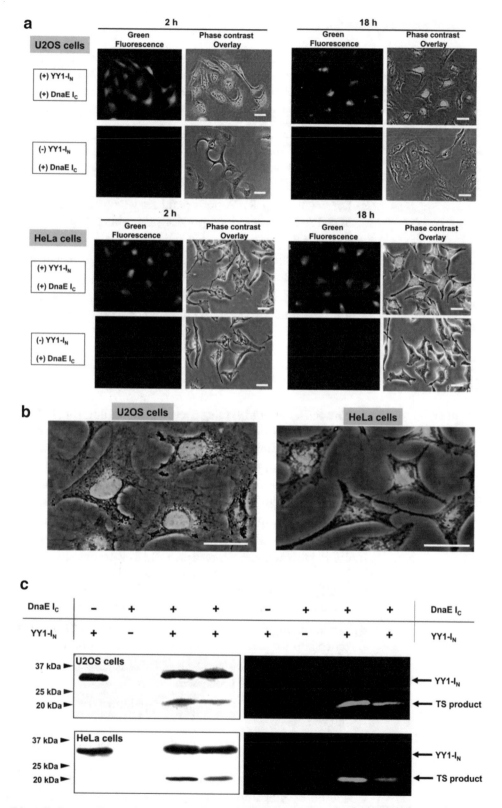

Fig. 5 In-cell site-specific labeling of YY1 DBD with a nuclear localization signal and concomitant fluorescence activation using protein trans-splicing. (**a**) U2OS (and HeLa) cells were first transiently transfected with a plasmid encoding YY1 -I$_N$ and with DnaE I$_C$ polypeptides as described. Cells were then extensively washed and examined by fluorescence microscopy. (**b**) Magnification of cells after 18 h of incubation showing the migration

2. Wash cells with PBS. Transfect peptide I_C (50 nM) using the Chariot protein delivery reagent (Active Motif) for 1 h as described in the manufacturer manual (*see* **Note 11**).

3. Wash transfected cells with full media two times and incubate for 1 h at 37 °C.

4. Wash cells three times with serum free media and observe fluorescence-labeled proteins in live cells using a fluorescence microscope (Fig. 5a, b) (*see* **Note 12**).

5. Lyse cells using RIPA buffer and analyze soluble cell lysate by SDS-PAGE as described previously (Subheading 3.7, **steps 5** through **9**).

6. Visualize fluoresceine-labeled protein on PAGE-gel using a molecular imaging system (Fig. 5c).

7. Use the same PAGE-gel to analyze the efficiency of the labeling reaction by western blotting as described previously (Subheading 3.7, **steps 10** through **14**) (Fig. 5c).

4   Notes

1. The buffer should be degassed before adding the TCEP. The buffer should be prepared fresh every time and not stored as TCEP has a limited stability at pH 7.0 in the presence of air.

2. We recommend to screen at least 3–4 different colonies for protein expression in BL21(DE3) cells to make sure that the polypeptide encoding DnaE I_N is expressed efficiently.

3. We recommend to screen at least 3–4 different colonies for protein expression in Rosetta(DE3) cells to make sure that the clone with highest expression yield is selected. Expression in Rosetta cells is recommended when the protein of interest (POI) contains rare codons in *E. coli*. Select the clone with the highest expression yield and proceed to the next section.

4. When plating the transformed cells with pET-YY1-I_N, it is better to aim for plates containing 200–300 colonies.

5. Cell pellets can be stored at −80 °C for no more than 2–3 weeks before being processed.

6. During sonication, be sure the temperature of the sample does not overheat.

7. A french press can be also used to lyse cells, depending on the availability.

8. Use the highest quality urea available. PBS containing urea should be prepared fresh and not stored for more than 1 day, as urea is not stable in buffers at pH above 7.

9. It is extremely import to check that the labeling/*trans*-splicing reaction works efficiently in vitro before trying it in live cells.

10. In this particular work the N-terminal of the DBD of YY1 contained a His-tag and therefore an anti-His IgG was used to detect the protein by western blotting.

11. Peptide transfection should be optimized for any particular case. In our hands, the best peptide transfection efficiency was obtained for 50 nM I_C using a molecular ratio I_C: Pep-1 of 1:20. Pep-1 is the name of the peptide used in the Chariot transfection system and its sequence is Ac-KETWWETWWTEWSQPKKKRKV-cysteamine. This peptide can be also obtained from different commercial sources (e.g., Anaspec).

12. The use of a FRET-quenched I_C polypeptide allows concomitant labeling of the protein with the corresponding fluorophore and activation of its fluorescence.

References

1. Xie XS, Yu J, Yang WY (2006) Living cells as test tubes. Science 312:228–230

2. Chen I, Ting AY (2005) Site-specific labeling of proteins with small molecules in live cells. Curr Opin Biotechnol 16:35–40

3. Miller LW, Cornish VW (2005) Selective chemical labeling of proteins in living cells. Curr Opin Chem Biol 9:56–61

4. Borra R, Dong D, Elnagar AY, Woldemariam GA, Camarero JA (2012) In-cell fluorescence activation and labeling of proteins mediated by FRET-quenched split inteins. J Am Chem Soc 134:6344–6353

5. Perler FB (2005) Protein splicing mechanisms and applications. IUBMB Life 57:469–476

6. Saleh L, Perler FB (2006) Protein splicing in cis and in trans. Chem Rec 6:183–193

7. Xu MQ, Evans TC Jr (2005) Recent advances in protein splicing: manipulating proteins in vitro and in vivo. Curr Opin Biotechnol 16:440–446

8. Muir TW (2003) Semisynthesis of proteins by expressed protein ligation. Annu Rev Biochem 72:249–289

9. Shi J, Muir TW (2005) Development of a tandem protein trans-splicing system based on native and engineered split inteins. J Am Chem Soc 127:6198–6206

10. Kwon Y, Coleman MA, Camarero JA (2006) Selective immobilization of proteins onto solid supports through split-intein-mediated protein trans-splicing. Angew Chem Int Ed 45:1726–1729

11. Kurpiers T, Mootz HD (2008) Site-specific chemical modification of proteins with a prelabelled cysteine tag using the artificially split Mxe GyrA intein. Chembiochem 9:2317–2325

12. Giriat I, Muir TW (2003) Protein semisynthesis in living cells. J Am Chem Soc 125:7180–7181

13. Woo YH, Stubbs L, Camarero JA (2008) In vivo protein labeling via protein trans-splicing. In: The 21st annual symposium of the protein society, San Diego, p 32

14. Dassa B, Amitai G, Caspi J, Schueler-Furman O, Pietrokovski S (2007) Trans protein splicing of cyanobacterial split inteins in endogenous and exogenous combinations. Biochemistry 46:322–330

15. Zettler J, Schutz V, Mootz HD (2009) The naturally split Npu DnaE intein exhibits an extraordinarily high rate in the protein trans-splicing reaction. FEBS Lett 583:909–914

16. Mootz HD, Blum ES, Tyszkiewicz AB, Muir TW (2003) Conditional protein splicing: a new tool to control protein structure and function in vitro and in vivo. J Am Chem Soc 125:10561–10569

17. Vila-Perello M, Hori Y, Ribo M, Muir TW (2008) Activation of protein splicing by protease- or light-triggered O to N acyl migration. Angew Chem Int Ed 47:7764–7767

18. Berrade L, Kwon Y, Camarero JA (2010) Photomodulation of protein trans-splicing through backbone photocaging of the DnaE split intein. Chembiochem 11:1368–1372

19. Binschik J, Zettler J, Mootz HD (2011) Photocontrol of protein activity mediated by the cleavage reaction of a split intein. Angew Chem Int Ed Engl 50:3249–3252

20. Wu H, Hu Z, Liu XQ (1998) Protein *trans*-splicing by a split intein encoded in a split DnaE gene of *Synechocystis* sp. PCC6803. Proc Natl Acad Sci U S A 95:9226–9231

21. Taki M, Sisido M (2007) Leucyl/phenylalanyl(L/F)-tRNA-protein transferase-mediated aminoacyl transfer of a nonnatural amino acid to the N-terminus of peptides and proteins and subsequent functionalization by bioorthogonal reactions. Biopolymers 88:263–271

22. Oeemig JS, Aranko AS, Djupsjobacka J, Heinamaki K, Iwai H (2009) Solution structure of DnaE intein from Nostoc punctiforme: structural basis for the design of a new split intein suitable for site-specific chemical modification. FEBS Lett 583:1451–1456

23. Wei XY, Sakr S, Li JH, Wang L, Chen WL, Zhang CC (2006) Expression of split dnaE genes and trans-splicing of DnaE intein in the developmental cyanobacterium Anabaena sp. PCC 7120. Res Microbiol 157:227–234

24. Iwai H, Zuger S, Jin J, Tam PH (2006) Highly efficient protein trans-splicing by a naturally split DnaE intein from Nostoc punctiforme. FEBS Lett 580:1853–1858

25. Wang CC, Chen JJ, Yang PC (2006) Multifunctional transcription factor YY1: a therapeutic target in human cancer? Expert Opin Ther Targets 10:253–266

26. Gronroos E, Terentiev AA, Punga T, Ericsson J (2004) YY1 inhibits the activation of the p53 tumor suppressor in response to genotoxic stress. Proc Natl Acad Sci U S A 101:12165–12170

27. Sui G, el Affar B, Shi Y, Brignone C, Wall NR, Yin P, Donohoe M, Luke MP, Calvo D, Grossman SR (2004) Yin Yang 1 is a negative regulator of p53. Cell 117:859–872

Chapter 9

Segmental Isotopic Labeling of Proteins for NMR Study Using Intein Technology

Dongsheng Liu and David Cowburn

Abstract

Segmental isotopic labeling of samples for NMR studies is attractive for large complex biomacromolecular systems, especially for studies of function-related protein–ligand interactions and protein dynamics (Goto and Kay, Curr Opin Struct Biol 10:585–592, 2000; Rosa et al., Molecules (Basel, Switzerland) 18:440, 2013; Hiroaki, Expert Opin Drug Discovery 8:523–536, 2013). Advantages of segmental isotopic labeling include selective examination of specific segment(s) within a protein by NMR, significantly reducing the spectral complexity for large proteins, and allowing for the application of a variety of solution-based NMR strategies. By utilizing intein techniques (Wood and Camarero, J Biol Chem 289:14512–14519, 2014; Paulus, Annu Rev Biochem 69:447–496, 2000), two related approaches can generally be used in the segmental isotopic labeling of proteins: expressed protein ligation (Muir, Annu Rev Biochem 72:249–289, 2003) and protein *trans*-splicing (Shah et al., J Am Chem Soc 134:11338–11341, 2012). Here, we describe general implementation and latest improvements of expressed protein ligation method for the production of segmental isotopic labeled NMR samples.

Key words Expressed protein ligation, Isotopic labeling, Intein, Structural biology, NMR

1 Introduction

Nuclear magnetic resonance (NMR) spectroscopy provides useful and unique information into the structure, dynamics of biological macromolecules, especially multi-domain proteins and their complexes [8]. As the molecular weight of an NMR sample increases, limitations become apparent because of slowed molecular motion leading to fast transverse relaxation, which is associated with line broadening, reduced sensitivity, and because of greater complexity of the spectra from the increased number of spins. A number of these challenges have been addressed through the development of sophisticated pulse schemes [9]. One important and useful approach for the NMR measurement of large protein systems involves sample preparation, using specific and selective isotopic labeling including deuteration. The combination of ^{15}N, ^{13}C

Henning D. Mootz (ed.), *Split Inteins: Methods and Protocols*, Methods in Molecular Biology, vol. 1495,
DOI 10.1007/978-1-4939-6451-2_9, © Springer Science+Business Media New York 2017

triple-resonance spectroscopy with deuteration holds the most promise for addressing the molecular weight limitations currently imposed on structure determination by solution NMR [9]. The segmental labeling of proteins, in which segments of the sequence with arbitrary isotopic composition are separately expressed and recombined, represents a significant step to reduce dramatically the spectral complexity and signal overlaps. Besides these benefits, segmental labeling of part of macromolecules known to be a major epitope for interactions with ligand in a fuller sequence context can significantly reduce the amount of assignment load, while retaining specificity for the complete target protein. Recent advances in the introduction of NMR unobservable solubility enhancement tags show that segmental labeling can provide another method to address poorly expressed or insoluble proteins [10, 11].

Three methods have been implemented to segmental labeling of proteins, EPL (expressed protein ligation) [12], PTS (protein *trans*-splicing) [13] and sortase-mediated ligation [10, 14]. EPL and PTS utilize intein technology, while sortase mediated protein ligation utilizes the sortase A enzyme which specifically recognizes a LPXTG motif on the N-terminal segment [15]. These methods are generally much more efficient and useful than prior splicing methods using reversal of proteolysis (e.g., ref. [16]). All of these segmental labeling methods rely on the expression of two separate protein segments, typically with specific isotope labeling (i.e., ^{15}N, ^{13}C, ^{2}H) on one and the other typically unlabeled, followed by linking the two chains to produce a native peptide bond forming the native protein or a minimally substituted analog. For PTS and sortase mediated ligation, longer linker region or recognition motif are required for *trans*-splicing or enzyme recognition, respectively, i.e., the splice is not 'traceless'. EPL is a protein engineering approach utilizing native chemical ligation (NCL) [17] that allows recombinant or synthetic polypeptides to be chemoselectively and regioselectively joined together. The NCL reaction can proceed in neutral aqueous conditions and internal cysteine residues within both segments are unreacted and thus permitted.

NCL reaction is catalyzed by in situ transthioesterification with thiol additives (Fig. 1). The most common thiol catalysts to date have been sodium 2-mercaptoethanesulfonate (MESNA), or 4-mercaptophenylacetic acid (MPAA). The first step of NCL is the chemoselective transthioesterification of an unprotected protein segment with C-terminal α-thioester and a second segment with an N-terminal Cysteine. The so-formed thioester spontaneously undergoes an S → N-acyl transfer to form a native peptide bond and the resulting peptide product is obtained. The thiol–thioester exchange reaction is reversible while the amide forming step is irreversible. Although the isolated α-thioester and N-terminal Cys segments can be ligated in the presence of many additives such as denaturants, chaotropes, or detergents,

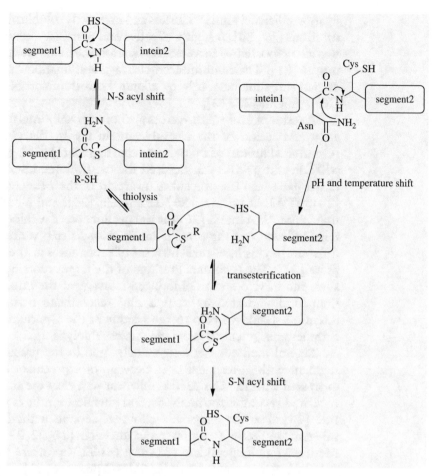

Fig. 1 Schematic drawings of segmental isotopic labeling using expressed protein ligation (EPL). Intein-based fusion approaches allow the straightforward introduction of α-thioester and N-terminal Cys. N-terminal Cys can also be made by cleavage of the leader sequence and chemical synthesis methods. Expressed protein ligation is based on native chemical ligation (NCL), in which one peptide with a C-terminal thioester group and the other peptide with an N-terminal cysteine are ligated in aqueous solution

additional attention should be paid since the reaction rate highly relies on the molar concentration of both precursors. For the segmental labeling of extremely large macromolecules, the ligation speed is generally slow because of the concentration limitations. Special attention should also be paid to the X position which will contain the α-thioester. Rapid ligation rates are observed when the α-thioester residue X is Gly, Cys, or His, while Val, Ile, and Pro represent bad choices for the α-thioester because of the slow ligation rates [18], presumably from conformational restriction. In additional to chemical synthesis of the C-terminal α-thioester peptide, a reactive thioester can also be generated when an appropriate thiol reagent is used to induce cleavage of the intein. MESNA is generally the chosen reagent since it provides higher ligation efficiency, and is odorless and extremely soluble in aqueous solutions [19, 20]. A highly effective and practical catalyst MPAA was also reported to increase the ligation speed an order of magnitude [21]. The thiol-induced cleavage and ligation can also be carried out simultaneously by adding both thiol and N-terminal Cys segments [22, 23].

A series of expression vectors are commercially available which allow the fusion of the cleavable intein tag to the C-terminus (C-terminal fusion) of a target protein (segment1) [24]. pTWIN1, pTXB1, and pTXB3 can be used for the in-frame insertion of a target gene into the polylinker upstream of the *Mxe* GyrA intein [20]. pTWIN2 utilizes *Mth* RIR1 intein [25], and pTYB1,2,3,4 utilize *Sce* VMA intein [26] for the production of α-thioester protein. Efficiency of the protein–intein thiolysis step varies because different inteins have their own set of preferences at the junction point [27]. The C-terminal residue of the target protein is critical since the in vivo and thiol-induced cleavage at the intein can be dramatically affected by it. It is thus recommended that special attention should be paid to the residue at the C-terminus of the α-thioester segment to avoid incomplete thiolysis.

Several methods can be generally used in the preparation of segment with N-terminal Cys: cleavage of a precursor sequence expressed with the Cys residue adjacent to a cleavage site; intein-mediated recombinant methods; and chemical synthesis of a peptide [28]. For the enzymatic cleavage, several methods can be utilized: (1) Factor Xa protease and thrombin [12, 22, 29, 30]; (2) Methionyl aminopeptidase [31–33]; (3) TEV protease [34, 35]; (4) Cyanogen bromide [36]; (5) SUMO protease [37]; (6) Leader peptidase [38]; (7) EnteroKinase [39].

There have been quite a few intein mediated segmental labeling studies of proteins and protein complexes, using EPL as well as PTS as previously reviewed [40, 41]. In particular, the development of in solution NMR techniques made it possible to extend NMR studies to structures in the size range 50–150 kDa and beyond. Recent examples of using NMR studies on segmental labeling techniques include the efficient reduction of spectral complexity in the solid-state NMR spectra by segmental labeling [42]; in an intrinsically disordered protein [43]; with amino acid-selective segmental isotope labeling of multidomain proteins [44]; in the inter-domain interaction and domain orientation studies [45–48]; and in NMR studies of difficult protein targets [49, 50].

The successful assignment of NMR resonances in solid-state NMR is generally hindered by spectral crowding and extreme resonance overlap, both of which increase rapidly with the number of residues. Segmental labeling of the protein has been employed in solid-state NMR sample preparation to facilitate sequence-specific resonance assignment. Tobias Schubeis et al. obtained unambiguous sequence-specific resonance assignments of a repetitive amyloid sequence and demonstrated that all repeats are most likely structurally equivalent [51]. They also showed that the segmentally ^{13}C, ^{15}N-labeled prion domain of HET-s exhibits significantly reduced spectral overlap while retaining the wild-type structure. Intramolecular unambiguous distance restraints can be identified without the need for preparing a dilute sample [42]. Intrinsically disordered proteins are difficult NMR targets because of the severe overlap of resonance arising from residues in the disordered segment in a narrow ^{1}H spectral region. Yuko Nabeshima et al. have produced segmentally isotope-labeled PQBP1 using the EPL method with non-labeled N-segment and $^{13}C/^{15}N$-labeled C-segment and detected a very weak intramolecular interaction between these two segments which might not be detectable if the IDP is completely disordered [43]. These results demonstrated segmental labeling could possibly identify very weak intramolecular interactions. Tandem protein domains and their interaction with their ligand can also be studied with the segmental labeling NMR method. The unambiguous assignment of the data collected from the RRM protein demonstrated that hnRNP A1 RRMs interact in solution, and the relative orientation of the two RRMs observed in solution is different from the crystal structure of free UP1 suggesting crystallization can sometimes confine protein domains in nonfunctional conformations [45]. Segmental isotope labeling is a method of choice to obtain reliable information, in solution, on the relative orientation and on the surface of interaction of different domains of multi-domain proteins [45]. The system was also studied by combining cell-free expression and ligation of the expressed proteins to produce selective amino acid-type segmental labeled multidomain proteins, which could potentially lead to structural investigations of even larger proteins with sizes up to several hundred kDa [44].

Here, we provide a general implementation protocol utilizing pTWIN1 expression vector which was designed for protein purification and for the preparation of proteins with an N-terminal cysteine and/or a C-terminal thioester. This system has been successfully used for segmental isotopic labeling of several different samples [44, 49, 50, 52]. pTWIN1 vector contains DNA of a modified *Ssp* DnaB intein as N-terminal fusion and *Mxe* GyrA intein as C-terminal fusion, which, as the same time, can serve as a solubility enhance tag to difficult protein targets. The fused CBD (chitin binding domain) tag on both N- and C- terminal can facilitate purification using chitin beads.

2   Materials

1. pTWIN1 plasmid (New England Biolabs).
2. Competent *E. coli* cells (DH5α, BL21(DE3), BL21(DE3) RIL, etc.).
3. Plasmid miniprep and PCR purification kits.

4. Proper restriction enzymes, DNAse, RNAse, lysozyme.

5. DNA polymerase.

6. dNTPs.

7. DNA ligase and DNA ligase buffer.

8. Chitin resin (New England Biolabs).

9. Ion exchange column (Q sepharose, SP sepharose).

10. Ni-NTA and/or TALON resin.

11. Dialysis cassette (various MWCO).

12. Milli-Q (Millipore) or double-distilled water.

13. pH meter and paper.

14. Ampicillin sodium sulfate stock solution (AMP): Dissolve ampicillin sodium sulfate to a concentration of 100 mg/mL in distilled water. Syringe filter and store at −80 °C.

15. LB broth and LB agar plates, supplemented with 100 μg/mL ampicillin from AMP stock solution.

16. M9 medium per liter: 6.8 g Na_2HPO_4, 3 g KH_2PO_4, 0.5 g NaCl was added to 950 mL deionized H_2O (for deuteration, D_2O is used). Autoclave to sterilize. Cool down to room temperature and add the following sterile (passed through 0.22 μM filter) ingredients: 2 mL $MgSO_4$(1 M), 1 mL ampicillin(100 mg/mL). 10 mL D-glucose (20% w/v), 5 mL ammonium chloride (20% w/v). Change the carbon and nitrogen source to ^{13}C or ^{15}N labeled feedstock if needed.

17. [U-^{13}C] D-glucose (^{13}C, 99%). Prepare 20% (w/v) stock solution. Sterilize by 0.22 μM filter.

18. Ammonium chloride (^{15}N, 99%). Prepare 20% (w/v) stock solution. Sterilize by 0.22 μM filter.

19. Deuterium oxide (D, 99%).

20. Buffer A: 50 mM Tris–HCl, 200 mM NaCl, pH 7.5.

21. 50 mM phosphate buffer (pH 8.0): Containing Na_2HPO_4 and NaH_2PO_4, pH 8.0.

22. 50 mM phosphate buffer (pH 6.0): Containing Na_2HPO_4 and NaH_2PO_4, pH 6.0.

23. Sodium azide (NaN_3, Highly toxic). The use of 0.04% sodium azide in NMR samples is highly recommended to prevent growth of microorganisms for samples used in multiday experiments. To make 10% stock solution of sodium azide, Dissolve 1 g of sodium azide in 10 mL of D_2O. Filter-sterilize and store at room temperature.

24. DSS(4,4-dimethyl-4-silapentane-1-sulfonate). 10 mM DSS stock solution can be made in D_2O. Final concentration of 10–20 μM were used as an internal chemical shift reference standard.

25. EDTA (ethylenediaminetetraacetic acid). Trace amount (10–100 μM) of EDTA can be added to the NMR buffer to prevent contaminating heavy metals from interference with the protein. Make 0.5 M stock solution, adjust to pH 8.0 with pellets of NaOH.

26. IPTG (isopropyl-β-d-thio-galactopyranoside). Make 1 M stock solution. Dissolve 2.38 g of IPTG in distilled water. Adjust the volume of the solution to 10 mL with distilled water and sterilize by filtration through a 0.22 μm disposable filter. Dispense the solution into 1-mL aliquots and store them at –20 °C.

27. DTT (1, 4-dithiothreitol). Make 1 M stock solution and store at –80 °C.

28. MESNA (sodium 2-mercaptoethanesulfonate). Make 1 M stock solution and store at –80 °C.

29. Electrophoresis system, precast gels, buffer strips, staining kits, and protein molecular weight markers.

30. Coomassie blue-staining solution: 0.25 % (w/v) Coomassie Brilliant Blue, 45 % (v/v) methanol, 10 % (v/v) acetic acid.

31. 5.0 mm standard NMR sample tube.

32. Software: NMRPipe (Delaglio, NIH), NMRViewJ (One Moon Scientific, Inc.), or similar.

2.1 Equipment

1. Thermal cycler for PCR.

2. Orbital shaker incubator.

3. French pressure cell press.

4. Refrigerated centrifuge (various speed).

5. Amicon Ultra centrifugal filter with appropriate molecular weight cutoff.

6. UV-Photometer (e.g., NanoDrop 2000, Thermo Scientific).

7. Water bath.

8. ÄKTA purifier chromatography system (GE Healthcare) or similar.

9. NMR spectrometer(s).

3   Methods

3.1 Cloning Segment1 into Expression Vector

1. Amplify the destination gene with polymerase chain reaction (PCR) using primers 5′-GGT GGT CAT ATG NNN… (segment1_intein2_fw) and 5′-GGT GGT TGC TCT TCC GCA NNN… (segment1_intein2_rev) with *NdeI* and *SapI* cleavage sites (The target gene sequence is represented by "NNN…"). Carry out the PCR reactions in 50 μL final volume with your

DNA polymerase of choice, 0.5 mM of all four dNTPs, 0.2 μM of each oligonucleotide, and 50 ng plasmid DNA using standard cycling conditions (e.g., 30 cycles with annealing at 60 °C and elongation at 72 °C).

2. Purify the PCR product with PCR purification kit.

3. Digest the PCR product as well as the pTWIN1 vector with *NdeI* and *SapI* restriction enzymes.

4. Ligate the purified digestion products overnight using DNA ligase.

5. Transform *E. coli* DH5α competent cells with the ligation product and select on LB agar plates supplemented with 100 μg/mL ampicillin.

6. Pick colonies, isolated plasmids and validate clone by DNA sequencing using T7 universal primer (5′-TAA TAC GAC TCA CTA TAG GG). The corresponding protein product was named segment1-intein2 (*see* **Note 1**).

3.2 Cloning Segment2 into Expression Vector

1. Amplify the destination gene with PCR using primers 5′-GGT GGT TGC TCT TCC AAC TGC NNN … (intein1_ segment2_fw) and 5′-GGT GGT GGA TCC TTA NNN … (intein1_segment2_rev) with *SapI* and *BamHI* cleavage sites (The target gene sequence is represented by "NNN…"). Carry out the PCR reactions in 50 μL final volume with your DNA polymerase of choice, 0.5 mM of all four dNTPs, 0.2 μM of each oligonucleotide, and 50 ng plasmid DNA using standard cycling conditions (e.g., 30 cycles with annealing at 60 °C and elongation at 72 °C).

2. Purify the PCR products with PCR purification kit.

3. Digest the PCR product as well as the pTWIN1 vector with *SapI* and *BamHI* restriction enzyme.

4. Ligate the purified digestion products overnight using DNA ligase.

5. Transform *E. coli* DH5α competent cells with the ligation product and select on LB agar plates supplemented with 100 μg/mL ampicillin.

6. Pick colonies, isolated plasmids and validate clone by DNA sequencing using *Ssp* DnaB forward primer (5′-ACT GGG ACT CCA TCG TTT CT) and T7 terminator reverse primer (5′-TAT GCT AGT TAT TGC TCA G). The corresponding protein product was named intein1-segment2.

3.3 Expression and Purification of the Protein Segments

1. For the expression of segment1-intein2 and intein1-segment2 proteins, transform the plasmids containing the designed DNA sequences into BL21-CodonPlus (DE3)-RIL competent cells (Stratagene). Inoculate a fresh colony harboring the plasmids in

20 mL of LB medium containing 100 µg/mL ampicillin and culture at 37 °C until the optical density (A_{600}) reaches 0.8.

2. Transfer the whole culture to 1 L of the same medium and shake at 37 °C until the optical density (A_{600}) reaches 0.6. Induce with 0.5 mL of 1.0 M IPTG at 30 °C for 4 h.

3. For the isotopic labeling, use M9 media with 0.1 % $^{15}NH_4Cl$ and/or 0.2 % [U-^{13}C] glucose.

4. Collect the cells of both cultures by centrifugation at $4000 \times g$ for 20 min at 4 °C, and then suspend the separate cell pellets in 35 mL of Buffer A and store at −80 °C for future use.

5. Thaw the cells on ice with gentle mixing. To break the cells, pass the cells containing the segment1-intein2 or intein1-segment2 protein through a French pressure cell twice. Transfer each cell lysate to 50 mL centrifuge tubes. Centrifuge the cell lysate at $12,000 \times g$ for 20 min at 4 °C in a conical fixed-angle rotor (*see* **Notes 2, 3** and **4**).

6. Load the clarified cell extract to 6 mL of chitin beads which was pre-equilibrated with Buffer A. Wash the column with 50 mL of Buffer A to remove the unbound protein.

7. For segment1-intein2 protein, induce cleavage of the intein-tag with 100 mM MESNA (the concentration may vary due to the different cleavage efficiency) in 50 mM phosphate buffer, pH 8.0, at room temperature for 24 h. Elute the cleaved protein from the column by 5 mL of Buffer A. Segment1 containing reactive α-thioester can be further purified by ion exchange column chromatography (optional, *see* **Notes 5** and **6**).

8. For intein1-segment2 protein, the cleavage of the intein-tag was induced by equilibrating the chitin beads with 50 mM phosphate buffer, pH 6.0, at room temperature for 24 h. Elute the cleaved protein from the column by 5 mL of Buffer A. Segment2 containing reactive N-Cys can be further purified by ion exchange chromatograpy (optional).

9. Perform SDS-PAGE analysis with all of the collected samples to assess the cleavage procedure and protein purity.

3.4 Protein Ligation and Purification

1. Concentrate the purified segment1 and segment2 proteins to 1 mM using Amicon Ultra centrifugal filter at 4 °C. Choose an appropriate molecular weight cutoff (MWCO) for the nominal molecular weight limit of the protein. Mix the concentrated segment1 and segment2 precursor protein and add 0.5 mM EDTA and 100 mM MESNA to the mixture and keep the ligation mixture at 4 °C for 24–72 h for the protein ligation, use SDS-PAGE to detect the ligation efficiency (for an example, *see* Fig. 2 and **Note 7**).

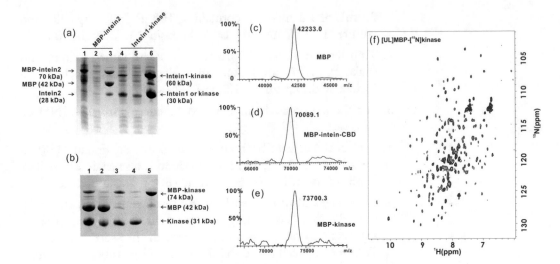

Fig. 2 Example of incorporation of NMR invisible solubility enhancement tag (MBP) to Csk kinase domain through expressed protein ligations. (**a**) and (**b**) SDS-PAGE to monitor the purification, cleavage and ligation procedure. (**c**) to (**e**) MALDI-TOF MASS to identify the right precursor and ligation product. (**f**) ¹H-¹⁵N HSQC spectrum of the ligation product of MBP-kinase, with kinase ¹⁵N labeled

2. The ligation mixture needs to be further purified to remove the unligated protein segments. Purification procedures like IMAC (Immobilized Metal Chelating Chromatography), IEC (Ion Exchange Chromatography), SEC (Size Exclusion Chromatography) on an Äkta system or similar may be utilized. If a protein sample of high purity is needed, the combination of the aforementioned purification steps may be necessary (Fig. 2).

3. Exchange to NMR buffer use Amicon Ultra centrifugal filter at 4 °C. Concentrate to the appropriate final volume of using a centrifugal concentrator, depending on protein solubility. Quantify protein concentration using absorbance at 280 nm (e.g., by using a NanoDrop 2000).

3.5 NMR Spectroscopy

1. Add 10% D₂O, 0.04% (w/v) NaN₃, and 10 μM DSS (4,4-dimethyl-4-silapentane-1-sulfonate) to the segmental labeled sample and transfer it to an NMR tube. NMR samples can be sealed in the NMR tube under a nitrogen atmosphere. Particularly readily oxidized materials may be prepared with solutions flushed with argon, and argon used in a sealed tube.

2. Lock, tune and shim the sample and take a normal 1D water suppression proton spectrum on appropriate spectrometer. Determine the proton spectral region to be used.

3. Acquire an initial ¹H-¹⁵N HSQC NMR spectrum at 25 °C. Pulsed-field gradient techniques with a WATERGATE pulse sequence can be used for H₂O samples, resulting in a good suppression of the solvent signal. In the 2D ¹H-¹⁵N

HSQC or TROSY experiments, 512 complex points can be collected in the ^{1}H dimension and 128 complex points in the ^{15}N dimension. ^{13}C and ^{15}N dimensions can be multiplied by a cosine-bell window function and zero-filled to 512 points using Topspin (Bruker) before Fourier transformation.

4. The ^{1}H chemical shifts can be referenced to internal DSS. The ^{15}N chemical shifts can be referenced indirectly using the ^{1}H/^{15}N frequency ratios of the zero point 0.101329118 (^{15}N) [53, 54].

5. TROSY spectra with sensitivity enhancement and water flip-back provide significant better resolution for larger protein products at 800 MHz or higher. Similarly, TROSY versions of HNCACB, HN(CO)CACB, HNCA, HN(CO)CA experiments are appropriate for assignment in larger systems.

6. Assign the peaks to the corresponding residues and determine chemical shift perturbations of residues for each peak computed from proton and nitrogen chemical shifts according to the following formula $\Delta\delta_{total} = \mathrm{sqrt}([\Delta\delta_H^2 + (0.154\,\Delta\delta_N)^2])$ [55].

4 Notes

1. The general procedure of performing segmental labeling using pTWIN1 vector is outlined in Fig. 3.

2. Insoluble expression of target protein (inclusion body) caused by the misfolding can be avoided by the following trials: decrease the temperature of the cell culture; decrease the amount of IPTG usage; fusion or co-expressing to molecular chaperones or solubility enhancement tags; refolding in vitro.

3. For proteins that are susceptible to in vivo degradation we recommend that the cell growth and protein purification steps be performed in the same day and 4 °C to improve yields.

4. Lyse the cells by methods other than French press will give similar results.

5. Protein purification procedure also depends on any additional fusion tag that is used, as different tags serve different purposes. In our laboratory, we generally apply HisX6 tag in addition to CBD tag.

6. The chitin resin can be regenerated by washing the chitin resin with 10 bed volumes 0.3 M NaOH and rinse with 20 bed volumes of water and stored at 4 °C.

7. Because high protein concentrations are needed for protein ligation, we typically concentrate proteins to a concentration of ~1 mM or higher, depending on protein solubility.

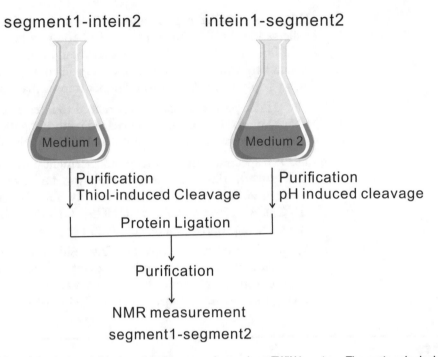

Fig. 3 Outline of the segmental isotope labeling procedure using pTWIN1 system. The system is designed for preparation of proteins with N-terminal cysteine as well as proteins with C-terminal thioester. Two precursors are expressed separately in different medium with different label scheme. After the release of segment1 with C-terminal α-thioester and segment2 with N-terminal Cys, the N-terminal cysteine of segment2 attacks the C-terminal thioester present at segment1. This results in a peptide bond at the site of ligation

Acknowledgments

DL acknowledges ShanghaiTech University and Shanghai Municipal Government for financial support; Supported by NIH grants GM47042 (DC).

References

1. Goto NK, Kay LE (2000) New developments in isotope labeling strategies for protein solution NMR spectroscopy. Curr Opin Struct Biol 10(5):585–592

2. De Rosa L, Russomanno A, Romanelli A, D'Andrea L (2013) Semi-synthesis of labeled proteins for spectroscopic applications. Molecules (Basel, Switzerland) 18(1):440

3. Hiroaki H (2013) Recent applications of isotopic labeling for protein NMR in drug discovery. Expert Opin Drug Discovery 8(5):523–536. doi:10.1517/17460441.2013.779665

4. Wood DW, Camarero JA (2014) Intein applications: from protein purification and labeling to metabolic control methods. J Biol Chem 289(21):14512–14519. doi:10.1074/jbc. R114.552653

5. Paulus H (2000) Protein splicing and related forms of protein autoprocessing. Annu Rev Biochem 69(1):447–496. doi:10.1146/ annurev.biochem.69.1.447

6. Muir TW (2003) Semisynthesis of proteins by expressed protein ligation. Annu Rev Biochem 72(1):249–289. doi:10.1146/annurev. biochem.72.121801.161900

7. Shah NH, Dann GP, Vila-Perello M, Liu ZH, Muir TW (2012) Ultrafast protein splicing is common among cyanobacterial

split inteins: implications for protein engineering. J Am Chem Soc 134(28):11338–11341. doi:10.1021/Ja303226x

8. Wüthrich K (2003) NMR studies of structure and function of biological macromolecules (Nobel Lecture). J Biomol NMR 27(1):13–39. doi:10.1023/a:1024733922459

9. Mittermaier A, Kay LE (2006) New tools provide new insights in NMR studies of protein dynamics. Science 312(5771):224–228. doi:10.1126/science.1124964

10. Kobashigawa Y, Kumeta H, Ogura K, Inagaki F (2009) Attachment of an NMR-invisible solubility enhancement tag using a sortase-mediated protein ligation method. J Biomol NMR 43(3):145–150. doi:10.1007/s10858-008-9296-5

11. Kobayashi H, Swapna GV, Wu KP, Afinogenova Y, Conover K, Mao B, Montelione GT, Inouye M (2012) Segmental isotope labeling of proteins for NMR structural study using a protein S tag for higher expression and solubility. J Biomol NMR 52(4):303–313. doi:10.1007/s10858-012-9610-0

12. Xu R, Ayers B, Cowburn D, Muir TW (1999) Chemical ligation of folded recombinant proteins: segmental isotopic labeling of domains for NMR studies. Proc Natl Acad Sci U S A 96(2):388–393

13. Yamazaki T, Otomo T, Oda N, Kyogoku Y, Uegaki K, Ito N, Ishino Y, Nakamura H (1998) Segmental isotope labeling for protein NMR using peptide splicing. J Am Chem Soc 120(22):5591–5592. doi:10.1021/ja980776o

14. Freiburger L, Sonntag M, Hennig J, Li J, Zou P, Sattler M (2015) Efficient segmental isotope labeling of multi-domain proteins using Sortase A. J Biomol NMR 63(1):1–8. doi:10.1007/s10858-015-9981-0

15. Ton-That H, Liu G, Mazmanian SK, Faull KF, Schneewind O (1999) Purification and characterization of sortase, the transpeptidase that cleaves surface proteins of Staphylococcus aureus at the LPXTG motif. Proc Natl Acad Sci 96(22):12424–12429

16. Kubiak T, Cowburn D (1986) Enzymatic semisynthesis of porcine despentapeptide (B26-30) insulin using unprotected desoctapeptide (B23-30) insulin as a substrate. Model studies. Int J Pept Protein Res 27(5):514

17. Dawson PE, Muir TW, Clark-Lewis I, Kent SB (1994) Synthesis of proteins by native chemical ligation. Science 266(5186):776–779

18. Hackeng TM, Griffin JH, Dawson PE (1999) Protein synthesis by native chemical ligation: expanded scope by using straightforward methodology. Proc Natl Acad Sci U S A 96(18):10068–10073

19. Ayers B, Blaschke UK, Camarero JA, Cotton GJ, Holford M, Muir TW (1999) Introduction of unnatural amino acids into proteins using expressed protein ligation. Biopolymers 51(5):343–354

20. Evans TC Jr, Benner J, Xu MQ (1998) Semisynthesis of cytotoxic proteins using a modified protein splicing element. Protein Sci 7(11):2256–2264. doi:10.1002/pro.5560071103

21. Johnson ECB, Kent SBH (2006) Insights into the mechanism and catalysis of the native chemical ligation reaction. J Am Chem Soc 128(20):6640–6646

22. Camarero JA, Shekhtman A, Campbell EA, Chlenov M, Gruber TM, Bryant DA, Darst SA, Cowburn D, Muir TW (2002) Autoregulation of a bacterial sigma factor explored by using segmental isotopic labeling and NMR. Proc Natl Acad Sci U S A 99(13):8536–8541. doi:10.1073/pnas.132033899

23. Vitali F, Henning A, Oberstrass FC, Hargous Y, Auweter SD, Erat M, Allain FHT (2006) Structure of the two most C-terminal RNA recognition motifs of PTB using segmental isotope labeling. EMBO J 25(1):150–162

24. Xu MQ, Evans TC Jr (2001) Intein-mediated ligation and cyclization of expressed proteins. Methods 24(3):257–277. doi:10.1006/meth.2001.1187

25. Evans TC, Benner J, Xu MQ (1999) The cyclization and polymerization of bacterially expressed proteins using modified self-splicing inteins. J Biol Chem 274(26):18359–18363

26. Chong SR, Xu MQ (1997) Protein splicing of the Saccharomyces cerevisiae VMA intein without the endonuclease motifs. J Biol Chem 272(25):15587–15590

27. Chong SR, Williams KS, Wotkowicz C, Xu MQ (1998) Modulation of protein splicing of the Saccharomyces cerevisiae vacuolar membrane ATPase intein. J Biol Chem 273(17):10567–10577

28. Muralidharan V, Dutta K, Cho J, Vila-Perello M, Raleigh DP, Cowburn D, Muir TW (2006) Solution structure and folding characteristics of the C-terminal SH3 domain of c-Crk-II. Biochemistry 45(29):8874–8884. doi:10.1021/bi060590z

29. Romanelli A, Shekhtman A, Cowburn D, Muir TW (2004) Semisynthesis of a segmental isotopically labeled protein splicing precursor: NMR evidence for an unusual peptide bond at the N-extein-intein junction. Proc Natl Acad Sci U S A 101(17):6397–6402. doi:10.1073/pnas.0306616101

30. Liu D, Xu R, Dutta K, Cowburn D (2008) N-terminal cysteinyl proteins can be prepared using thrombin cleavage. FEBS Lett 582(7):1163–1167. doi:10.1016/j.febslet.2008.02.078

31. Iwai H, Pluckthun A (1999) Circular beta-lactamase: stability enhancement by cyclizing the backbone. FEBS Lett 459(2):166–172

32. Camarero JA, Fushman D, Sato S, Giriat I, Cowburn D, Raleigh DP, Muir TW (2001) Rescuing a destabilized protein fold through backbone cyclization. J Mol Biol 308(5):1045–1062

33. Gentle IE, De Souza DP, Baca M (2004) Direct production of proteins with N-terminal cysteine for site-specific conjugation. Bioconjug Chem 15(3):658–663. doi:10.1021/bc049965o

34. Tolbert TJ, Franke D, Wong CH (2005) A new strategy for glycoprotein synthesis: ligation of synthetic glycopeptides with truncated proteins expressed in E-coli as TEV protease cleavable fusion protein. Bioorg Med Chem 13(3):909–915

35. Tolbert TJ, Wong CH (2004) Conjugation of glycopeptide thioesters to expressed protein fragments: semisynthesis of glycosylated interleukin-2. Methods Mol Biol 283:255–266. doi:10.1385/1-59259-813-7:255

36. Macmillan D, Arham L (2004) Cyanogen bromide cleavage generates fragments suitable for expressed protein and glycoprotein ligation. J Am Chem Soc 126(31):9530–9531. doi:10.1021/ja047855m

37. Weeks SD, Drinker M, Loll PJ (2007) Ligation independent cloning vectors for expression of SUMO fusions. Protein Expr Purif 53(1):40–50. doi:10.1016/j.pep.2006.12.006

38. Hauser PS, Ryan RO (2007) Expressed protein ligation using an N-terminal cysteine containing fragment generated in vivo from a pelB fusion protein. Protein Expr Purif 54(2):227–233

39. Hosfield T, Lu Q (1999) Influence of the amino acid residue downstream of (Asp)(4)Lys on enterokinase cleavage of a fusion protein. Anal Biochem 269(1):10–16

40. Cowburn D, Muir TW (2001) Segmental isotopic labeling using expressed protein ligation. Methods Enzymol 339:41–54

41. Liu D, Xu R, Cowburn D (2009) Segmental isotopic labeling of proteins for nuclear magnetic resonance. Methods Enzymol 462:151–175. doi:10.1016/S0076-6879(09)62008-5

42. Schubeis T, Luhrs T, Ritter C (2015) Unambiguous assignment of short- and long-range structural restraints by solid-state NMR spectroscopy with segmental isotope labeling. Chembiochem 16(1):51–54. doi:10.1002/cbic.201402446

43. Nabeshima Y, Mizuguchi M, Kajiyama A, Okazawa H (2014) Segmental isotope-labeling of the intrinsically disordered protein PQBP1. FEBS Lett 588(24):4583–4589. doi:10.1016/j.febslet.2014.10.028

44. Michel E, Skrisovska L, Wuthrich K, Allain FH (2013) Amino acid-selective segmental isotope labeling of multidomain proteins for structural biology. Chembiochem 14(4):457–466. doi:10.1002/cbic.201200732

45. Barraud P, Allain FH (2013) Solution structure of the two RNA recognition motifs of hnRNP A1 using segmental isotope labeling: how the relative orientation between RRMs influences the nucleic acid binding topology. J Biomol NMR 55(1):119–138. doi:10.1007/s10858-012-9696-4

46. Xue J, Burz DS, Shekhtman A (2012) Segmental labeling to study multidomain proteins. Adv Exp Med Biol 992:17–33. doi:10.1007/978-94-007-4954-2_2

47. Cho JH, Muralidharan V, Vila-Perello M, Raleigh DP, Muir TW, Palmer AG III (2011) Tuning protein autoinhibition by domain destabilization. Nat Struct Mol Biol 18(5):550–555. doi:10.1038/nsmb.2039

48. Chen J, Wang J (2011) A segmental labeling strategy for unambiguous determination of domain-domain interactions of large multi-domain proteins. J Biomol NMR 50(4):403–410. doi:10.1007/s10858-011-9526-0

49. Minato Y, Ueda T, Machiyama A, Shimada I, Iwai H (2012) Segmental isotopic labeling of a 140 kDa dimeric multi-domain protein CheA from Escherichia coli by expressed protein ligation and protein trans-splicing. J Biomol NMR 53(3):191–207. doi:10.1007/s10858-012-9628-3

50. Xu R, Liu D, Cowburn D (2012) Abl kinase constructs expressed in bacteria: facilitation of structural and functional studies including segmental labeling by expressed protein ligation. Mol Biosyst 8(7):1878–1885. doi:10.1039/c2mb25051a

51. Schubeis T, Yuan P, Ahmed M, Nagaraj M, van Rossum BJ, Ritter C (2015) Untangling a repetitive amyloid sequence: correlating biofilm-derived and segmentally labeled curli fimbriae by solid-state NMR spectroscopy. Angew Chem Int Ed Engl 54:14669–14672. doi:10.1002/anie.201506772

52. Clerico EM, Zhuravleva A, Smock RG, Gierasch LM (2010) Segmental isotopic labeling of the Hsp70 molecular chaperone DnaK using expressed protein ligation. Biopolymers 94(6):742–752. doi:10.1002/bip.21426

53. Live DH, Davis DG, Agosta WC, Cowburn D (1984) Long-range hydrogen-bond mediated effects in peptides - N-15 Nmr-study of gramicidin-S in water and organic-solvents. J Am Chem Soc 106(7):1939–1941

54. Wishart DS, Bigam CG, Yao J, Abildgaard F, Dyson HJ, Oldfield E, Markley JL, Sykes BD (1995) 1H, 13C and 15N chemical shift referencing in biomolecular NMR. J Biomol NMR 6(2):135–140

55. Mulder FA, Schipper D, Bott R, Boelens R (1999) Altered flexibility in the substrate-binding site of related native and engineered high-alkaline Bacillus subtilisins. J Mol Biol 292(1):111–123

Chapter 10

Segmental Isotope Labeling of Insoluble Proteins for Solid-State NMR by Protein Trans-Splicing

Tobias Schubeis, Madhu Nagaraj, and Christiane Ritter

Abstract

Solid-state NMR spectroscopy (ssNMR) is uniquely suited for atomic-resolution structural investigations of large protein assemblies, which are notoriously difficult to study due to their insoluble and non-crystalline nature. However, assignment ambiguities because of limited resolution and spectral crowding are currently major hurdles that quickly increase with the length of the polypeptide chain. The line widths of ssNMR signals are independent of proteins size, making segmental isotope labeling a powerful approach to overcome this limitation. It allows a scalable reduction of signal overlap, aids the assignment of repetitive amino acid sequences, and can be easily combined with other selective isotope labeling strategies. Here we present a detailed protocol for segmental isotope labeling of insoluble proteins using protein *trans*-splicing. Our protocol exploits the ability of many insoluble proteins, such as amyloid fibrils, to fold correctly under in vitro conditions. In combination with the robust *trans*-splicing efficiency of the intein DnaE from *Nostoc punctiforme*, this allows for high yields of segmentally labeled protein required for ssNMR analysis.

Key words Amyloid, Intein, Protein complexes, Protein *trans*-splicing, Isotopic labeling, Protein structures, Solid-state NMR spectroscopy

1 Introduction

Magic-angle spinning solid-state nuclear magnetic resonance spectroscopy (ssNMR) has evolved into a powerful tool for investigating the structure and dynamics of large proteins and protein assemblies such as amyloid fibrils or membrane proteins [1–4]. Large macromolecular assemblies are typically inaccessible to X-ray crystallography or solution NMR due to their insoluble nature. However, despite the impressive progress made in ssNMR instrumentation and experimental techniques over the recent years, sensitivity, resolution and spectral crowding remain serious challenges for the sequence-specific resonance assignment and for the unambiguous identification of structurally relevant cross-peaks. The problem of spectral overlap quickly increases with increasing molecular weight of the protein (and hence the number of amino

Henning D. Mootz (ed.), *Split Inteins: Methods and Protocols*, Methods in Molecular Biology, vol. 1495, DOI 10.1007/978-1-4939-6451-2_10, © Springer Science+Business Media New York 2017

acid residues to be studied). This is a limiting factor for uniformly $^{13}C,^{15}N$-labeled samples, which intrinsically would contain the maximum amount of information. Therefore, it has become common practice to employ selective labeling with specific amino acids or sparse labeling with metabolic precursors such as 2-^{13}C-glycerol or 1-^{13}C-glucose [5]. However, an even more rigorous, and more controllable, reduction of spectral crowding can be achieved by labeling only a specific segment within a full-length protein [6]. This is achieved by ligation of an isotope labeled with an unlabeled part of the protein of interest by expressed protein ligation (EPL), protein *trans*-splicing (PTS), or sortase-mediated ligation [7, 8]. Segmental labeling techniques preserve the native size and biophysical characteristics of the protein. To simplify the spectral information even further, segmental and specific labeling strategies can be easily combined, as both approaches rely on the recombinant expression of proteins in *Escherichia coli* grown in minimal media. Especially for β-solenoid type amyloid fibrils, which assemble via a repetitive stacking of β-strands, an additional advantage of segmental isotope labeling is the facilitated differentiation between inter- and intramolecular cross-peaks.

The natural split intein DnaE from *Nostoc punctiforme (Npu)* has been shown to provide decisive advantages for the segmental labeling of proteins for solution NMR [9]. These include a general acceptance of residues adjacent to the splicing junction, enabling an almost traceless reaction. In addition, this intein has a very high reaction rate even in the presence of high concentrations of chaotropic agents [10, 11]. These features make *Npu* DnaE ideally suited for the segmental labeling of insoluble proteins that are purified from *E. coli* as unstructured precursors or under denaturing conditions [12, 13]. This is typically the case for amyloidogenic proteins, and efficient protocols for the purification from inclusion bodies and subsequent refolding have also been presented for other proteins, such as different classes of membrane proteins [14, 15]. Once a refolding protocol is established, expression and purification in inclusion bodies can actually be beneficial since very high yields can be obtained and degradation by proteases is almost entirely suppressed under denaturing conditions. Denaturation and refolding furthermore facilitates amid proton back exchange in case of deuteration. The protocol presented here can be applied to all proteins with an established in vitro folding/refolding protocol. The simplicity of the method in fact relies on repetitive folding and denaturation. It needs to be noted that high-quality, homogenously folded samples are of prime importance for ssNMR spectra. For amyloidogenic proteins, it often helps to add preformed fibrils as seeds, but detailed protocols need to be established for every protein of interest individually.

Here we focus on labeling of either the N- or the C- terminal fragments, but labeling of a central fragment has also been demonstrated [16]. The decision where to split the protein of interest (POI) is usually guided either by the interest in the structure, dynamics, or interactions of a specific protein domain, or by aiming at stepwise coverage of the full protein to complete resonance assignments. A traceless reaction can be achieved if a natural cysteine is chosen as the splicing junction. Otherwise, a specific residue needs to be selected that will be replaced by the catalytic cysteine of the C-terminal intein fragment (defined as the IntC +1 position). Ideal are amino acids with similar size and charge like serine or threonine, but a single cysteine mutation is generally well tolerated, especially on the protein surface or if a loop region has been selected. In addition, the IntC +2 position can have a strong influence on the splicing efficiency of *Npu*DnaE (Fig. 1a). While phenylalanine is the native residue, splicing activity has been demonstrated for the majority of +2 amino acids, albeit with slower kinetics or over all reduced efficiency. Good choices are other aromatic (Trp, His) or bulky (Met) amino acids. Proline and glycine should be avoided [10, 17]. In practice the analysis of the splicing reaction as well as the product purification is simplified if the intein fusion proteins (termed POI-n-IntN and IntC-POI-c) and the splicing product all differ in molecular weight. A short protocol for generating traceless expression plasmids by blunt end cloning is provided in this chapter.

Many intein fusion constructs of insoluble proteins will form inclusion bodies during protein expression in *E. coli*. We thus recommend protein purification under denaturing conditions using hexa-histidine affinity tags (Fig. 1b). The His_6-tags should be placed on the C-terminus of the IntN domain and N-terminus of the IntC domain to generate a tag free splicing product, which forms the basis of our purification strategy (Fig. 1b).

Elaborate refolding of the purified intein fusion constructs is not required, as the split intein domains are almost entirely unstructured before binding to each other [18]. Thus, the splicing reaction can be directly initiated by adjusting the denaturant concentration low enough for maximal *Npu* DnaE activity, but high enough to keep the fusion constructs soluble. Purification of the correctly spliced, segmentally labeled POI from the reaction mix is straightforward and independent of the aggregation state of all proteins involved (Fig. 1c) as it employs reversed affinity chromatography under denaturing conditions. Because of the placement of the His_6-tags, all intein-containing educts, reaction by-products or products are removed from the tag-free POI. A final size exclusion chromatography under denaturing conditions is recommended to further improve ssNMR sample quality.

a   -1
IntN   X (CLSYETEILTVEYGLLPIGKIVEKRIECTVYSVDNNGNIYTQPVAQWHDRG
          EQEVFEYCLEDGSLIRATKDHKFMTVDGQMLPIDEIFERELDLMRVDNLPN)

                                                          +1+2
b   IntC   (MIKIATRKYLGKQNVYDIGVERDHNFALKNGFIASN) C X

1.) Intein Fusion Protein
    - Expression
     - Cell lysis under denaturing conditions
     - Purification by IMAC
     - Refolding
2.) *In vitro* Protein *trans-* Splicing

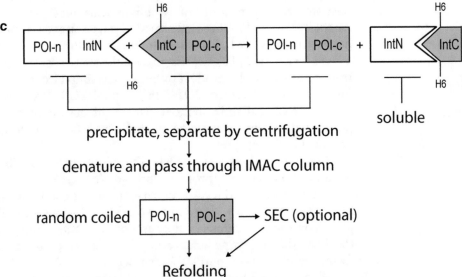

Fig. 1 Primary structure of the split intein DnaE of *Nostoc punctiforme* and the adjacent residues that can have an influence on the splicing efficiency (**a**), Overview of the preparative steps towards the in vitro protein *trans*-splicing reaction (**b**) and graphical representation of the splicing and purification strategy used for insoluble proteins (**c**)

2   Materials

2.1   Cloning of Intein Fusion Protein Expression Vectors

1. A DNA template containing the gene encoding the protein of interest (POIn-c).

2. DNA templates containing the genes encoding the *Npu* DnaE split intein (IntN/IntC) (e.g., available at Addgene).

3. A target vector suitable for recombinant protein expression, here pET28a (Novagen).

4. A total of eight primers are needed for two expression vectors coding for POI-n-IntN and IntC-POI-c, respectively, which

should be designed as follows: A POI-n forward primer that introduces an NcoI restriction site and a POI-n blunt end reverse primer that is 5′ phosphorylated. An IntN blunt end forward primer and an IntN reverse primer that introduces an XhoI restriction site. Furthermore an IntC forward primer that introduces an NdeI restriction site and an IntC blunt end reverse primer that is 5′ phosphorylated. Finally a POI-c blunt end forward primer and a POI-c reverse primer that introduces a stop codon followed by a restriction site (e.g., XhoI).

5. A high fidelity DNA polymerase, typically Pfx or Phusion (ThermoFisher), and its specific buffers and dNTP mix.

6. Standard restriction enzymes and their related buffers, here NdeI, NcoI, and XhoI.

7. T4 DNA ligase and its related buffer.

8. *E. coli* DH5α chemically competent cells (or another standard cloning strain), prepared in house or purchased.

9. LB medium, 1% tryptone, 0.5% yeast extract, 1% NaCl, pH 7.4, and LB agar plates, both supplemented with 50 μg/mL kanamycin.

10. A robust nucleic acid extraction and purification kit, here the NucleoSpin kit (Macherey-Nagel).

11. Standard equipment, consumables and chemicals for routine molecular biology techniques including PCR amplification of DNA fragments, DNA separation, visualization, and concentration determination as well as *E. coli* culturing.

2.2 Expression of Intein Fusion Proteins

1. *E. coli* BL21(DE3) chemically competent cells (or another standard protein expression strain), prepared in house or purchased.

2. LB medium, 1% tryptone, 0.5% yeast extract, 1% NaCl, pH 7.4, and LB agar plates, both supplemented with 50 μg/mL kanamycin.

3. 100× MEM vitamin mix, 100 mg/L choline chloride, 100 mg/L d-calcium pantothenate, 100 mg/L folic acid, 100 mg/L nicotinamide, 100 mg/L pyridoxal hydrochloride, 10 mg/L riboflavin, 100 mg/L thiamine hydrochloride, 200 mg/L i-inositol, 8.5 g/L sodium chloride. Sterile filtered or purchased as sterile solution.

4. 5000× trace metal solution, 50 mM $FeCl_3$, 20 mM $CaCl_2$, 10 mM $MnCl_2$, 10 mM $ZnSO_4$, 2 mM $CoCl_2$, 2 mM $CuCl_2$, 2 mM $NiCl_2$, 2 mM Na_2MoO_4, 2 mM Na_2SeO_3, 2 mM H_3BO_3, 60 mM HCl. Sterile filtered.

5. Minimal medium CN040, 50 mM Na_2HPO_4, 50 mM KH_2PO_4, 5 mM Na_2SO_4, 2 mM $MgSO_4$; after autoclaving add 1× MEM vitamin mix, 1× trace metal solution, 4 g/L $^{13}C_6$-glucose,

1 g/L $^{15}NH_4Cl$ (or other isotopes as needed), 50 µg/mL kanamycin. Equipment for *E. coli* culturing (baffled 3 L Fernbach flasks, shaking incubator), OD_{600} measurements, harvesting by centrifugation.

2.3 Fusion Protein Purification and Refolding

1. Buffer A, 6 M guanidine hydrochloride (GuHCl), 50 mM potassium phosphate, pH 7.4.

2. Buffer B, 6 M GuHCl, 50 mM potassium phosphate, 20 mM imidazole, pH 7.4.

3. Buffer C, 6 M GuHCl, 50 mM potassium phosphate, 300 mM imidazole, pH 7.4.

4. Ni-NTA gravity flow resin.

5. Empty gravity flow columns (e.g., EconoPac from Bio-Rad).

2.4 Protein *trans*-Splicing and Purification of the Splicing Product

1. Splicing buffer. 30 mM Tris-HCl, 150 mM NaCl, 0.5 M urea (freshly prepared), 1 mM EDTA, pH 7.8.

2. Dialysis tubing, with a molecular weight cutoff well below (approx. twofold) the molecular weight of the fusion constructs.

3. 100 mM Tris(2-carboxyethyl)phosphine (TCEP), in water, pH adjusted with 1 M NaOH to 7.2–8.0.

4. Buffer D, 50 mM potassium phosphate, pH 7.4.

5. Superdex 200 prep grade (e.g., GE Healthcare) size exclusion chromatography column.

6. FPLC chromatography system, e.g., Äkta series (GE Healthcare) or BioLogic series (Bio-Rad).

7. Centrifugal concentrators (e.g., VivaSpin, Vivaproducts).

2.5 Refolding and NMR Sample Preparation

1. HiPrep 26/10 desalting column (e.g., GE Healthcare) or similar.

2. Buffer D, 50 mM potassium phosphate, pH 7.4.

3. Ultracentrifuge with swing bucket rotor.

4. ZrO_2 Magic Angle Spinning rotor (e.g., Cortecnet).

5. Rotor filling device for ultracentrifuge (e.g., Giotto Biotech).

3   Methods

3.1 Cloning of Plasmids for the Expression of POI-Intein Fusion Constructs

To obtain native-like segmentally labeled protein, the intein domains need to be directly fused to the protein of interest without introducing additional residues. Traditional cloning therefore involves a blunt end ligation step to not introduce additional bases between the fusion protein coding sequences. Since it is probably still the most widely used technique, we included a short section on this topic. Other options are plasmid preparation by ligation independent cloning or direct gene synthesis.

1. Amplify the genes for POI-n, POI-c, IntN, and IntC by polymerase chain reaction (PCR) using the designated primers under standard conditions and purify the product.

2. To analyze the result run 1–2 μL on an agarose gel and estimate the concentration of the PCR product by comparison to a marker band.

3. Perform a DNA ligation in a total volume of 20 μL containing 50–100 ng PCR products POI-n + IntN and POI-c + IntC, respectively, 2 μL 10× ligation buffer, and 0.5 μL T4 DNA ligase for 1 h at room temperature. Heat-inactivate the ligase (65 °C, 10 min).

4. Use 1 μL of the ligation mix as template for a second PCR to amplify specifically the correct gene length using the primers POI-n forward/IntN reverse and IntC forward/POI-c reverse. Purify the PCR product and check for correct size and purity on an agarose gel.

5. Digest the POI-n/IntN PCR product with NcoI and XhoI. Digest the IntC/POI-c PCR product with NdeI and XhoI. Digest two aliquots of the destination vector (here pET28a) with the respective enzyme pairs.

6. Perform two standard sticky end ligations using 25 ng vector DNA and 22 ng PCR product (for a 1000 base pair insert).

7. Transform 50 μL chemically competent DH5α with max. 5 μL ligation mix by applying a heat shock (42 °C for 45 s), add 450 μL LB medium and incubate at 37 °C for 1 h. Pellet the cells for 1 min at 3000×g and discard 400 μL of the supernatant. Resuspend the cells in the remaining 100 μL and spread on LB-agar plates containing 50 μg/mL Kanamycin. Incubate at 37 °C overnight.

8. Screen for positive colonies using PCR.

9. Grow 5 mL cultures of positive colonies in LB medium containing 50 μg/mL kanamycin at 37 °C overnight.

10. Extract the plasmid DNA and verify the cloning success by sequencing.

3.2 Intein Fusion Protein Expression and Purification

All the protein purification steps mentioned below are concerned with amyloidogenic and other hydrophobic, aggregation-prone proteins and are carried out at room temperature unless stated otherwise. The steps explained below involve isotope labeling of IntN-fusion with protein of interest (POI-n-IntN) and natural abundance labeling of IntC-fusion with protein of interest (POI-c-IntC).

1. The POI-n-IntN and IntC-POI-c fusion construct plasmids are freshly transformed into *E. coli* BL21 (DE3) cells and grown at 37 °C for 14–16 h on LB agar plates containing 50 μg/mL kanamycin. To make a preculture, inoculate LB liquid medium with a single colony and incubate at 37 °C overnight.

2. Optimize the conditions for highest yields of protein expression in inclusion bodies. Always perform the optimization with the medium you are going to use (i.e., both LB and CN040). A critical parameter is the culture temperature post induction. We routinely test 37, 30, and 20 °C. Other parameters that can have an influence are the IPTG concentration and the length of induction. As an initial test inoculate three small cultures (100 mL) with the preculture to a starting OD_{600} of 0.05–0.1. A starting OD_{600} of 0.1 is recommended for minimal medium. Incubate at 37 °C to an OD_{600} of 0.9. Keep one culture at 37 °C bring the other two to 30 and 20 °C. Induce the cultures at 37 and 30 °C with 500 μM IPTG and the 20 °C culture with 200 μM IPTG. Monitor the OD_{600} and take several samples for SDS-PAGE analysis (*see* **Notes 1** and **2**).

3. Grow and induce the main cultures under optimal conditions. Use CN040 minimal medium for isotope enrichment (*see* **Note 3**). Harvest cells by centrifugation at (6000×*g*, 15 min, 20 °C), resuspend in Buffer A and stir overnight. To increase the cell lysis efficiency, overnight stirring is followed by pulsed sonication for 30 min, 70 % amplitude and continued stirring for 1–4 h (*see* **Notes 4** and **5**).

4. Centrifuge at 36,000×*g* for 30 min to remove the cell debris and incubate the supernatant at least for 1 h with Ni-NTA beads equilibrated with Buffer A (*see* **Note 6**). Wash with 20 column volumes Buffer B. Pour the suspension into a gravity flow column and elute the His-tagged protein with 5×1 column volume Buffer C (*see* **Notes 7** and **8**). Check the flow-through for residual POI by SDS PAGE and if necessary incubate once more with Ni-NTA beads, wash and elute again (*see* **Note 9**).

5. Refold the purified intein fusion constructs by removing the denaturant and imidazole through dialysis against 100 times the volume of splicing buffer (*see* **Note 10**).

6. Ultracentrifuge the protein at 60,000×*g* for 30 min at 4 °C to remove any aggregated protein. Filtration through 0.22 μm might be sufficient if none or only minor aggregation is visible and the filter does not get clogged by unfolded POI. Pool the clear supernatant and store on ice before setting up for the protein *trans*-splicing reaction.

3.3 **In Vitro Protein** *Trans*-**Splicing**

1. Mix the POI-n-IntN and IntC-POI-c fusion proteins at equimolar concentration. Adjust the protein concentration to 5–15 μM by dilution with dialysis buffer to avoid nonspecific aggregation of the fusion proteins (*see* **Note 11**).

2. Initiate protein *trans*-splicing by the addition of TCEP to a final concentration of 0.5 mM (*see* **Note 12**).

3. Take samples for SDS-PAGE at several time points (e.g., 10 min, 30 min, 1 h, 3 h, 6 h, 16 h) to analyze the reaction progress (*see* **Note 13**). Incubate at room temperature with gentle agitation (avoid frothing) for up to 48 h to allow sufficient splicing as well as aggregation of the spliced POI. At an interval of e.g., 24 h spin down the splicing mixture to remove the spliced or any other aggregated proteins by spinning at $6000 \times g$ for 15 min (*see* **Note 14**) and replenish the supernatant with 0.5 mM TCEP. This may help to further improve the yield. An SDS-PAGE analysis of the *trans*-splicing reaction and the purification of the product is exemplified in Fig. 2.

4. After a steady state has been reached, collect all aggregated protein by centrifugation. The obtained protein pellet usually contains the splicing product as well as aggregated intein fusion proteins and splicing by-products. Check for completeness of the precipitation by SDS PAGE (*see* **Notes 15** and **16**).

3.4 Purification of Spliced Protein

1. Wash the protein pellet with 50 mM potassium phosphate buffer, pH 7.4, by resuspending and spinning down the pellet at $6000 \times g$ for 15 min.

2. To purify the splicing product, dissolve the pellet in a sufficient volume of Buffer A until no aggregates are visible and adjust the TCEP concentration to 0.5 mM (*see* **Note 17**).

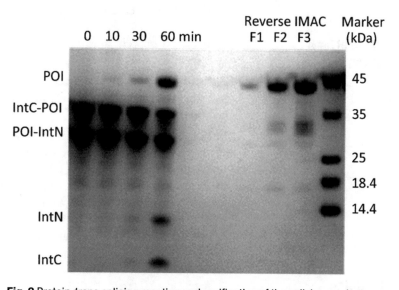

Fig. 2 Protein *trans*-splicing reaction and purification of the splicing product analyzed by SDS-PAGE. Samples were taken at several points during the reaction. Appearance of three additional bands over time (one product, and two Intein bands) is a criteria to evaluate the splicing efficiency. The splicing product was purified by passage through a Ni²⁺ charged IMAC column connected to an FPLC system using continues fractionation (F1, 2, 3)

3. To remove the intein and non-reacted educts, incubate the solution with equilibrated Ni-NTA beads, pass through a gravity flow column and wash with one column volume Buffer A. The tag-free splicing product can be found in the flow-through. Repeat if the purity needs improvement.

4. To remove preformed aggregates or smaller splicing by-products perform a size exclusion chromatography under denaturing conditions, e.g., in Buffer A containing 0.5 mM TCEP at room temperature. Pure and monomeric splicing product should be obtained in denaturing conditions after this step.

3.5 Protein Refolding

A refolding protocol needs to be established for every protein of interest individually. The steps described here are a simple starting point for amyloidogenic proteins.

1. The protein concentration after the size exclusion column may be rather dilute. If necessary concentrate the pooled fractions of monomeric splicing product (*see* **Note 18**).

2. Desalt over a Sephadex G-25 column into 50 mM potassium phosphate buffer, pH 7.4, containing 0.005% sodium azide. Use a 26/10 column for loading volumes up to 15 mL. Allow to form aggregates with gentle agitation for at least 2 days at room temperature. For amyloids, adding a preformed fibrillar seed often helps to achieve higher sample homogeneity. If you need to load you samples in multiple runs, it is sufficient to add seed to the first eluate.

3.6 NMR Rotor Packing

1. Collect the protein aggregates by centrifugation ($14,000 \times g$) and wash the pellet several times with pure water.

2. For efficient packing, transfer aggregates into an MAS rotor using an ultracentrifuge device in a swing out centrifuge rotor at $100,000 \times g$ (*see* **Note 19**).

3. Perform regular solid-state NMR experiments to validate the sample homogeneity and the successful implementation of the segmental labeling scheme [19]. Most commonly used experiments are one-dimensional ^{13}C spectra and two dimensional ^{13}C-^{13}C correlation spectra. Figure 3 shows extracts of the Cα region of 2D spectra recorded on uniformly and segmentally labeled HET-s (218–289) amyloid fibrils.

4   Notes

1. A high expression yield is especially important for the protein that is planned to be labeled. Expression levels in minimal medium are usually lower than in full medium. Perform a Western Blot analysis with an anti His-tag antibody in case no

overexpression is detectable with regular SDS-PAGE. If the expression levels of *Npu* IntC fusion proteins turn out to be low, the addition of an N-terminal solubility tag in the expression vector should be considered (e.g., Trx, Smt3). For higher yields of unlabeled protein, you may try terrific broth (TB: 12 g tryptone, 24 g yeast extract, 4 mL glycerol, 2.31 g KH_2PO_4, and 12.54 g K_2HPO_4 in 1 L H_2O) instead of LB.

2. We strongly advice to always work with freshly transformed *E. coli* expression strains, and to provide optimal aeration during growth. Should the protein be difficult to express, it may even be considered to avoid any interruption of growth, i.e., to grow the starter culture during the day (about 4–6 h), directly followed by the main culture and induction (often overnight is ok here).

3. If the intein fusion constructs expression is low then often the cell density before induction can be increased by growing the bacteria in unlabeled medium, harvesting them gently by centrifugation (10 min, $5000 \times g$, 20 °C) at on OD_{600} of 0.9, and resuspension in a quarter of the original volume isotope labeled CN040. This helps to start induction at high OD in the minimal medium with cells at the log phase.

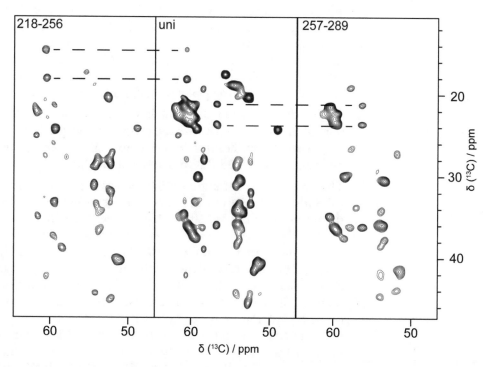

Fig. 3 Extracts of the Cα region of 2D ^{13}C-^{13}C solid-state NMR spectra recorded on uniformly (middle) and segmentally labeled HET-s (218–289) amyloid fibrils

4. A certain amount of intein fusion protein might be expressed as soluble protein in the cytoplasm. This strategy guaranties the purification of the entire expression yield. Cell lysis under native conditions followed by differential centrifugation should be applied in case of solely soluble or insoluble expression.

5. A common alternative to 6 M GuHCl are 8 M urea buffers. Prolonged incubation of proteins in urea, or the use of old urea stock solutions, might lead to irreversible modifications which can interfere with the refolding process and protein function. Furthermore the denaturing power of urea might be too low for certain amyloid forming proteins. In cases of severely insoluble proteins, 8 M GuHCl (keep at room temperature) or 5 M guanidine thiocyanate may be tried.

6. For optimal purity, avoid excessive amounts of Ni-NTA resin. Try 3–5 mL for well-expressing proteins.

7. The Ni-NTA binding is carried out in a 50 mL conical screw-cap centrifuge tube (e.g., Falcon) which is then assembled on a rotating wheel for incubation. It is then centrifuged at $2000 \times g$ for 2 min to pellet the beads. The pellet is washed and centrifuged again using the abovementioned conditions. Centrifugation should always be carried out at maximum acceleration and the lowest deceleration.

8. Keep each elution fraction separated initially, and test for protein content by SDS PAGE. GuHCl would cause poor electrophoretic performance, and needs to be removed by precipitation of the protein with ten sample volumes 50 % ice cold ethanol for at least 1 h, better overnight, at –20 °C. Spin down for 10 min in a tabletop centrifuge, wash once with ice cold ethanol, and resuspend the pellet in SDS PAGE sample buffer.

9. The binding efficiency might be lower in 6 M GuHCl. A substantial amount of recombinant protein could get lost when only a single purification round is applied. Check purity of both elutions by SDS PAGE before combining.

10. The protein concentration during dialysis should be kept low (<50 µM). The dialysis buffer is stirred slowly to slow down the diffusion of the initially highly concentrated GuHCl from the sample across the dialysis membrane, which might lead to nonspecific aggregation. Care should be taken that the dialysis tubing is filled no more than half the maximal volume, because during dialysis the volume of the protein solution will almost double. Alternatives to dialysis are fast buffer exchange using a desalting column or refolding on the Ni-NTA column. The most suitable method needs to be determined for each POI individually. The folding of the POI domains at this point is not relevant. The only requirement is to reestablish the intein

activity, which requires the fusion constructs to be soluble. Should aggregation of the precursor fusion constructs occur, the urea concentration during refolding can be further increased (we typically found 0.5–1 M urea to be sufficient).

11. The reaction is not concentration depended. Concentrations between 1 and 30 μM should work equally well. Nevertheless the stability of the fusion proteins is decisive for the splicing yield, which is usually increased at low concentrations.

12. The reaction is autocatalytic, no further energy source like ATP is required. The reduction of cysteins which are involved in the reaction is the only requirement. TCEP is the reagent of choice because of its stability and non-nucleophilic character. Other reducing agents like DTT or β-mercaptoethanol work as well but can lead to a high degree of incorrect intein cleavage by-products.

13. The optimal reaction length needs to be adjusted. Under optimal conditions the reaction is quantitative and finalized within 1 h, but may well take longer for suboptimal residues at IntC +2 position. In addition, the reduction of cysteins might destabilize the intein fusion proteins so that aggregation events compete with the splicing reaction.

14. Some proteins require higher g-forces to precipitate, e.g., thin filaments. You may have to use ultracentrifugation in this case.

15. Many amyloidogenic proteins aggregate after splicing, making them easy to separate from the rest of the intein proteins by centrifugation.

16. If the splicing product remains in solution, move on to the purification as soon as a steady state has been reached.

17. Low concentrations of TCEP usually do not reduce Ni^{2+} loaded IMAC columns. Reducing condition at this step are used to avoid unwanted disulfide bridge formation between the splicing product and precursors and the consecutive binding to the column.

18. A PES membrane concentrator is used for concentrating the flow-through from the reverse column and pooled sample from gel filtration. 6 M GuHCl is highly viscous, so the concentrating process is carried out at room temperature to fasten the concentrating process.

19. Using a packing tool improves sensitivity by getting more labeled protein in to the NMR rotor and also helps in uniform packing of the hydrated protein pellet/fibers.

References

1. Comellas G, Rienstra CM (2013) Protein structure determination by magic-angle spinning solid-state NMR, and insights into the formation, structure, and stability of amyloid fibrils. Annu Rev Biophys 42:515–536. doi:10.1146/annurev-biophys-083012-130356

2. Schütz AK, Vagt T, Huber M et al (2015) Atomic-resolution three-dimensional structure of amyloid β fibrils bearing the Osaka mutation. Angew Chem Int Ed Engl 54:331–335. doi:10.1002/anie.201408598

3. Wang S, Munro RA, Shi L et al (2013) Solid-state NMR spectroscopy structure determination of a lipid-embedded heptahelical membrane protein. Nat Methods 10:1007–1012. doi:10.1038/nmeth.2635

4. Loquet A, Sgourakis NG, Gupta R et al (2012) Atomic model of the type III secretion system needle. Nature 486:276–279. doi:10.1038/nature11079

5. Verardi R, Traaseth NJ, Masterson LR et al (2012) Isotope labeling for solution and solid-state NMR spectroscopy of membrane proteins. Adv Exp Med Biol 992:35–62. doi:10.1007/978-94-007-4954-2_3

6. Volkmann G, Iwaï H (2010) Protein trans-splicing and its use in structural biology: opportunities and limitations. Mol Biosyst 6:2110–2121. doi:10.1039/c0mb00034e

7. Minato Y, Ueda T, Machiyama A et al (2012) Segmental isotopic labeling of a 140 kDa dimeric multi-domain protein CheA from Escherichia coli by expressed protein ligation and protein trans-splicing. J Biomol NMR 53:191–207. doi:10.1007/s10858-012-9628-3

8. Freiburger L, Sonntag M, Hennig J et al (2015) Efficient segmental isotope labeling of multi-domain proteins using Sortase A. J Biomol NMR 63:1–8. doi:10.1007/s10858-015-9981-0

9. Muona M, Aranko AS, Raulinaitis V, Iwaï H (2010) Segmental isotopic labeling of multi-domain and fusion proteins by protein trans-splicing in vivo and in vitro. Nat Protoc 5:574–587. doi:10.1038/nprot.2009.240

10. Iwai H, Züger S, Jin J, Tam P-H (2006) Highly efficient protein trans-splicing by a naturally split DnaE intein from Nostoc punctiforme. FEBS Lett 580:1853–1858. doi:10.1016/j.febslet.2006.02.045

11. Zettler J, Schütz V, Mootz HD (2009) The naturally split Npu DnaE intein exhibits an extraordinarily high rate in the protein trans-splicing reaction. FEBS Lett 583:909–914. doi:10.1016/j.febslet.2009.02.003

12. Schubeis T, Lührs T, Ritter C (2015) Unambiguous assignment of short- and long-range structural restraints by solid-state NMR spectroscopy with segmental isotope labeling. Chembiochem 16:51–54. doi:10.1002/cbic.201402446

13. Schubeis T, Yuan P, Ahmed M et al (2015) Untangling a repetitive amyloid sequence: correlating biofilm-derived and segmentally labeled curli fimbriae by solid-state nmr spectroscopy. Angew Chem Int Ed Engl 54:14669–72. doi:10.1002/anie.201506772

14. Park SH, Casagrande F, Chu M et al (2012) Optimization of purification and refolding of the human chemokine receptor CXCR1 improves the stability of proteoliposomes for structure determination. Biochim Biophys Acta 1818:584–591. doi:10.1016/j.bbamem.2011.10.008

15. Michalke K, Huyghe C, Lichière J et al (2010) Mammalian G protein-coupled receptor expression in Escherichia coli: II. Refolding and biophysical characterization of mouse cannabinoid receptor 1 and human parathyroid hormone receptor 1. Anal Biochem 401:74–80. doi:10.1016/j.ab.2010.02.017

16. Otomo T, Ito N, Kyogoku Y, Yamazaki T (1999) NMR observation of selected segments in a larger protein: central-segment isotope labeling through intein-mediated ligation. Biochemistry 38:16040–16044

17. Cheriyan M, Pedamallu CS, Tori K, Perler F (2013) Faster protein splicing with the Nostoc punctiforme DnaE intein using non-native extein residues. J Biol Chem 288:6202–6211. doi:10.1074/jbc.M112.433094

18. Shah NH, Eryilmaz E, Cowburn D, Muir TW (2013) Naturally split inteins assemble through a "capture and collapse" mechanism. J Am Chem Soc 135:18673–18681. doi:10.1021/ja4104364

19. Shi L, Ladizhansky V (2012) Magic angle spinning solid-state NMR experiments for structural characterization of proteins. Methods Mol Biol 895:153–165. doi:10.1007/978-1-61779-927-3_12

Chapter 11

Split-Intein Triggered Protein Hydrogels

Miguel A. Ramirez and Zhilei Chen

Abstract

Proteins are nature's building blocks and indispensable in living organisms. Protein-based hydrogels have a wide variety of applications in research and biotechnology. In this chapter, we describe an intein-mediated protein hydrogel that utilizes two synthetic soluble protein block copolymers, each containing a subunit of a trimeric protein that serves as a cross-linker and one half of the naturally split DnaE intein from *Nostoc punctiforme*. Mixing of these two protein block copolymers initiates an intein *trans*-splicing reaction that constitutes a self-assembling polypeptide flanked by cross-linkers, triggering protein hydrogel formation. The generated hydrogels are highly stable under both acidic and basic conditions, and at temperatures up to 50 °C. In addition, these hydrogels are able to undergo rapid reassembly after shear-induced rupture. Incorporation of an appropriate binding motif into the protein block copolymers enables the convenient site-specific incorporation of functional globular proteins into the hydrogel network.

Key words Split-intein, Self-assembly, Shear-thinning, Enzyme, Immobilization

1 Introduction

Protein-based hydrogels have emerged as an important biomaterial in scientific research and biotechnology with applications including tissue engineering, drug delivery, and biotransformation [1–4]. Protein hydrogels composed of designed artificial proteins programed to form molecular networks offer important advantages in the engineering of dynamic biomaterials systems [1, 3]. Hydrogel cross-links can be either chemical or physical. Although chemical hydrogels are typically more stable, most chemical cross-linkers are toxic and give rise to non-injectable gels. Physical hydrogels self-assemble via non-covalent interactions and many are shear-thinning and injectable, an important criterion for noninvasive cell and drug delivery. Physical hydrogels include mixing-induced two-component protein hydrogels in which gelation is triggered by an interaction/reaction between two artificial proteins, forming a cross-linked protein network or hydrogel [5–7]. Unfortunately, physical hydrogels often suffer from poor solution stability due to the non-covalent interactions [8, 9].

Henning D. Mootz (ed.), *Split Inteins: Methods and Protocols*, Methods in Molecular Biology, vol. 1495,
DOI 10.1007/978-1-4939-6451-2_11, © Springer Science+Business Media New York 2017

Inteins are protein splicing elements that are able to excise themselves out of precursor proteins via the protein splicing reaction and have emerged as an important protein engineering tool in numerous and diverse biotechnological applications. Split inteins have the unique ability to join two separate polypeptides into a single functional protein. In this work, we describe a split-intein-catalyzed protein hydrogel as a general scaffold for enzyme immobilization [10]. Two synthetic protein block copolymers, CutA-NpuN (**N**) and NpuC-S-CutA (**C**) (Fig. 1, Table 1) are constructed. NpuN and NpuC are the N- and C-fragment of the naturally split DnaE intein from *Nostoc punctiforme* (Npu) [11, 12] (*see* **Note 1**). CutA is a small trimeric protein (12 kDa) from *Pyrococcus horikoshii* [13, 14] and functions as the hydrogel cross-linker (*see* **Note 2**). The S fragment is a flexible polyanionic linker consisting of the amino acid sequence $[(AG)_3PEG]_{10}$ [15, 16] and was incorporated as the midblock for water retention. Mixing of the

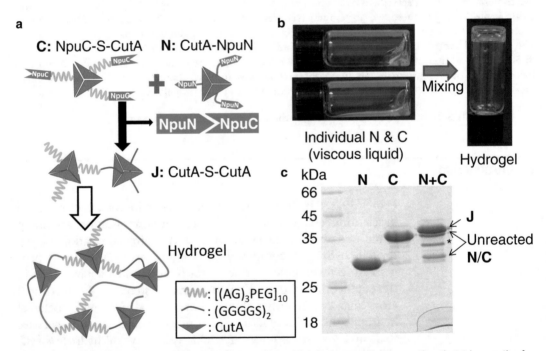

Fig. 1 Intein-mediated protein hydrogel. (**a**) Schematics of intein *trans*-splicing reaction that triggers the formation of an extended protein chain (J) with cross-linker proteins at both termini. Cross-linker proteins from multiple J protein chains non-covalently associate and, upon intein-mediated protein ligation, induce the formation of a highly cross-linked protein network with hydrogel properties. NpuN/C: intein N-/C-fragment. (**b**) Mixing of purified **N** and **C** (8.3 % w/v) yields the formation of a highly cross-linked hydrogel network (1.6 mM J). (**c**) SDS-PAGE analysis of purified **N** and **C** building blocks before and after mixing. "N + C" corresponds to a sample taken directly from a 1.6 mM hydrogel. *Asterisk* denotes an intein C-terminal cleavage side reaction product. Intensity of each band was quantified using the Trace module in the Quantity One software. Band intensity was divided by the protein molecular weight to obtain the molar equivalent. *Trans*-splicing efficiency (~80 %) was calculated from the amount of the product J and the unreacted N/C in the same lane. Reprinted with permission from the Journal of American Chemical Society (DOI 10.1021/ja401075s)

Table 1
Protein constructs used in this study

Short name	Protein sequence	MW (kDa)
CutA-NpuN (**N**)	*CutA*-EAC-(GGGGS)$_2$-AS-*NpuN*-6xH	26.3
NpuC-S-CutA (**C**)	*NpuC*-CFNKLYRDPMG-[(AG)$_3$PEG]$_{10}$-ARMPYV-*CutA*-6xH	26.1
NpuC-S-SH3$_{lig}$-CutA (**C-SH3**$_{lig}$)	*NpuC*-CFNKLYRDPMG-[(AG)$_3$PEG]$_{10}$-ARMPYVGS-PPPALPPKRRR-(GGGGS)$_2$-AS-*CutA*-6xH	28.3
SH3-GFP	*SH3*-KL-(GGGGS)$_2$-AS-*GFP*-6xH	34.5

In "Protein sequence column": *Italics*—Names of genes; <u>*Underlined*</u>—SH3$_{lig}$ amino acid sequence; all others: amino acid one-letter codes

protein block copolymers **N** and **C** under reducing conditions enables *trans*-splicing between NpuN and NpuC and generates a longer protein chain (**J**) containing the cross-linker protein CutA at both termini. Cross-linkers from multiple **J** interact with each other, forming a highly cross-linked network or hydrogel (Fig. 1a).

Our intein-triggered protein hydrogel exhibits high solution-stability with minimal erosion observed following immersion in saline buffer at room temperature for 3 months and over a wide range of pH (6–10) and temperatures (4–50 °C). These hydrogels can also rapidly regain their structural integrity after shear-thinning, making them injectable. We further demonstrated the facile immobilization of functional enzymes onto the hydrogel via the insertion of a "docking-station peptide" within a hydrogel building block and subsequent non-covalent docking of "docking-peptide"-fused model protein (GFP) onto the docking-station peptide. In this chapter, we provide protocols for the expression and purification of the intein-containing protein building blocks, formation of the protein hydrogel and its characterization by rheology.

2   Materials

All buffers should be prepared using double-deionized water (ddH$_2$O, for example prepared by purifying deionized water through a MilliQ purification system).

2.1 Protein Expression Components

1. Expression host: *E. coli* BL21 (DE3) chemically competent cells in 50 μL aliquots stored at –80 °C.

2. Luria–Bertani (LB) broth: 25 g LB powder per liter. Autoclaved.

3. 50 mg/mL kanamycin (Kan): Filtered through 0.45 μm sterile filter, aliquoted at 1 mL per tube, and stored at –20 °C.

4. LB-agar plates supplemented with 50 μg/mL Kan.

5. Plasmids (Table 1): pNpuC-S-CutA; pCutA-NpuN; pNpuC-S-SH3$_{lig}$-CutA and pSH3-GFP encoding for proteins NpuC-S-CutA (**S**); CutA-NpuN (**N**); NpuC-S-SH3$_{lig}$-CutA (**C**-SH3$_{lig}$) and SH3-GFP, respectively (*see* **Note 3**).

6. 1 M isopropyl-β-d-1-thiogalactopyranoside (IPTG), filtered through a 0.45 μm sterile filter, aliquoted at 1 mL per tube and stored at −20 °C.

2.2 Protein Purification

1. Sonicator to disrupt *E. coli* cells.

2. Dulbecco's phosphate-buffered saline (DPBS), pH 7.4.

3. DPBSD buffer: DPBS buffer supplemented with 2 mM DTT.

4. 1 M Dithiothreitol (DTT): Filter through a 0.22 μm filter, aliquot at 100 μL per tube and store at −20 °C.

5. Buffer A: 500 mM NaCl, 10 mM Tris–HCl, pH 8.0.

6. Buffer DA: 500 mM NaCl, 8 M urea, 10 mM Tris–HCl, pH 8.0.

7. Buffer CA: 500 mM NaCl, 50 mM sodium phosphate, pH 6.0.

8. HisTrap HP column, 5 mL column volume (GE Healthcare Life Sciences).

9. Low-pessure liquid chromatography system (e.g., BioLogic, Bio-Rad).

10. Amicon Ultra 0.5-mL centrifugal filters (10,000 NMWL).

11. Amicon Ultra 15-mL centrifugal filters (10,000 NMWL).

12. 2× SDS sample buffer: 0.5 M Tris–HCl, pH 6.8, 20% glycerol, 10% SDS, 0.1% w/v bromophenol blue, 2% β-mercaptoethanol.

2.3 Hydrogel Formation

1. 5% NaN$_3$ solution in water (*see* **Note 4**).

2. 2 mL clear glass vials with caps.

3. Dulbecco's phosphate-buffered saline (DPBS) pH 7.4.

4. 1 M DTT.

2.4 Hydrogel Characterization

1. Paar-Physica MCR-300 parallel plate rheometer (Anton Paar, Ashland, VA) with a 25 mm plate fixture (PP25) (Graz, Austria).

2. Mineral Oil.

3 Methods

3.1 Protein Expression

1. For each fusion protein (Table 1), transform 50 μL chemically competent BL21 (DE3) cells with 1 ng of plasmid, plate the cells on LB agar plates containing 50 μg/mL Kan.

2. Incubate plates at 37 °C for 12–15 h (*see* **Note 5**).

3. Add 10 mL of LB medium to the plate and thoroughly resuspend all the colonies by repeatedly pipetting up and down.

4. Transfer the suspension of colonies to a flask containing 1 L of LB broth supplemented with 50 µg/mL Kan and incubate the culture at 37 °C with shaking at 250 rpm.

5. Monitor the growth of the culture. When the culture reaches $OD_{600} \sim 0.8$, remove the flask from incubator. Rapidly cool the culture to ~18 °C in an ice–water bath (*see* **Note 6**), and then add IPTG to a final concentration of 1 mM. Incubate the culture at 18 °C, 250 rpm for 14 h.

6. Harvest cells by centrifugation at $6000 \times g$ for 20 min. The bacterial pellets can be stored for several months at −80 °C.

3.2 Protein Purification

1. Thaw *E. coli* pellets on ice and resuspend the pellet in 10 mL buffer per gram of wet pellet. Use Buffer A for **N** and SH3-GFP, and Buffer CA for **C** and C-SH3$_{lig}$ (*see* **Note 7**).

2. Lyse cells in an ice–water bath by sonication. We used the following settings with a microtip probe: Amp 10, 1-s pulse and 5-s pause for 3 min.

3. Centrifuge the lysate at $16,000 \times g$ for 20 min at 4 °C.

4. For the purification of **N** (in the insoluble fraction): Discard soluble fraction and resuspend the pellet in Buffer DA. Carry out an additional round of centrifugation at $16,000 \times g$ for 20 min at 4 °C. Skip this step for the purification of **C**, C-SH3$_{lig}$ and SH3-GFP.

5. Filter the supernatant through a 0.45-µm filter. Load the filtrate onto a 5-mL HisTrap HP column previously equilibrated with the appropriate buffer using the BioLogic Low-Pressure Liquid Chromatography system (*see* **Note 8**). Use Buffer DA for **N**, Buffer A for SH3-GFP and Buffer CA for **C** and C-SH3$_{lig}$.

6. Wash column with ten column volumes (CV) of the appropriate buffers containing 45 mM imidazole. Use Buffer DA for **N**, Buffer A for SH3-GFP and Buffer CA for **C** and C-SH3$_{lig}$.

7. Elute proteins using 4 CV of buffer containing 150 mM imidazole. Use Buffer DA for **N** and Buffer A for all the other proteins.

8. Determine protein purity via SDS-PAGE. To facilitate gel analysis, the concentration of NaCl in purified **N**, **C** and C-SH3$_{lig}$ needs to be reduced to <1 mM (*see* **Note 9**). All buffers contain 500 mM NaCl. To reduce the NaCl concentration, dilute the purified samples (10–20 µg) in 500 µL of ddH$_2$O and transfer the solution to an Amicon Ultra 0.5-mL Centrifugal Filter (10,000 NMWL). Centrifuge at $13,000 \times g$ until the protein volume is 10–20 µL (typically 5 min). Refill the column with ddH$_2$O and centrifuge again until the volume is 10–20 µL. Two rounds of this dilution/centrifugation is sufficient to reduce the NaCl concentration in the sample to <1 mM.

9. Exchange the buffer for the remaining protein to DPBSD buffer and concentrate the protein using an Amicon Ultra 15-mL Centrifugal Filter (**10,000 NMWL**) (*see* **Note 10**). To completely exchange the buffer (remove urea and imidazole), we typically perform four rounds of concentration/dilution in which the protein sample (15 mL) is first supplemented with 2 mM DTT, concentrated to ~1 mL by centrifugation at $2800 \times g$ and 4 °C, and then diluted with ~14 mL of DPBSD (*see* **Note 11**). During the last round of centrifugation, the protein is concentrated to ~100 mg/mL (*see* **Note 12**) and stored at –80 °C in 100 µL aliquots.

3.3 Hydrogel Synthesis

Gelation is triggered by the *trans*-splicing reaction between NpuN and NpuC in blocks **N** and **C**. Purified and concentrated **N** and **C** are viscous liquids (Fig. 1b) and mixing of **N** and **C** triggers the formation of a protein hydrogel. The minimum concentration of each building block protein needed for hydrogel formation is 0.8 mM (4.2% w/v). The hydrogel appears soft and semitransparent (Fig. 1b).

For stable hydrogel formation it is critical that the two hydrogel building blocks are mixed at a 1:1 M ratio. Any excess building blocks will affect the hydrogel structural integrity. Since each batch of protein will have a different purity, the volumetric ratio of the building blocks needs to be determined empirically as follows.

1. Calculate the theoretical total protein concentration by the absorbance at 280 nm and the predicted extinction coefficient (**N**: 1.7; **C**: 1.5; **C**-SH3lig: 1.4; SH3-GFP: 1.0).

2. Mix 10 µL of **C** (100 µM) with different amounts of **N** in DPBS supplemented with 5 mM DTT and adjust the reaction volume to 20 µL. The molar ratio of N to C in each tube should range from 0.5:1 to 2:1. Incubate the reactions at room temperature overnight.

3. The next day, analyze the reactions via SDS-PAGE. The optimal ratio should result in a similar amount of unreacted **N** and **C**.

4. Inside a 2 mL glass vial, mix the appropriate amount of **C** with **N** in DPBS supplemented with NaN_3 (final 0.25%) and DTT (final 2 mM) (*see* **Notes 13** and **14**). Hydrogels with total protein concentrations as low as 0.8 mM (42 mg/mL) can be obtained using this technique. The volume of the hydrogel needs to be at least 100 µL to completely cover the bottom of a 2 mL glass vial.

5. Mix thoroughly using a pipette tip or toothpick. Do not pipette up and down since the protein mixture rapidly becomes viscous upon mixing. Close the lid tightly to minimize the evaporation of DTT.

6. Remove air bubbles by briefly centrifuging the tube at $2800 \times g$. The intein *trans*-splicing reactions will proceed at room

temperature overnight (between 12 and 16 h), yielding a stable hydrogel the following day.

7. To confirm gel formation, perform a "vial gravity test" by turning the glass vial upside down. Hydrogels will not flow down the wall (Fig. 1b).

3.4 Hydrogel Rheology Characterization

Rheological studies are performed to elucidate the mechanical properties of the hydrogel. We used a Paar-Physica MCR-300 rheometer (Anton Paar, Ashland, VA) with a 25 mm parallel plate fixture (PP25).

1. Prepare sufficient hydrogel building blocks (protein **N** and **C**) for the synthesis of a 250-µL hydrogel at 1.6 mM of each building block (~83 mg/mL or 8.3% w/v). Calculate the volume needed for each component including **N**, **C**, NaN$_3$, DTT and DPBS buffer.

2. Add the appropriate amount of NaN$_3$ and DTT to DPBS solution and transfer the mixture to the center of the plate. The final concentration of NaN$_3$ and DTT in the solution, after accounting for the volume of **N** and **C**, should be 0.25% and 2 mM, respectively (*see* **Note 15**).

3. Add the appropriate amount of **C** and **N** to the rheometer plate. Mix well using a toothpick in a swirling motion (1–2 min).

4. Lower the top plate fixture (PP25) to a measuring gap of 0.2 mm.

5. Carefully remove excess protein around the fixture using a pipette tip avoiding physical contact with the plate.

6. Carefully seal the edge of the fixture with 1 mL mineral oil to minimize drying of the hydrogel.

7. Monitor the gelation process by measuring the G' and G" of the protein mixture at 1% strain and 10 rad/s frequency at room temperature. The gel typically fully cures after 24 h.

8. Upon gelation, the rheological properties of the hydrogel can be determined (Fig. 2c) (*see* **Note 16**).

4 Notes

1. We chose the Npu intein because at the time of our study, this intein exhibited the fastest reaction kinetics ($t_{1/2} = 63$ s) and very high *trans*-splicing yield (75–85%) [11]. Since our study, other ultra-fast split inteins have been reported [17, 18] and they can potentially replace the Npu intein for hydrogel synthesis.

2. In principle. any multimeric protein may be used as the hydrogel cross-linker. CutA was selected as the hydrogel cross-linker for our study because there is a large number of intra- and

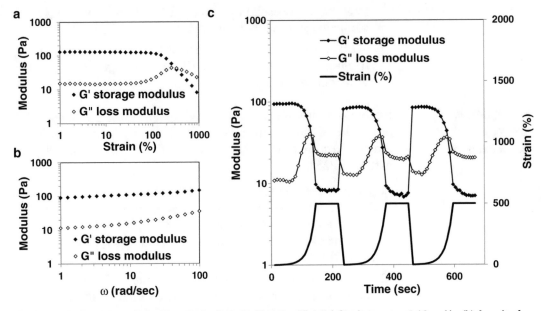

Fig. 2 Rheological characterization of a hydrogel with 1.6 mM J. (**a**) Strain sweep at 10 rad/s. (**b**) Angular frequency sweep at 10 % strain. (**c**) LAOS cycles at 10 rad/s. Reprinted with permission from the Journal of American Chemical Society (DOI 10.1021/ja401075s)

inter-subunit ionic pairs within a CutA trimer, yielding an extensive ion-pair network and contributing to the ultrahigh stability of the CutA trimer. The CutA trimer has an extremely high denaturation temperature of almost 150 °C and retains its quaternary structure in solutions with as much as 5 M GuHCl. This property of CutA confers CutA-mediated protein hydrogels with tremendous stability, as CutA-linked hydrogels are much less likely to undergo erosion caused by intermolecular closed loop formation at the surface of hydrogel. Proteins in closed loops quickly disengage from the hydrogel bulk and diffuse into the solution [8, 16].

3. The plasmids depicted in Table 1: CutA-NpuN (**N**); NpuC-S-CutA (**C**); NpuC-S-SH3$_{lig}$-CutA (**C**-SH3$_{lig}$); SH3-GFP are available upon request from the author Zhilei Chen's lab.

4. NaN$_3$ is a highly toxic compound. We typically prepare a 5 % w/v NaN$_3$ stock solution in water.

5. Plate different dilutions of the culture and use a plate containing ~100 colonies for the subsequent protein expression. Overcrowding the plate and prolonged incubation (>16 h) may reduce the protein expression yield.

6. To ensure even and rapid cooling, swirl the culture every 2 min. For a 1-L culture, it typically takes 6–8 min for the culture to cool from 37 °C to ~18 °C. Expression of proteins **C** and **C**-SH3$_{lig}$ can also be carried out with 1 mM IPTG at 37 °C for 4 h without affecting the protein yield.

7. We use Buffer CA (pH 6.0) instead of Buffer A (pH 8.0) for the purification of all NpuC-fusion proteins to minimize the proteolytic degradation of NpuC fusions by cellular proteases. After purification, the proteins are eluted in Buffer A.

8. If FPLC is not available, Ni-NTA agarose beads can be used. In that case, incubate the appropriate amount of beads with lysate at room temperature for 20 min or 4 °C overnight. Dispense the slurry into a gravity column and proceed with the purification using the same buffers.

9. CutA is a thermostable trimeric protein that is able to mostly maintain its trimeric form in an SDS-PAGE gel even after boiling in SDS-containing buffer. The CutA trimer in our fusion proteins appears to be less stable than the native CutA, yielding a mixture of trimers, dimers and monomers on a SDS-PAGE gel after boiling in SDS buffer, making it difficult to assess the sample purity. We found that reducing the NaCl concentration in the protein sample to <1 mM can abolish most of the CutA trimers in our fusion proteins after boiling in SDS-containing buffer, facilitating gel analysis.

10. The reducing agent DTT is needed to prevent inter-protein disulfide bond formation during buffer exchange. In the absence of DTT, purified **N** and **C** interlocks during buffer exchange in the Amicon column, forming a hydrogel-like material which clogs the centrifugal filters.

11. Buffer exchange may also be performed using conventional dialysis tubing with the same molecular-weight cutoffs.

12. The time needed for each centrifugation step varies depending on the protein concentration. It typically takes ~45 min to concentrate 15 mL of 4 mg/mL protein to 60 mg/mL, and ~30 min to concentrate 1 mL of 60 mg/mL protein to 100 mg/mL.

13. If a SH3-tagged protein (e.g., SH3-GFP) is to be immobilized, first mix SH3-GFP with C-SH3$_{lig}$ at a 1:1 M ratio by thorough pipetting and incubate the mixture at room temperature for 30 min to allow interaction between SH3 and SH3$_{lig}$. After incubation, transfer this mixture to a glass vial containing DPBS, DTT and NaN$_3$, and then add the appropriate amount of **N**.

14. The immobilization of a multimeric protein fused to SH3 may further increase the hydrogel storage modulus due to its ability to bridge different hydrogel building blocks.

15. Since overnight incubation is typically required for the hydrogel to fully cure, we included NaN$_3$ in the solution to minimize microbial contamination which can degrade the hydrogel backbone and prevent gelation.

16. Rheological studies can be used to monitor the viscoelastic properties of the hydrogel. For a hydrogel, the storage modulus (G') is higher than the loss modulus (G''). A viscous liquid has a higher G'' than G'. Figure 2a represents a strain sweep at constant frequency of 10 rad/s. This study shows that at values from 0 to ~200% strain, the material behaves like a gel. At 500% strain, the gel loses its elasticity and behaves like a viscous liquid. Figure 2b is a frequency sweep in which a constant 10% strain is applied. Figure 2c is the large-amplitude oscillatory shear analysis (LAOS) in which G' and G'' are measured at 10 rad/s frequency and 0–500% strain. The gel loses its elasticity and behaves like a liquid at high strain, but is able to instantaneously recover elasticity upon removal of the strain, a property characteristic of injectable materials.

Acknowledgements

This work was supported in part by the National Science Foundation CAREER award, US Air force YIP and Norman Hackman Advanced Research Program. We would like to acknowledge Dr. David Tirrell (Caltech) for his kind gift of the plasmid pQE9 AC10Atrp [16], Dr. Takehisa Matsuda (Kanazawa Institute of Technology, Hakusan, Ishikawa, Japan) for his kind gift of the plasmid pET30-CutA-Tip1 [19], Dr. Tom Muir (Princeton University) for his kind gift of the plasmid KanR-IntRBS-NpuNC-CFN [20], and Dr. Jay D. Keasling (UC Berkley) for his kind gift of the plasmid pJD757 [21].

References

1. Langer R, Tirrell DA (2004) Designing materials for biology and medicine. Nature 428(6982):487–492

2. Kopeček J, Yang J (2012) Smart self-assembled hybrid hydrogel biomaterials. Angew Chem Int Ed 51(30):7396–7417. doi:10.1002/anie.201201040

3. Banta S, Wheeldon IR, Blenner M (2010) Protein engineering in the development of functional hydrogels. Annu Rev Biomed Eng 12:167–186

4. Ramirez M, Guan D, Ugaz V, Chen Z (2013) Intein-triggered artificial protein hydrogels that support the immobilization of bioactive proteins. J Am Chem Soc 135(14):5290–5293

5. Sun F, Zhang WB, Mahdavi A, Arnold FH, Tirrell DA (2014) Synthesis of bioactive protein hydrogels by genetically encoded SpyTag-SpyCatcher chemistry. Proc Natl Acad Sci U S A 111(31):11269–11274. doi:10.1073/pnas.1401291111

6. Wong Po Foo CT, Lee JS, Mulyasasmita W, Parisi-Amon A, Heilshorn SC (2009) Two-component protein-engineered physical hydrogels for cell encapsulation. Proc Natl Acad Sci U S A 106(52):22067–22072. doi:10.1073/pnas.0904851106

7. Guan D, Ramirez M, Shao L, Jacobsen D, Barrera I, Lutkenhaus J, Chen Z (2013) Two-component protein hydrogels assembled using an engineered disulfide-forming protein-ligand pair. Biomacromolecules 14(8):2909–2916. doi:10.1021/bm400814u

8. Shen W, Zhang K, Kornfield JA, Tirrell DA (2006) Tuning the erosion rate of artificial protein hydrogels through control of network topology. Nat Mater 5(2):153–158. doi:10.1038/nmat1573, nmat1573 [pii]

9. Shen W, Kornfield JA, Tirrell DA (2007) Dynamic properties of artificial protein

hydrogels assembled through aggregation of leucine zipper peptide domains. Macromolecules 40(3):689–692

10. Muralidharan V, Muir TW (2006) Protein ligation: an enabling technology for the biophysical analysis of proteins. Nat Methods 3(6):429–438

11. Zettler J, Schütz V, Mootz HD (2009) The naturally split Npu DnaE intein exhibits an extraordinarily high rate in the protein trans-splicing reaction. FEBS Lett 583(5):909–914

12. Iwai H, Züger S, Jin J, Tam P-H (2006) Highly efficient protein trans-splicing by a naturally split DnaE intein from Nostoc punctiforme. FEBS Lett 580(7):1853–1858

13. Tanaka Y, Tsumoto K, Nakanishi T, Yasutake Y, Sakai N, Yao M, Tanaka I, Kumagai I (2004) Structural implications for heavy metal-induced reversible assembly and aggregation of a protein: the case of Pyrococcus horikoshii CutA. FEBS Lett 556(1):167–174

14. Sawano M, Yamamoto H, Ogasahara K, Kidokoro S, Katoh S, Ohnuma T, Katoh E, Yokoyama S, Yutani K (2008) Thermodynamic basis for the stabilities of three CutA1s from Pyrococcus horikoshii, Thermus thermophilus, and Oryza sativa, with unusually high denaturation temperatures. Biochemistry 47(2):721–730

15. McGrath KP, Fournier MJ, Mason TL, Tirrell DA (1992) Genetically directed syntheses of new polymeric materials. Expression of artificial genes encoding proteins with repeating-(AlaGly) 3ProGluGly-elements. J Am Chem Soc 114(2):727–733

16. Shen W, Lammertink RGH, Sakata JK, Kornfield JA, Tirrell DA (2005) Assembly of an artificial protein hydrogel through leucine zipper aggregation and disulfide bond formation. Macromolecules 38(9):3909–3916

17. Shah NH, Dann GP, Vila-Perello M, Liu Z, Muir TW (2012) Ultrafast protein splicing is common among cyanobacterial split inteins: implications for protein engineering. J Am Chem Soc 134(28):11338–11341. doi:10.1021/ja303226x

18. Carvajal-Vallejos P, Pallisse R, Mootz HD, Schmidt SR (2012) Unprecedented rates and efficiencies revealed for new natural split inteins from metagenomic sources. J Biol Chem 287(34):28686–28696. doi:10.1074/jbc.M112.372680

19. Ito F, Usui K, Kawahara D, Suenaga A, Maki T, Kidoaki S, Suzuki H, Taiji M, Itoh M, Hayashizaki Y, Matsuda T (2010) Reversible hydrogel formation driven by protein-peptide-specific interaction and chondrocyte entrapment. Biomaterials 31(1):58–66. doi:10.1016/j.biomaterials.2009.09.026, S0142-9612(09)00956-9 [pii]

20. Lockless SW, Muir TW (2009) Traceless protein splicing utilizing evolved split inteins. Proc Natl Acad Sci U S A 106(27):10999–11004. doi:10.1073/pnas.0902964106

21. Dueber JE, Wu GC, Malmirchegini GR, Moon TS, Petzold CJ, Ullal AV, Prather KLJ, Keasling JD (2009) Synthetic protein scaffolds provide modular control over metabolic flux. Nat Biotechnol 27(8):753–759. doi:10.1038/Nbt.1557

Chapter 12

A Recessive Pollination Control System for Wheat Based on Intein-Mediated Protein Splicing

Mario Gils

Abstract

A transgene-expression system for wheat that relies on the complementation of inactive precursor protein fragments through a split-intein system is described. The N- and C-terminal fragments of a barnase gene from *Bacillus amyloliquifaciens* were fused to intein sequences from *Synechocystis* sp. and transformed into wheat plants. Upon translation, both barnase fragments are assembled by an autocatalytic intein-mediated *trans*-splicing reaction, thus forming a cytotoxic enzyme. This chapter focuses on the use of introns and flexible polypeptide linkers to foster the expression of a split-barnase expression system in plants. The methods and protocols that were employed with the objective to test the effects of such genetic elements on transgene expression and to find the optimal design of expression vectors for use in wheat are provided. Split-inteins can be used to form an agriculturally important trait (male sterility) in wheat plants. The use of this principle for the production of hybrid wheat seed is described. The suggested toolbox will hopefully be a valuable contribution to future optimization strategies in this commercially important crop.

Key words Split-inteins, Protein *trans*-splicing, *Triticum aestivum*, Intron-mediated Gene enhancement, Hybrid wheat, Pollination control

1 Introduction

Come together, right now

Over me!

John Lennon, 1969

(LP Abbey Road, Apple Rec. C062 04 243)

Functional proteins can be produced *in planta* through intein-mediated fusion of precursor peptides that are expressed by independent DNA sequences [1]. As "intervening protein elements", inteins are autocatalytically excised from precursor molecules and covalently ligate the flanking protein sequences (exteins) via a process termed protein splicing [2]. Notably, inteins catalyze such "*cis*-splicing" reactions but are also able to catalyze "*trans*-splicing" events between independent, non-covalently linked protein

Henning D. Mootz (ed.), *Split Inteins: Methods and Protocols*, Methods in Molecular Biology, vol. 1495,
DOI 10.1007/978-1-4939-6451-2_12, © Springer Science+Business Media New York 2017

fragments [3]. In recent years, numerous applications of intein-mediated *trans*-splicing in transgenic expression systems have been suggested. For example, the posttranslational intein-mediated assembly of precursor molecules represents a tool to circumvent the problems of gene and mRNA instability that are encountered in the synthesis of highly repetitive recombinant polypeptides, such as spider silk proteins [4]. Furthermore, the application of intein-mediated protein splicing allows for the production of active proteins by bringing together complementary gene fragments that are nonfunctional when being detached. Such combinations of gene fragments can be achieved by the sexual crossing of plants. In turn, active protein production may be prevented when the fragments are physically separated (e.g., by chromosomal segregation or maternal inheritance). As a result, such fundamental biological principles in combination with split-gene technology allow metabolic pathways to be activated or traits to be formed within a designated generation, tissue, or developmental stage [5].

Intein-mediated assembly of cytotoxic proteins, such as barnase (RNAse from *Bacillus amyloliquifaciens*), in male reproductive tissues can result in the abortion of pollen development and, consequently, male sterility. This technology was suggested as a tool for numerous technical solutions in correlation with efficient germplasm or trait control of transgene flow, thereby enabling the generation of more biologically contained transgenic plants [6]. In the technology described in this chapter, the barnase is encoded by two nonoverlapping sequences and specifically expressed in the anther tissue tapetum that is essential for pollen development. After translation, inactive barnase fragments are assembled into an active phytotoxic enzyme via a fused *Synechocystis* sp. *DnaB* intein system. As a result, the tapetum is destroyed, the formation of active pollen is prevented and the plant is male-sterile. This principle was suggested for the development of a hybrid seed production technology for plants [7–9]. In this method, the two inactive barnase loci are inserted in an identical (allelic) position on two homologous chromosomes, and thus are "linked in repulsion".

Despite the potential of this technology, it is apparent that the higher complexity of split gene approaches results in a reduced efficiency compared to systems based on continuous coding regions. Thus, for applications in plants with complex genomes, such as wheat, there is an obvious demand for stimulation of the expression level of the split-barnase gene.

In a previous chapter of this series, the fundamental mechanisms of split-transgene expression in plants were described [9]. In the present chapter, the results of subsequent studies that aimed at optimizing the expression of split transgenes in wheat to allow for a practical agricultural application of the technology are provided. Protocols are given for analyzing the effects of intron-mediated enhancement (IME) and of flexible polypeptide linker sequences on split-transgene expression. Methods of evaluating the genetic modifications are described stepwise in (a) rapid transient assays, (b) the model species *Arabidopsis thaliana* and (c) eventually by the analysis of transgenic wheat plants. Furthermore, it is demonstrated how careful selection of introns and flexible spacers enables the production of male sterility from a single copy of the barnase transgene. This represents a breakthrough towards the applicability of split-transgene systems because multiple-copy insertions result in unwanted complications for breeding programs.

Finally, this chapter describes how the developed high-expressing transgenic wheat plants can be used to establish a hybrid seed production system [10]. The main advantages of hybrid versus line varieties are increases in trait values due to the exploitation of heterosis (hybrid vigor). They include larger yield stability (especially in marginal environments), sturdiness against biotic and abiotic stresses, the ease of stacking dominant major genes and a larger return of investment for seed companies due to a built-in plant variety protection by inbreeding depression [11]. Currently, a major bottleneck in breeding hybrid wheat is, however, the lack of an efficient pollination control (sterility) system. This split-intein-based system might help to circumvent this problem and thus, to exploit the heterosis effect in the staple food wheat.

2   Materials

2.1   DNA Vectors and Cloning Procedures

1. Vectors for transient GUS-assays:
   pHWGUS, pHWGUS-N, and pHWGUS-

2. Vectors for transient barnase assays:
   pHW21, pHW611, pHW631, pHW651, pHW701, and pICH24612

3. Vectors for stable plant transformation:
   pHW451, pHW461, pHW471, pHW481, pHW491, pHW501, pHW561, pHW681, pHW771, pHW781, pHW821, pHW831, pHW841, and pHW851

2.2   Plant Material

1. Eight-week-old *Nicotiana benthamiana* plants grown under greenhouse conditions with 16 h of light at 20 °C and 8 h of darkness at 18 °C (60% relative humidity, circulating air).

2. *Arabidopsis thaliana* plants grown under light chamber conditions with 16 h of light at 23 °C and 8 h of darkness at 20 °C.

3. Spring wheat (*Triticum aestivum* L., cultivar 'Bobwhite') grown under greenhouse conditions with 16 h of light at 20 °C and 8 h of darkness at 16 °C (60–70% relative humidity, circulating air).

2.3 Transient Assays Using Agroinfiltration

1. *Agrobacterium tumefaciens* strain GV3101::pMP90I grown in LB medium.

2. LB medium containing 100 mg/L rifampicin, 25 mg/L gentamycin, and an antibiotic agent for selecting the binary vector (depends on the selection marker that is used; 50 mg/L in case of kanamycin or carbenicillin).

3. 10 mM 2-(N-morpholino) ethanesulfonic acid (MES) buffer: 10 mM MES, pH 5.5, 10 mM $MgSO_4$, 100 µM acetosyringone.

4. Infiltration buffer: 10 mM MES, pH 5.5, 10 mM magnesium sulfate.

2.4 Analysis of GUS-Expression Level

1. Extraction buffer: 50 mM $NaHPO_4$, pH 7.0, 10 mM beta-mercapto-ethanol, 10 mM EDTA, 0.1 % laurylsarcosine, 0.1 % Triton X-1000.

2. Reaction buffer: Extraction buffer + 2 mM MUG-Hydrate (*see* **Note 1**).

3. Stop solution: 0.2 M Na_2CO_3.

4. Fluorimeter (e.g., Versafluor-Flurometer, BIO-RAD)

Calibrate the fluorimeter with 4-Methylumbelliferone (MU) in the following concentrations:

100 µM, 1 µM, 0.5 µM, 0.1 µM, 0.05 µM, 0 µM (blank).

2.5 Analysis of Barnase Expression via Quantitative Reverse Transcription PCR (qRT-PCR)

1. RNeasy Plant Mini Kit (Qiagen).

2. Spectrophotometer (e.g., NanoDrop ND-1000 fromPeqlab Biotechnology).

3. Maxima First Strand cDNA Synthesis Kit (e.g., from Thermo Scientific).

4. Real-Time PCR System (e.g., 7900HT from Applied Biosystems Inc).

5. qRT-PCR Kit (e.g., Maxima SYBR Green/ROX qPCR Master Mix from Thermo Scientific or GoTaq RT-qPCR from Promega), according to the manufacturer's instructions.

6. Primers for detection of amount of barnase-C transcript: (5′–3′).
   FW (TCCGGCTTCCGCAACTC)
   Rev (TGGTCGGTGGTCTTGTAGATGA).

2.6 Transformation of Arabidopsis by Floral Dipping

1. Liquid LB-Medium for growing the transformed *Agrobacterium tumefaciens* strains.

2. Silwet L-77 solution: 0.05 % (500 µL/L, w/v in water).

3. 0.5 MS medium (Sigma).

4. 70 % ethanol and 7 % NaClO (w/v) for surface sterilization.

2.7 Transformation of Wheat Plants

1. Biolistic Particle Delivery System for transformation of wheat (e.g., PDS-1000/He from BIO-RAD).

2. Macrocarriers, stopping screens, and 900 psi rupture disks (BIO-RAD).

3. Plasmid-DNA-purification kit (e.g., from Qiagen or Macherey-Nagel).

4. For preparation of the Micron gold suspension, sterilize 60 mg of gold particles (0.6 μm gold; BIO-RAD) with 1 mL of 70% ethanol in a microcentrifuge tube by shaking briefly and incubating for 15 min. Sediment the particles by centrifugation at $13,000 \times g$ for 5 min, carefully remove supernatant and add 1 mL of sterile water. Centrifuge the tubes and discard the supernatant as before. Repeat the last washing step twice. After the final washing step, resuspend the particles in 1 mL of 50% (v/v) glycerol.

5. For particle coating, quickly mix 50 μL of the Micron gold suspension, 10 μg of plasmid DNA (maximal volume 10 μL; isolate DNA with plasmid DNA purification kit according the manufactures instructions), 50 μL of 2.5 M $CaCl_2$ and 20 μL of 0.1 M spermidine. Shake for 2 min, incubate at room temperature for 10 min, centrifuge at $13,000 \times g$ for 20 s, wash once with 70% ethanol and once with 99.5% ethanol and resuspend in 60 μL of 99.5% ethanol.

6. MS culture medium containing MS inorganic salts including vitamins (Duchefa, M0222), 30 g/L sucrose, 1 g/L N-Z-Amine A, 1 mL/L copper (II) sulfate pentahydrate stock solution, 2 mL/L 2,4-D stock solution, 5.2 g/L Phytagel (for solidification), pH 5.8–5.9.

7. Pre-transformation medium is MS culture medium containing 100 g/L sucrose.

8. Post-transformation medium is MS culture medium containing 60 g/L sucrose.

9. Prepare a 1% ethylenediaminetetraacetic acid, ferric sodium salt (FeNa-EDTA) solution and store at 4 °C.

10. Callus selection medium contains MS inorganic salts including vitamins (Duchefa, M0222), 30 g/L sucrose, 1 g/L N-Z-Amine A, 1 mL/L copper (II) sulfate pentahydrate stock solution, 1 g/L 2-(N-morpholino)ethanesulfonic acid (MES), 7 mL/L FeNa-EDTA solution, 2 mL/L 2,4-D stock solution, 5.2 g/L Phytagel, pH 6.0. After autoclaving, add 150 mg/L Hygromycin B (Duchefa).

11. For preparation of the kinetin solution, dilute to 10 g/L in DMSO and store at –20 °C.

12. For preparation of the zeatin riboside solution, dilute to 100 g/L in DMSO and store at –20 °C.

13. MS regeneration medium contains MS inorganic salts including vitamins, 20 g/L maltose, 1 g/L N-Z-Amine A, 1 mL/L copper (II) sulfate pentahydrate stock solution, 100 µL/L kinetin solution, 5.2 g/L Phytagel, pH 5.8–5.9. After autoclaving, add 70 µL/L zeatin of the riboside solution.

14. For preparation of the B1-myo-inositol solution, dilute 1 g/L of thiamine HCl and 100 g/L of myo-inositol in distilled water and store at −20 °C.

15. Plant selection medium contains half-strength MS inorganic salts including vitamins (Duchefa, M0222), 15 g/L sucrose, 0.1 g/L N-Z-Amine A, 1 mL/L copper (II) sulfate pentahydrate stock solution, 0.5 mL/L B1-myo-inositol solution, 5.2 g/L Phytagel, pH 5.8–5.9. After autoclaving, add 50 mg/L Hygromycin B (Duchefa).

2.8 Molecular Analysis of Transgenic Plants

2.8.1 DNA Isolation

1. Mill for disruption of plant tissue, e.g., TissueLyser from Qiagen.

2.8.2 Detection and Analysis of Transgenic Events by PCR

1. Primer list for PCR-analysis
   The following polymerase chain reaction (PCR) primers were used (5′–3′):

   (a) Detection of the N-terminal *barnase* sequence and to produce the Bar-N probe for DNA gel blot analysis,
   Bar-N-Fw (GCATCGATATGGCCCAAGTG)
   dnaB-intN-Rev (GAGCTGGAGGGAGGAGGATTCG)

   (b) Detection of the C-terminal *barnase* sequence and to produce the Bar-C probe for DNA gel blot analysis,
   Bar-C-Rev (GATCTTGGTGAAGGTCTGGTAG)
   dnaB-intC-Fw (GGGACTCCATCGTGTCCATC)

2.8.3 DNA Gel Blot Analysis

1. Nylon membranes (e.g., Biodyne B from Pall).

2. Depurination buffer: 0.25 N HCl. Mix 25 mL 37 % HCl and 985 mL dH_2O.

3. Denaturation buffer: 1.5 M NaCl, 0.5 M NaOH.

4. Neutralization buffer: 0.5 M Tris–HCl, 3.0 M NaCl, pH 7.0.

5. 20× SSC: 0.3 M Tri-Na-citrate, 3 M NaCl, pH 7.0.

6. Stratalinker (UV Crosslinker from Stratagene or similar instruments from other companies).

2.8.4 Hybridization with Radiolabeled Probe

1. Church-buffer: 0.5 M NaPi (18.96 g/L NaH_2PO_4 + 48.56 g/L Na_2HPO_4), 1 mM EDTA, 1 % BSA (w/v), 7 % SDS (w/v).

2. 20 % SDS stock solution: 200 g SDS/L, pH 7.2.

3. Washing solution I: 2× SSC, 0.5% SDS.

4. Washing solution II: 1× SSC, 0.1% SDS.

3   Methods

3.1   General Design of Split-Gene Expression Vectors

See Fig. 1 for a schematic illustration of the concept. The chemical nature of the splice site junction determines the efficiency of the splicing reaction. The rules were previously described in detail [9, 12, 13] (*see* **Note 2**) and can be summarized as follows:

1. The splicing junction within the protein of interest must not be located within important regions or functional domains.

2. Most inteins contain three highly conserved splice junction residues, which include a Cys or Ser at the beginning of the intein and a His-Asn at the end of the intein. An additional key amino acid for *trans*-splicing is a Ser, Thr or Cys as the first amino acid of the C-extein [3].

3. Short extein stretches derived from the intein host protein located at the splice junction may be beneficial for the efficiency of the *trans*-splicing reaction [12] (*see* **Note 3**).

4. The insertion of flexible GGGGS sequences, which space the fusion domains of the assembled barnase protein, increases the system's efficiency in plants. Within a certain range, the effect of split-gen stimulation is positively correlated with the length of the spacer [13].

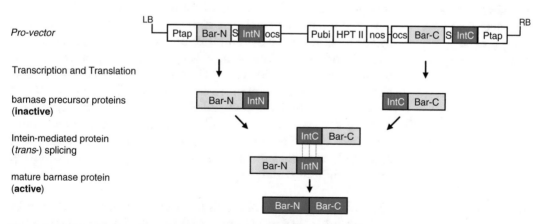

Fig. 1 Assembly of active barnase cytotoxin intein-mediated by *trans*-splicing of two precursor molecules. Note that the pro-vector harbors *attP* and *attB* sequences as targets for site-specific recombination (not shown). Only the T-DNA part of the vector is illustrated (not drawn to scale). Abbreviations: Bar-N and Bar-C, N- and C-terminal gene fragments of the *Bacillus amyloliquefaciens* barnase gene; IntN and IntC, N- and C-terminal intein sequences from the DnaB gene of *Synechocystis* sp.; Ptap, tapetum-specific *osg6B* promoter from rice; HPTII, hygromycin phosphotransferase selection marker gene; Pubi, maize ubiquitin 1 promoter; S, (GGGGS)$_3$ flexible peptide linker; ocs, octopine synthase terminator; nos, nopaline synthase terminator; LB and RB, T-DNA *left* and *right* borders

Introns have been reported to stimulate gene expression in different eukaryotic systems, including plants [14]. Although the molecular nature of this so-called "intron mediated gene enhancement" (IME) is still not fully understood, several common characteristics have been identified among introns capable of inducing IME. Taking this into account, compliance with the following rules is suggested when IME is intended to be the strategy for stimulating transgene expression.

1. A careful selection of the type of intron is necessary. Most introns that induce IME are the first intron of the native gene, of which many are present in the 5′-untranslated region (UTR). Other introns show no or even a negative effect.

2. Unlike "classical" transcriptional enhancers, an intron involved in IME must be located within the transcription unit [14–16].

3. The intron should be inserted in its "natural," i.e., the correct, orientation [17, 18].

4. The insertion site of the intron should be in close proximity to the start of the gene (not more than 1 kb from the promoter [14]).

5. Not all exon-intron boundaries are suitable for splicing. GeneSplicer (http://cbcb.umd.edu/software/GeneSplicer/) and NetGene (http://www.cbs.dtu.dk/services/NetGene2/) are valuable tools to validate the suitability of using the exon-intron boundaries as splice sites.

3.2 Analysis of GUS-Expression from Split-Genes Using a Rapid Infiltration Assay

Rapid transient assays can be used to monitor the efficiencies of split gene systems. Such assays are particularly helpful when a newly designed intein-mediated split-gene system (according to the rules listed in Subheadings 3.1 and 3.2) needs to be evaluated (Fig. 2). In leafs of *N. benthamiana*, a vector containing a continuous β-glucuronidase (GUS) gene (*pHWGUS*) or vectors containing N- and C-terminal GUS-intein fusions (*pHWGUS-N*, *pHWGUS-C*) were delivered by infiltrated and the amounts of active GUS protein were measured. The results indicate that the used intein-mediated split-transgene expression system is functional but apparently less efficient than the analogous system that is relying on a continuous coding region.

3.2.1 Agroinfiltration

1. Collect *Agrobacterium* cells from an overnight culture (5 mL) by centrifugation (10 min, $4500 \times g$, 4 °C) and resuspend in infiltration buffer.

2. Adjust bacterial suspension to a final OD_{600} of 0.8. In case of delivery of several constructs, *Agrobacteria* suspensions of different (split-gene-) constructs were mixed in equal volumes before infiltration.

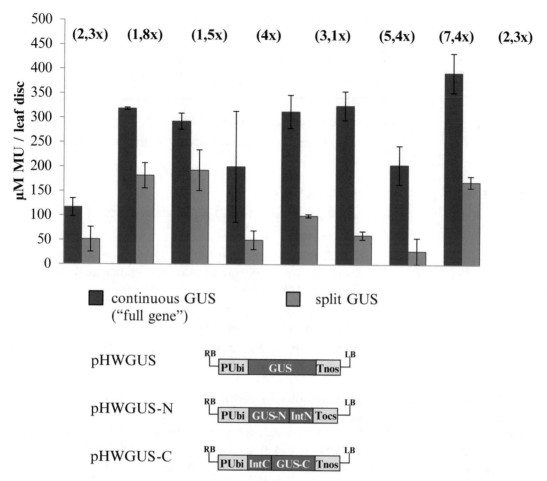

Fig. 2 GUS activities resulting from a split or a "full" GUS gene Two samples of eight infiltrated *N. benthamiana* leaves at different developmental stages were examined. The standard errors obtained from the two replicates. In all experiments, the quantity of GUS protein produced by the split GUS gene system was significantly lower (up to sevenfold) than that of the "full" GUS gene

3. Conduct agroinfiltration on near fully expanded leaves that are still attached to the intact plant. Slightly scratch the lower side of the leaf with a blade and infiltrate with a 5 mL syringe (*see* **Note 4**).

4. By infiltrating approximately 100 μL of bacterial suspension into each location (typically 3–4 cm² in infiltrated area), 8–16 locations separated by veins could be arranged in a single tobacco leaf.

5. After infiltration, plants were further grown under greenhouse conditions at 22 °C with 16 h of light.

3.2.2 Tissue Extraction   Sample leaf tissue from *N. benthamiana* that had been infiltrated with *Agrobacterium* harboring GUS-vectors 24 h after infiltration. Carry out measurements after a reaction with 4-methylumbelliferyl.

1. Harvest leaf tissue in Eppendorf tubes and add two steel balls (4 mm).

2. Homogenize in a TissueLyser (Qiagen) for 60 s.

3. Immediately add 100 µL extraction buffer.

4. Vortex briefly.

5. Centrifuge at 4500 ×g for 1 min at 4 °C.

6. Transfer the supernatant in a new Eppendorf tube and store on ice.

3.2.3 GUS Assay

1. Use a glass test tube.

2. Mix 60 µL of extract + 240 µL of reaction buffer.

3. Incubate at 37 °C.

4. After 10 min, 30 min and 1 h of incubation: take out 100 µL of solution and terminate the reaction immediately by adding 900 µL of Stop solution.

5. Measure fluorescence (in Versafluor-Fluorometer, BIO-RAD) in different dilutions (*see* **Note 5**).

6. Set-up of filters: 365 nm Excitation/455 nm Emission (*see* **Note 6**).

3.3 Transient Plant Assays for the Analysis of Split-Barnase Vector Functionality

The efficiency of split-barnase expression in plant tissue can be measured by infiltration of leaves of *N. benthamiana* with Agrobacteria carrying expression vectors. In the example shown in Fig. 3, different mixtures of Agrobacterium carrying T-DNAs for constitutive N- and C-terminal barnase expression (according to Subheading 3.2.1) were used.

3.3.1 Assessment of Split-Transgene Expression via Quantitative Reverse Transcription

PCR (qRT-PCR)

qRT-PCRs were performed to determine whether the observed intron-mediated enhancement in barnase protein activity could be explained by increases in the levels of C- and N-terminal barnase mRNA transcripts (Fig. 3b).

From the summary of results shown in Fig. 3, it can be concluded that the chosen introns can facilitate the production of barnase from a split-transgene and that the introns increased the accumulation of barnase transcripts.

1. Harvest 100–105 mg of infiltrated leaf material (*see* Subheading 3.2.1) 24 h after infiltration with Agrobacteria.

Fig. 3 (continued) were used. The mean expression level of plants transformed with pHW21 and pHW24612 was assigned a value of 1. The indicated positions of the introns are relative to the ATG of the N- or C-terminal transcription units. Abbreviations: *P35S*, cauliflower mosaic virus (CaMV) *35S* promoter; *Pact*, *Arabidopsis actin* 2 promoter; *Tocs*, octopine synthase terminator. For other abbreviations, *see* Fig. 1

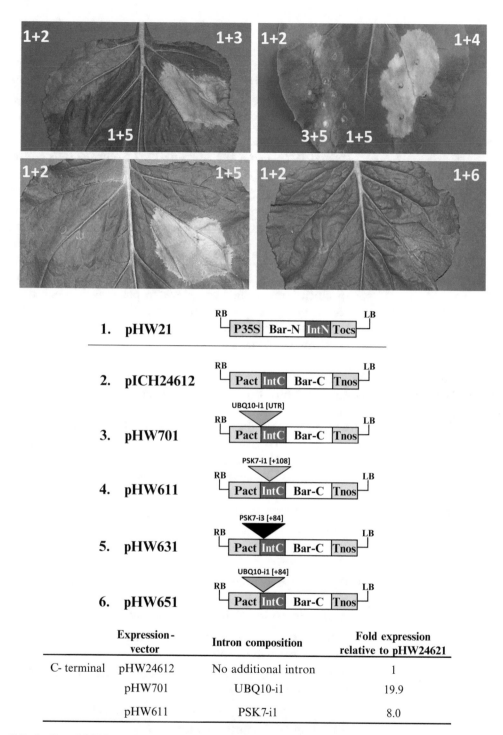

Expression-vector		Intron composition	Fold expression relative to pHW24621
C- terminal	pHW24612	No additional intron	1
	pHW701	UBQ10-i1	19.9
	pHW611	PSK7-i1	8.0

Fig. 3 Evaluation of IME through transient assays and qRT-PCR (**a**) For enhancing the barnase expression, three introns (PSK7-i1 and PSK7-i3 from *Petunia hybrida* and UBQ10-i1 from *Arabidopsis* [20] were cloned into the C-terminal transcription units. The relative degree of necrosis at a defined time (photographs show leafs 4 days after infiltration) was considered indicative for the barnase expression level (13). (**b**) qRT-PCR was conducted with tissue extracts from leaves of which one part was infiltrated with the N-terminal vector pHW21 and the C-terminal vectors pHW611 (intron psk7-i1) or pHW701 (intron UBQ10-i1) and the other part was infiltrated with the N-terminal vector pHW21 and pICH24612 (no intron). Compared to the intron-less vector, a considerable higher accumulation of C-terminal barnase mRNA was found when intron-containing vectors

2. Isolate total RNA using the RNeasy Plant Mini Kit (Qiagen). For quality control, analyze the RNA with 1.8 % agarose gels.

3. Measure the RNA concentration using a spectrophotometer, e.g., NanoDrop ND-1000 (Peqlab Biotechnology).

4. Perform first strand cDNA synthesis on 0.5 μg of DNase-treated RNA using the Maxima First Strand cDNA Synthesis Kit from Thermo Scientific.

5. Perform qPCR (e.g., with a 7900HT Fast Real-Time PCR System from Applied Biosystems Inc). In the described experiments, the reaction was analyzed with a Maxima SYBR Green/ROX qPCR Master Mix (Thermo Scientific) according to the manufacturer's instructions. Primer binding specificity was analyzed with dissociation curve analysis and PCR product specificity was verified by gel electrophoresis. Three biological replicates (RNA preparations from three different infiltrated leaves) and three technical replicates per experiment (standard deviation ≤ 0.4) were executed.

3.4 Analysis of the Functionality of Intein-Mediated Trans-Splicing in Transgenic *Arabidopsis thaliana*

The genetic transformation of wheat is laborious and time-consuming. Thus, prior to their delivery into wheat, it is recommended to test the genetic constructs in an easy-to-transform model species. In the example shown in Fig. 4, the intron-containing split-barnase-vectors were tested for their functionality in *Arabidopsis thaliana*. As a result, all the tested vectors turned out to be functional after stable transformation.

3.4.1 Transformation of Arabidopsis thaliana

1. Grow plants 4–6 days under long-day conditions until early flowering. Cut primary shoots to induce secondary shoots. Secondary shoots will be ready for transformation 5–7 days after cutting.

2. Prepare *Agrobacterium tumefaciens* strains harboring the vectors of interest. Grow 2 L of liquid culture at 28 °C in LB with antibiotics to select for the binary plasmid.

3. Spin down *Agrobacterium*, resuspend to OD600 = 0.6–1.0 with 5 % saccharose (w/v).

4. Before dipping, add Silwet L-77 to a concentration of 0.05 % (v/v; = 500 μL/L) and mix well.

5. Dip aboveground parts of the plant in *Agrobacterium* solution for 1 min. with gentle agitation.

6. After dipping, cover the plants with bags for 16–24 h to maintain high humidity. During growth to maturity, water and grow plants normally under long-day conditions. Stop watering as the seeds become mature and bag the plants for easy seed harvesting.

7. Harvest dry seeds and surface sterilize for 1 min with 70 % ethanol and for 1 min in 7 % NaClO (w/v).

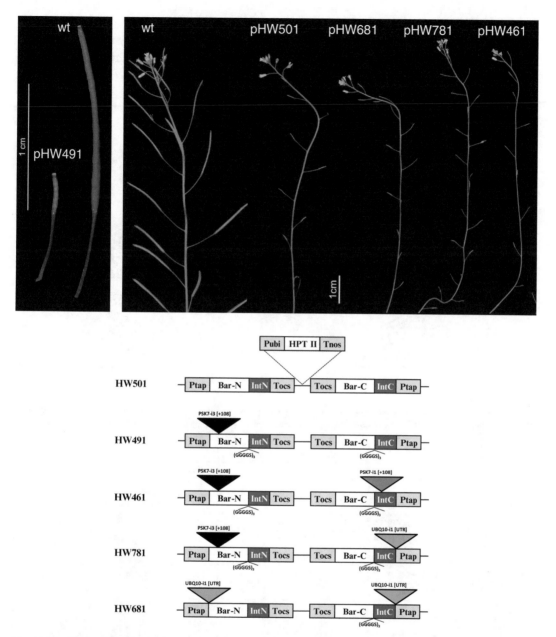

Fig. 4 Test of the intron-containing split-barnase-vectors in *Arabidopsis thaliana* Without exception, all *Arabidopsis* plants carrying split barnase vectors displayed a male-sterile phenotype. *Wt*, non-transgenic control plant

8. Sterilize seeds and transfer to 0.5 MS medium supplemented with selective antibiotics (e.g., 50 μg/mL kanamycin).

9. For fast selection: Stratification: 2 days, 4 °C in the dark, 6 h 22 °C light, 2 days 4 °C in the dark and 1 day 22 °C light.

10. Select for transformants using an antibiotic or herbicide selectable marker gene.

11. Transgenic plants display light green, open-shaped and long cotyledons.

3.5 Analysis of the Effects of Intron Insertion on Barnase Expression in Transgenic Wheat Plants: Transformation of Wheat Plants

In total 1491 independent primary transformant spring wheat plants harboring different versions of the split barnase constructs were produced (Fig. 5). The transformation of the vectors was carried out by biolistic bombardment [22].

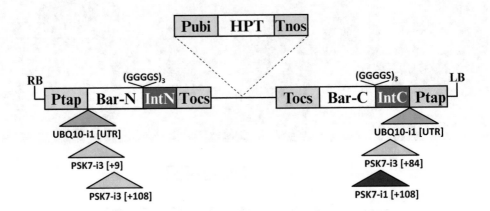

Transformation vector	N-terminal modification		C-terminal modification		No. plants	% male sterile plants
	intron	Linker (GGGGS)$_3$	intron	Linker (GGGGS)$_3$		
Non-transgenic regen.					845	3%
pHW501 ("control")	-	-	-	+	71	11%
pHW481	PSK7-i3^{+108}	-	-	+	41	28%
pHW491	PSK7-i3^{+108}	+	-	+	265	48%
pHW781	PSK7-i3^{+108}	+	UBQ-10-i1^{UTR}	+	301	54%
pHW461	PSK7-i3^{+108}	+	PSK7-i1^{+108}	+	81	19%
pHW451	PSK7-i3^{+108}	-	PSK7-i1^{+108}	+	59	19%
pHW831	PSK7-i3^{+9}	+	-	+	86	5%
pHW561	PSK7-i3^{+9}	-	PSK7-i3^{+84}	+	58	24%
pHW851	-	-	UBQ-10-i1^{UTR}	+	130	41%
pHW681	UBQ-10-i1^{UTR}	-	UBQ-10-i1^{UTR}	+	117	46%
pHW821	UBQ-10-i1^{UTR}	+	UBQ-10-i1^{UTR}	+	121	47%
pHW841	UBQ-10-i1^{UTR}	-	-	+	65	58%
pHW771	UBQ-10-i1^{UTR}	-	PSK7-i1^{+108}	+	51	6%
pHW471	-	-	PSK7-i1^{+108}	+	45	11%

Fig. 5 Intron-mediated gene enhancement in stably transformed wheat plants Mature plants were assayed for seed formation. 3% of non-sterile "escapes" exhibited a male-sterile phenotype. The intron-less vector pHW501 served as a control and 11% of the transformants displayed male sterility. From the summary of the results is can be concluded that different combinations of introns and flexible linkers have different effects on the efficiency of barnase-expression level and trait formation (sterility)

3.5.1 Isolation of Wheat Embryos and Callus Culture Maintenance

1. Cut spikes of wheat plants 14–18 days after anthesis (8–10-week-old plants) and carefully remove the immature caryopses by hand.

2. Perform a surface sterilization by the successive immersion of seeds in 70 % ethanol for 1–2 min and freshly prepared 2.5 % sodium hypochlorite/0.01 % SDS solution for 15 min.

3. Wash the seeds three times in sterile, distilled water.

4. Aseptically excise the immature embryos (1.0–1.5 mm in length, semitransparent) with forceps and a scalpel using a stereomicroscope and place them, scutellum side up, on MS culture medium.

5. Allow the embryogenic callus to develop in the dark at 25 °C for 3–4 weeks.

3.5.2 Ballistic Transformation of Wheat Plants

1. Pretreat the calli for 6 h on pre-transformation medium at 22 °C in the dark.

2. For each bombardment, place approximately 50 callus cultures in the middle of a plate in a circular-like formation (radius of about 1 cm). In that position, the probability of hits is high.

3. Use the Biolistic PDS-1000/He Particle Delivery System (Bio-Rad) and follow the manufactures instructions. Use the following settings for bombardment: 1 cm distance between the rupture disk and macrocarrier, 6 cm distance between the stopping screen and target tissue.

4. Sterilize the chamber and all components with 70 % ethanol.

5. Sterilize the macrocarrier holders, macrocarriers, stopping screens, and rupture disks by dipping them in absolute ethanol and allow it to evaporating completely.

6. Briefly vortex the Micron gold suspension (coated gold particles), place 6 μL onto the macrocarrier membrane and allow for complete drying.

7. Load a rupture disk (900 psi) into the rupture disk retaining cap and firmly tighten the screw into place.

8. Place a stopping screen into the microcarrier launch assembly.

9. Invert and position the macrocarrier holder containing the macrocarrier and coated gold particles over the stopping screen.

10. Place a sample on the target stage, draw a vacuum and fire the gun.

11. Transfer the bombarded cultures to 22 °C in the dark overnight.

12. Incubate the cultures on post-transformation medium for 2 weeks at 25 °C in the dark.

13. Transfer regenerating plantlets to jars with half strength hormone-free MS selection-medium containing 50 mg/L hygromycin B.

14. Transfer embryogenic calli to callus selection medium for 7 days, in the dark at 22 °C. Perform 5–6 successive callus selection steps (total time: 4–6 months).

15. Subculture callus in MS regeneration medium for about 4–6 weeks with 16 h of light at 24 °C and 8 h of dark at 15 °C. Change the medium every 2–3 weeks.

16. Transfer regenerating plantlets were transferred to jars with plant selection medium under light conditions for 4 weeks.

17. Acclimate fully developed plantlets 7–10 days at room temperature in liquid medium containing fourfold diluted MS salts.

18. Transfer plants with developed roots into soil and grown them under greenhouse conditions to maturity.

3.6 Molecular Analysis of Transgenic Plants for Gene Identification and Copy-Number Estimation via PCR and Southern-Blot Analysis

An overview about the genetic structure of the transgenic locus and the strategy of for copy-number analysis is given in Fig. 6. After copy-number analysis, selected lines with low-copy-numbers of transgenic events were assayed for trait formation (sterility; Fig. 7). The results indicate that the insertion of introns into the intein-barnase-transcription unit facilitates more efficient production of functional low-copy-events for practical applications.

3.6.1 DNA Isolation

1. For total DNA isolation [19], harvest leaf segments and freeze in liquid nitrogen.

2. Homogenize using a TissueLyser™ from Qiagen. The material can be stored at –80 °C.

3.6.2 PCR Analysis (See Fig. 6)

1. Perform PCR on the isolated DNA with an initial denaturation step at 95 °C for 3 min, followed by 35 amplification cycles of 94 °C for 30 s, 60 °C for 30 s and 72 °C for 1–2 min.

2. Analyze the amplified fragments by agarose gel electrophoresis.

3.6.3 DNA Gel Blot Analysis (See Fig. 6)

1. To estimate the copy number of the transgenes, digest the DNA (Subheading 3.6.1) with enzymes that release fragments containing

    (a) vector sequences homologous to the labeled probe along with

    (b) a genomic DNA stretch of unpredictable length.

The latter is expected to be different for every individually integrated vector sequence, which allows the transgene copy number to be estimated by counting the radiolabeled fragments on the autoradiogram.

2. Separate DNA fragments on a 0.8 % agarose gel (*see* **Note 7**).

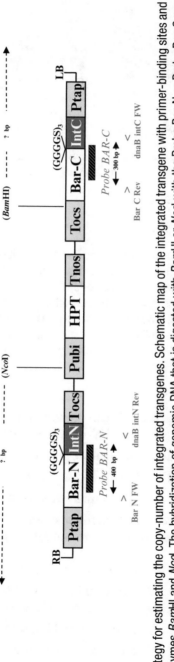

Fig. 6 Strategy for estimating the copy-number of integrated transgenes. Schematic map of the integrated transgene with primer-binding sites and recognition sites for the enzymes *Bam*HI and *Nco*I. The hybridization of genomic DNA that is digested with *Bam*HI or *Nco*I with the Probe Bar-N or Probe Bar-C results in fragments of unpredictable sizes that can be scored for copy-number estimation

190   Mario Gils

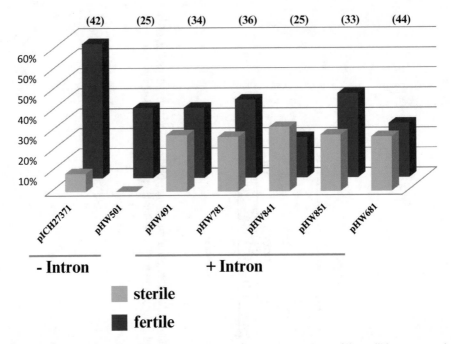

Fig. 7 Ratio of sterile versus fertile wheat plants harboring low-copy numbers of the split barnase vectors Note: All of the analyzed plants carry one of two copies of the transgene, as estimated by Southern blot analyses. The intron-less vector pICH27371 is part of separate study (12). The number of analyzed plants is given in *brackets*. Gene copy-numbers were detected by Southern blot analysis. None of the plants containing a low copy number of the intron-less vector pHW501 and 9 % of the plants carrying a low copy number of the intron-less vector pICH27371 displayed male-sterility. On the contrary, the populations carrying one or two copies of pHW491, pHW781, pHW841, pHW851 and pHW681 displayed a higher proportion of male-sterile plants (27–32 %)

3. Stain the gel with an ethidium bromide solution and analyze the gel under UV-light (*see* **Note 8**).

4. Incubate the agarose gel for 5 min in depurination buffer.

5. Incubate for 2×15 min in neutralization solution.

6. Blot on a nylon membrane with 20× SSC overnight.

7. Crosslink the membrane using a Stratalinker© (UV Crosslinker from Stratagene) according the manufactory's instruction.

3.6.4 Hybridization of Membrane with Radiolabeled Probe

1. Incubate in Church buffer for a minimum of 4 h, 65 °C.

2. Add radiolabeled probe ([^{32}P]-labeled DNA-fragments depicted as "probes" in Fig. 6), incubate over night at 50–60 °C.

3. Wash 2×30 min in washing solution I and II, 50–65 °C.

4. Seal the membrane and expose to an X-ray film.

3.7 Crossing of Winter Wheat Plants to Create a Male-Sterile Female Line and Hybrid F₁ Progeny

The establishment of an intein-based hybridization system for wheat relies on several crossing steps (*see* Fig. 8).

1. Transfer seeds for a period of 3–4 days into climate chambers (4 °C) to break dormancy.

2. Transfer seeds into greenhouse conditions (16 h of light and 8 h of darkness at 16 °C) for 1 week to stimulate germination.

3. Germinated plants are vernalized for 8 weeks at 4 °C with 10 h of light and 14 h of darkness.

4. Pick out one plant per pot (13 cm) in peat-rich soil (Klasmann substrate 2) and fertilize weekly with NPK (8:12:24) (*see* **Note 9**).

5. Grow plants to maturity under greenhouse conditions with 16 h of light and 8 h of darkness at 16 °C.

3.7.1 Cultivation of Plants

3.7.2 Demasculinization

1. Wheat is a self-pollinating species. Therefore, for crossing two different spikes, the removal of the male reproductive organs (anthers) is necessary to avoid self-fertilization. Before you start to demasculinize, it must be ensured that there is a male parent ready to donate pollen (*see* **Note 10**).

2. Only the outermost flowers from a spikelet should be used for cross-pollination. The middle floret should be removed from the spike.

3. Choose spikes with anthers that are beginning to yellow but are still at the green stage (pistils should not be fully developed).

4. To access the anthers, open the florets with forceps and tease apart the anthers with fine forceps (*see* **Note 11**).

5. After removing of the anthers, cut the top half of every spikelet to expose the ovaries (*see* **Note 12**).

6. Reexamine/double-check the cut flowers for remaining anthers.

7. To protect the flowers from injuries or pollen, bag the spike and label the bag, fix with a paper clip.

8. The demasculinated plant is reproductively immature and not yet ready to be pollinated. Leave the spike for 2–3 days under the bag before fertilization (*see* **Note 13**).

3.7.3 Cross-Pollination

1. After 2–3 days, the female plant is ready for cross-pollination.

2. Cross-pollinate in the morning, if possible, from 7 to 11 am.

3. Because no pollination took place, the florets usually slightly open and the stigmas appear larger and "fuzzy".

4. For pollination, choose spikes with yellow anthers that shed pollen (*see* **Note 14**).

5. If the male has long awns (depending on the variety), cut them short.

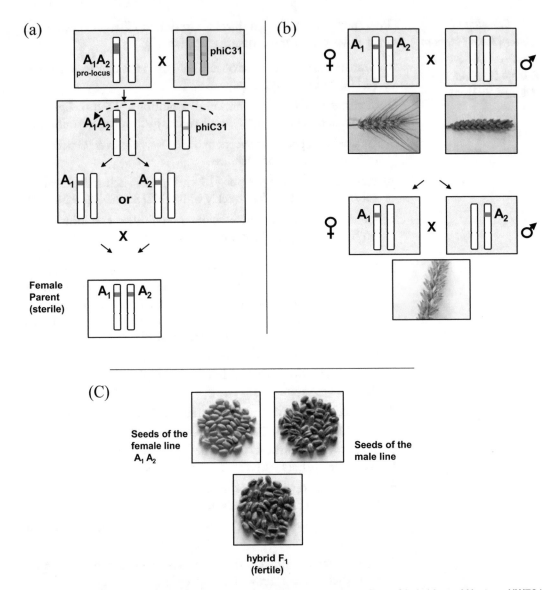

Fig. 8 Design of an intein-mediated split-gene system for the production of hybrid seed Vector pHW781 (Fig. 5) was used to develop the system. (**a**) system development: After transforming the "pro-vector" pHW781, the primary transformants are expected to be male-sterile because of barnase expression. In the following step, they are crossed with plants expressing a phiC31 site-specific recombinase. During progeny plant development, the integrase may catalyze site-specific deletions between the *attP* and *attB* sites of the pro-locus (recombination, target sites; not shown). Either of the two alternative reactions can lead to the deletion of the C- terminal or the N-terminal part of the locus [21, 22], which results in the formation of the derivative loci designated A₁ or A₂. The loci will naturally reside at exactly the same genetic position on two homologous chromosomes (because they originated from one pro-locus). Plants carrying only A₁ or A₂ should be fertile because no complete barnase protein is produced. The A₁ and A₂ lines are then crossed and a portion of the resulting progeny (~25 %) is expected to carry both loci. These heterozygous genetic segregants are expected to be male-sterile because of the co-expression of barnase from the A₁ and A₂ loci and the functional complementation of the N- and C-terminal barnase fragments.

6. In the optimal case, two male spikes should be used as pollen donors for one female spike.

7. To open up the female and male flowers, stimulate them by gently rubbing the spikes from top to bottom between the fingers of one hand (*see* **Note 15**).

8. Collect 1–3 anthers with the forceps and carefully dab the anthers onto every demasculinated flower of the female crossing partner (*see* **Note 16**).

9. Bag the pollinated spikes.

10. Note: Alternatively, the female and male crossing partners can be located adjacent in the greenhouse and the demasculinated spike can be bagged together with the spike of the pollen-donor. Flip the bag several times for pollen release and keep the bag closed until the seed set.

4   Notes

1. Before mixing the reaction buffer, dilute MUG in a small volume of DMSO and then add extraction buffer. Store at 4 °C, but not for more than a few weeks.

2. Useful information about inteins can also be obtained from *InBase*, The Intein Database (www.inteins.com).

3. Consider a codon usage optimized for monocots.

4. It is necessary to water *N. benthamiana* plants 15 min before infiltration. Use full-grown leaves, but not the oldest.

5. Use immediately or store at –70 °C.

6. Do not forget to "warm up" the device 15 min prior to the measurement.

7. The recommended voltage depends on the distance between the electrodes. Do not use too high voltage, as this will result in "fuzzy" signals on the autoradiogram (e.g., use not more than 25 V if the distance between the electrodes is 15 cm). For copy-number-detection in the hexaploid species wheat, distinct signals are particularly important.

8. It is recommended to mark the gel to define the orientation throughout further handling (for instance by cutting one defined corner).

Fig. 8 (continued) (**b**) Hybrid-cross: After crossing heterozygote segregants with pollen donors, the allelic position of the complementary barnase fragments enforces 100 % segregation during meiosis. This results in male fertility and seed set in the hybrid progeny, as all segregants carry only an inactive barnase fragment (either A₁ or A₂). As an example, hybridization with parental lines carrying a different seed color (females *bright* and males *purple*) is demonstrated (**c**)

9. Adapt temperature, light, watering and fertilization to the actual growth stage and appearance of the plant (within the temperature range of 14–22 °C and 10–16 h of light).

10. The flowering time of the female crossing partner needs to be synchronized with the flowering time of the male crossing partner. In the case of genotypes with different flowering times, it is recommended to plant the males/females in staggered intervals to ensure that plants with overlapping flowering times will be available.

11. To ensure that no anther remains in the flower deposit extracted anthers on your hand for counting.

12. Take the utmost care not to injure the stigma of the flower during the demasculinization procedure.

13. As a guide, the appropriate stage is reached when the base of the spike level is at the base of the flag leaf.

14. Maturation of wheat starts at the center of the spike and moves outwards. The middle spikelets are more mature and have possibly already exposed the anthers. If they are white, they usually have no more pollen.

15. The male spike can be cut and placed into water to stimulate anther extrusion.

16. Sterilize forceps with 70 % Et-OH if different males are used.

Acknowledgements

The critical comments of Angelika Gils and Dr. Heike Gnad (Saaten-Union Biotec GmbH) on the manuscript are gratefully acknowledged. The author thanks Dr. Volker Lein (Saaten-Union Recherche, France) for providing the purple-colored spring wheat variety that was used in the hybrid cross.

The majority of experiments described here were performed within a joint project between the Nordsaat GmbH and the Leibniz Institute of Plant Genetics and Crop Plant Research (IPK) Gatersleben and between Nordsaat GmbH and Saaten-Union Biotec GmbH under funding of the Bundesministerium für Bildung und Forschung (Grants FZK0315889 and 031B0030; "WEIZEN 2.0").

References

1. Yang J, Fox GC Jr, Henry-Smith TV (2003) Intein-mediated assembly of a functional beta-glucuronidase in transgenic plants. Proc Natl Acad Sci USA 100(6):3513–3518. doi:10.1073/pnas.0635899100

2. Perler FB (1998) Protein splicing of inteins and hedgehog autoproteolysis: structure, function, and evolution. Cell 92(1):1–4

3. Saleh L, Perler FB (2006) Protein splicing in cis and in trans. Chem Rec 6(4):183–193. doi:10.1002/tcr.20082

4. Hauptmann V, Weichert N, Menzel M, Knoch D, Paege N, Scheller J, Spohn U, Conrad U, Gils M (2013) Native-sized spider silk proteins synthesized in planta via intein-based multimerization. Transgenic Res 22:369–377

5. Evans TC Jr, Xu MQ, Pradhan S (2005) Protein splicing elements and plants: from transgene containment to protein purification. Annu Rev Plant Biol 56:375–392

6. Gleba Y, Marillonnet S, Klimyuk V (2004) Design of safe and biologically contained transgenic plants: tools and technologies for controlled transgene flow and expression. Biotechnol Genet Eng Rev 21:325–567

7. Gils M, Marillonnet S, Werner S, Grutzner R, Giritch A, Engler C, Schachschneider R, Klimyuk V, Gleba Y (2008) A novel hybrid seed system for plants. Plant Biotechnol J 6:226–235

8. Kempe K, Gils M (2011) Pollination control technologies for hybrid breeding. Mol Breed 27:417–437

9. Gils M, Rubtsova M, Kempe K (2012) Split-transgene expression in wheat. In: Dunwell JM, Wetten AC (eds) Transgenic plants: methods and protocols, vol 847, Methods Mol Biol. Humana, New York, pp 123–135

10. Kempe K, Rubtsova M, Gils M (2014) A split gene system for hybrid wheat production. Proc Natl Acad Sci U S A 111(25):9097–9102

11. Longin CF, Mühleisen J, Maurer HP, Zhang H, Gowda M, Reif JC (2012) Hybrid breeding in autogamous cereals. Theor Appl Genet 125(6):1087–1096

12. Kempe K, Myroslava R, Mario G (2009) Intein-mediated protein assembly in transgenic wheat: production of active barnase and acetolactate synthase from split genes. Plant Biotechnol J 7(3):283–297

13. Kempe K, Rubtsova M, Riewe D, Gils M (2013) The production of male-sterile wheat plants through split barnase expression is promoted by the insertion of introns and flexible peptide linkers. Transgenic Res 22(6):1089–1105

14. Rose AB, Elfersi T, Parra G, Korf I (2008) Promoter-proximal introns in Arabidopsis thaliana are enriched in dispersed signals that elevate gene expression. Plant Cell 20:543–551

15. Callis J, Fromm M, Walbot V (1987) Introns increase gene expression in cultured maize cells. Genes Dev 1:1183–1200

16. Mascarenhas D, Mettler IJ, Pierce DA, Lowe HW (1990) Intron-mediated enhancement of heterologous gene expression in maize. Plant Mol Biol 15:913–920

17. Clancy M, Vasil V, Curtis Hannah L, Vasil IK (1994) Maize Shrunken-1 intron and exon regions increase gene expression in maize protoplasts. Plant Sci 8:151–161

18. Curi GC, Chan RL, Gonzalez DH (2005) The leader intron of Arabidopsis thaliana genes encoding cytochrome c oxidase subunit 5c promotes high-level expression by increasing transcript abundance and translation efficiency. J Exp Bot 56:2563–2571

19. Dellaporta SL, Wood J, Hicks JB (1983) A plant DNA minipreparation: version II. Plant Mol Biol Report 1:19–29

20. Norris SR, Meyer SE, Callis J (1993) The intron of Arabidopsis thaliana polyubiquitin genes is conserved in location and is a quantitative determinant of chimeric gene expression. Plant Mol Biol 21:895–906

21. Kempe K, Rubtsova M, Berger C, Kumlehn J, Schollmeier C, Gils M (2010) Transgene excision from wheat chromosomes by phage phiC31 integrase. Plant Mol Biol 72(6):673–687

22. Rubtsova M, Kempe K, Gils A, Ismagul A, Weyen J, Gils M (2008) Expression of active Streptomyces phage phiC31 integrase in transgenic wheat plants. Plant Cell Rep 27:1821–1831

Chapter 13

Conditional Toxin Splicing Using a Split Intein System

Spencer C. Alford, Connor O'Sullivan, and Perry L. Howard

Abstract

Protein toxin splicing mediated by split inteins can be used as a strategy for conditional cell ablation. The approach requires artificial fragmentation of a potent protein toxin and tethering each toxin fragment to a split intein fragment. The toxin–intein fragments are, in turn, fused to dimerization domains, such that addition of a dimerizing agent reconstitutes the split intein. These chimeric toxin–intein fusions remain nontoxic until the dimerizer is added, resulting in activation of intein splicing and ligation of toxin fragments to form an active toxin. Considerations for the engineering and implementation of conditional toxin splicing (CTS) systems include: choice of toxin split site, split site (extein) chemistry, and temperature sensitivity. The following method outlines design criteria and implementation notes for CTS using a previously engineered system for splicing a toxin called sarcin, as well as for developing alternative CTS systems.

Key words Protein splicing, Split intein, Selective cell ablation, Sarcin, Dimerization domain, VMA intein, Rapamycin, FKBP12, FRB

1 Introduction

The ability to selectively eliminate populations of cells in multicellular model organisms is a highly desirable tool. For example, selective elimination of cells at key stages of development has been used to address questions related to cell autonomy and lineage [1–4]. Most commonly, tissue specific or inducible expression of a suicide gene is used to ablate desired cell populations at specific points during development. However, the time required for transcription and translation, which is on the scale of several hours, limits this approach. An alternative strategy employs prodrugs; here, nontoxic molecules are introduced at critical developmental time points and are converted to a toxic form in specific cell populations through tissue-specific expression of a prodrug converting enzyme [5, 6]. While this approach may exhibit rapid activation, it is limited by the number of unique pro-drugs, off-target toxicity, and inability to kill non-dividing cells [6].

Henning D. Mootz (ed.), *Split Inteins: Methods and Protocols*, Methods in Molecular Biology, vol. 1495,
DOI 10.1007/978-1-4939-6451-2_13, © Springer Science+Business Media New York 2017

The approach of selective cell ablation has also been utilized for anticancer strategies, where toxic molecules are selectively delivered or activated in cancer cells [7–9], as well as for tissue transplantation, where the ability to selectively eliminate transplanted cells is used as a failsafe measure to mitigate the potential for harm from unintended or aberrant cell behavior [10, 11]. Taken together, the above examples highlight a strong unmet need for alternative and efficient methods of selective cell killing. To address this need, we determined protein splicing can be employed effectively in an alternative selective cell ablation approach [12]. Specifically, we tested and demonstrated the ability of intein splicing to conditionally activate a ribotoxin called α-sarcin (Figs. 1 and 2), which is cytotoxic to many cell types [13–15]. The strategy relies on regulating the cytotoxicity of α-sarcin, such that addition of a dimerizing agent rapidly activates the toxin. Briefly, using rational design, we recombinantly split α-sarcin into nontoxic fragments and engineered a system to splice the fragments together upon addition of rapamycin [12]. By necessity, this strategy relies on the phenomenon of intein splicing.

Inteins are (self)-excised from larger precursors, and the flanking protein sequences (called exteins) are spliced together to form a mature functional protein [16]. Splicing typically requires two sequence determinants: a cysteine or serine within the intein sequence at the N-terminal extein–intein splice junction, and a histidine–asparagine dipeptide sequence at the C-terminal intein–extein splice junction, immediately followed by a cysteine, serine, or

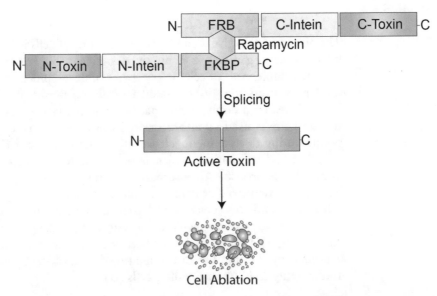

Fig. 1 Conditional toxin splicing strategy. A protein toxin is split into N-terminal and C-terminal fragments, which should render the toxin neutral. Each fragment is fused to a split intein fragment, which are in turn fused to heterodimerization domains, FKBP and FRB. Splicing is activated upon addition of a dimerizing agent (rapamycin; *grey hexagon*), which produces a full-length spliced toxin that mediates cell ablation

a

N-Sarcin   VMA^N

AVTWTCLNDQKNPKTNKYETKRLLYNQCFAKGTNVLMADGSIECIENIEVGNK...

Q27G

b

VMA^C   C-Sarcin

...DYYGITLSDDSDHQFLLANQVVVHNNKAESNSHHAPLSDGKTGSSYPHWFT...

N28C

c

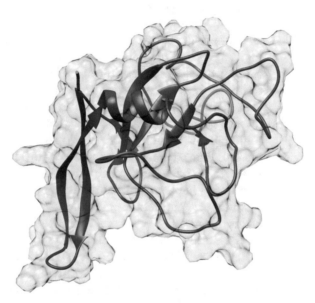

Fig. 2 Sequence constraints in generating a toxin (extein)-intein junction using α-sarcin and VMA intein. (**a**) The VMA^N intein contains a cysteine at the N-terminal extein–intein splice junction. It was found that substituting glutamine-27 with glycine within N-Sarcin at the −1 position of the extein–intein junction improved splicing [12]. (**b**) The VMA^C intein contains a histidine-asparagine dipeptide at the C-terminal intein–extein splice junction. Asparagine-28 within C-Sarcin (+1 position) was substituted with cysteine in order to satisfy splicing requirements. (**c**) α-sarcin was split into two fragments based on its structure, with N-sarcin composed of a β-hairpin (*blue*), and C-sarcin containing the catalytic fragment (*red*)

threonine as the first C-terminal extein residue [17]. Flanking residues within the N-terminal and C-terminal extein sequences also influence splicing efficiency [18–20], and many inteins can accommodate a variety of extein sequences and still splice effectively [21, 22]. Through implementation of a variety of engineered intein systems, protein splicing has been used to regulate protein function in live cells with rapid kinetic resolution (e.g., [23, 24]).

For selective cell ablation applications, the intein must exhibit stringent activation with no background activity. *Cis*-splicing (intramolecular) and *trans*-splicing (intermolecular) inteins have been developed for conditional activation with triggers such as hormones, changes in pH or temperature, light, or redox potential [25–32]. Since *cis*-systems rely on intramolecular reactions, they may have the inherent potential for higher background splicing and for protein fragment complementation that may activate the toxin in the absence of a trigger. Alternatively, *trans*-splicing intermolecular systems utilize split dimerization domains fused to split intein fragments, which reassemble to activate splicing upon addition of a small molecule dimerizer. Two such varieties of conditional protein splicing have been described for splicing proteins in cultured mammalian cells: an estrogen receptor-hydroxytamoxifen system [26, 27] and an FKBP–rapamycin–FRB system [33, 34]. In addition to choosing appropriate dimerization domains, the implementation of intein fragments that exhibit low affinity for self-association is an absolute requirement for low-background conditional *trans*-splicing. Almost all split intein systems are high affinity and exhibit spontaneous self-association (e.g., [35, 36]). However, the artificially split intein *S. cerevisiae* VMA intein does not self-associate and is the most appropriate intein for implementing CTS [34].

Despite a strong grasp of the mechanism and chemistry of protein splicing, examples of functional spliced proteins in live cell systems have been largely limited (though not exclusively) to reporter proteins (e.g., luciferase or GFP) [33, 34, 37–40]. Several limitations have slowed the broad applicability of protein splicing, including a minimum requirement for a naturally occurring (or artificially installed) cysteine, serine, or threonine at the C-terminal intein–extein splice junction, as well as the possible requirement of installing splicing-permissive extein mutations that may destabilize or inactivate the spliced target protein [19, 41]. Further, different inteins exhibit different functioning temperature optima, with some inteins better suited to applications in organisms at cooler temperatures [22, 24] (e.g., in *D. melanogaster* or *D. rerio*) (*see* **Note 1**). Careful consideration should be given to (1) either designing CTS systems to match organism and intein temperatures, or (2) changing the temperature of the system to permit efficient splicing. In addition to the various splicing parameters requiring optimization for CTS designs, many options exist for integrating different protein toxins into these systems.

Protein toxins produced by plants, bacteria, and fungi comprise some of the most deadly molecules known. They exhibit a broad range of function including: pore-forming toxins, which damage cell membranes; protein synthesis inhibitors, which inhibit translation by covalently modifying elongation factors or ribosomal RNA (rRNA); activators of second messengers, which target Rho-GTPases or G-proteins by covalent modification to disrupt cell signaling events; and protease toxins, which target essential proteins for destruction. In particular, toxins that inactivate protein synthesis are attractive candidates for conditional cell ablation strategies. They may function as ribonucleases, ADP-ribosyl transferases, or N-glycosylases [42] and due to their catalytic efficiency, only a few toxin molecules may be required to elicit cell death. Indeed, it is has been reported that a single molecule of ricin, a potent translation inhibitor, may be sufficient to kill HeLa cells [13]. While the ribonuclease α-sarcin was implemented for the initial CTS system described [12], alternative CTS systems may be engineered using different toxins, such as those listed above.

The ability to inducibly trigger toxin activity is valuable in genetic, developmental, and therapeutic applications where one wants to eliminate specific cells with precise temporal and spatial resolution. Here we describe the general design principles for developing a conditional toxin splicing (CTS) system, with considerations for toxin split site selection and modification, and temperature sensitivity. We provide the detailed design criteria and methods for α-sarcin CTS as a guide on using this strategy for selective cell ablation.

2   Materials

2.1   Reagents and Supplies

2.1.1   Genes Encoding CTS Constructs, GFP Splicing Constructs, and Diagnostic Constructs (^{N}MBP and ^{C}His)

Required plasmids are described in Table 1, and where indicated, are available through Addgene. *See* **Note 2** for considerations when generating constructs. Sequences for deposited constructs are available through Addgene.

Table 1
Constructs for designing conditional toxin splicing system

Construct components	Availability	Abbreviation
MBP-VMA^N-FKBP	[33, 34]	^{N}MBP
FRB-VMA^C-PolyHis	[33, 34]	^{C}His
3×FLAG-NSarcin(Q27G)-VMA^N-FKBP	Addgene (Plasmid ID 70229)	3×F-^{N}Sar
NSarcin(Q27G)-VMA^N-FKBP	Addgene (Plasmid ID 70230)	^{N}Sar
FRB-VMA^C-CSarcin(N28C)	Addgene (Plasmid ID 70223)	^{C}Sar
NGFP-VMA^N-FKBP	Addgene (Plasmid ID 70225)	^{N}GFP
FRB-VMA^C-CGFP	Addgene (Plasmid ID 70226)	^{C}GFP
N-Toxin-VMA^N-FKBP	User-designed construct	^{N}Tox
FRB-VMA^C-C-Toxin	User-designed construct	^{C}Tox

2.1.2 Rapamycin Solution (See Note 3)	1. Rapamycin (Sigma-Aldrich). 2. Dimethyl sulfoxide (DMSO). 3. Working rapamycin stock solution (10 µM Rapamycin in DMSO). Store working aliquots at –20 °C and thaw in your fingers just prior to use.

2.1.3 Transfection Reagent

1. Jet Prime transfection reagent (Polyplus) (*see* **Note 4**).

2.1.4 Cell Culture Media and Reagents

1. Dulbecco's High Glucose Modified Eagle's Medium (DMEM).

2. Penicillin–streptomycin solution (Pen-Strep).

3. Fetal bovine serum (FBS).

4. Trypsin–EDTA: 0.5 % trypsin, 913 µM EDTA.

5. Phosphate buffered saline (PBS).

6. Bovine serum albumin (BSA).

7. 6-well tissue culture plates.

8. 15 mL polypropylene conical tubes.

9. Hank's Balanced Salt Solution (HBSS) with HEPES.

2.1.5 Molecular Biology Reagents

1. Plasmid Maxi Kit.

2.1.6 Western Blot Reagents

1. 1× SDS sample buffer: 62.5 mM Tris–HCl, pH 6.8, 2 % (w/v) SDS, 75 mM DTT, 7.5 % glycerol, 0.02 % bromophenol blue.

2. Protease inhibitor cocktail.

3. 100 mM PMSF (phenylmethylsulfonyl fluoride).

4. Cell scrapers.

5. 27 gauge needles.

6. 1 mL syringes.

7. 4× Resolving gel buffer: 1.5 M Tris–HCl, pH 8.8, 4 % SDS.

8. 4× stacking gel buffer: 1 M Tris–HCl, pH 6.8, 4 % SDS.

9. SDS-PAGE gel. For 10 % separation gel 8 mL total volume: 3.25 mL dH$_2$O, 2.67 mL 30 % acrylamide, 2 mL 4× resolving gel buffer, 80 µL 10 % ammonium persulfate, 8 µL TEMED. For 3.5 % stacking gel 3 mL total volume: 1.74 mL dH$_2$O, 0.51 mL 30 % acrylamide, 0.75 mL 4× stacking gel buffer, 30 µL 10 % APS, 3 µL TEMED.

10. 1× running buffer: 25 mM Tris–HCl, pH 8.3, 192 mM glycine, 3.5 mM SDS.

11. Nitrocellulose membrane, 0.45 µm.

12. 1× transfer buffer: 25 mM Tris–HCl, pH 8.3, 192 mM glycine, 20 % (w/v) methanol.

13. Washing buffer: 1× TBST: 50 mM Tris–HCl, pH 7.5, 150 mM NaCl, 0.1 % Tween 20.

14. Blocking buffer: 5 % nonfat dry milk or 5 % bovine serum albumin (BSA), as indicated in primary antibody datasheet, dissolved in 1× TBST.

15. Antibodies: anti-GFP (e.g.,Roche), anti-FLAG (e.g.,Sigma-Aldrich), anti-MBP (e.g.,New England Biolabs), anti-β-actin (e.g.,Sigma-Aldrich), anti-6×His tag antibody (e.g.,Abcam). Goat anti-mouse IgG HRP-conjugated (e.g.,R&D Systems).

16. Western HRP substrate (e.g.,Luminata Forte, Cedarlane).

17. X-ray film.

2.1.7 Apoptosis Assay

1. FITC Annexin V Apoptosis Detection Kit I (BD Pharmingen).

2.1.8 Equipment

1. Inverted fluorescent microscope (e.g., Leica DMIRE2) (*see* **Note 5**).

2. Flow cytometer (e.g.,BDFACSCalibur, BD Biosciences) (*see* **Note 6**).

3. Centrifuges (e.g., Eppendorf centrifuge 5417C, Beckman Coulter Allegra X-15R centrifuge).

4. Electrophoresis Power Supply.

5. Western blot transfer apparatus.

6. X-ray film developer.

2.1.9 Flow Cytometry

1. Acquisition software for flow cytometer (e.g.,Cell Quest for BDFACSCalibur, BD Biosciences).

2. FlowJo analysis software (optional) (*see* **Note 7**).

3. Falcon round bottom polystyrene tubes (FACS tubes).

4. Falcon 70 μm cell strainers.

2.2 Cell Culture and Media Preparation

1. HeLa cells (ATCC®-CCL-2™).

2. DMEM supplemented with 10 % FBS and 1 % Pen-Strep. Store at 4 °C after preparation.

3. Activation media: DMEM supplemented with 10 % FBS, 1 % Pen-Strep and 10 nM rapamycin (or desired concentration) (*see* **Note 8**).

3 Methods

3.1 Gene Assembly of Splicing Constructs

1. Examine the protein sequence and the available 3D crystal structure (see RCSB database) of a desired toxin to identify sites for recombinant fragmentation (*see* **Note 9**).

2. Acquire the gene encoding the desired toxin, or order synthesized gene fragments (e.g., gBlocks from IDT).

3. Using standard molecular biology techniques (PCR, overlap extension PCR or Gibson assembly, etc.) assemble appropriate N-toxin (^{N}Tox) and C-toxin (^{C}Tox) CTS fusions (*see* Fig. 1) from existing constructs (Table 1) (*see* **Notes 2** and **10**). As an illustrative example, the methods described herein outline the implementation of α-sarcin CTS described in Alford et al. [12] with more generalized descriptive notes on generating new toxin CTS systems.

3.2 Biochemical Confirmation of In Vitro Splicing Conditions (See Notes 11 and 12)

1. First perform a pilot experiment to confirm the in vitro splicing conditions and detection methods are appropriate. This experiment recapitulates results described in [12, 34]. Aseptically prepare plasmids encoding ^{N}MBP and ^{C}His in sterile TE buffer using a Qiagen Maxi prep kit (*see* **Notes 13** and **14**). When not in use store plasmids at –20 °C. Plasmid solutions should be at least 500 μg/mL.

2. Distribute 3.0×10^5 HeLa cells per well into 6-well tissue culture plates in cell culture media and grow in a 37 °C, 5 % CO_2 incubator overnight. Cells should be at ~70–90 % confluence the next day for optimal transfection.

3. Co-transfect HeLa cells using a 1:1 ratio of ^{N}MBP–^{C}His plasmids (500 ng each). Perform the transfection according to the manufacturer's recommendations. Other equivalent transfection methods can be used (*see* **Note 4**). Incubate the cells in a 37 °C, 5 % CO_2 incubator overnight.

4. On the day of the experiment, thaw aliquots of 10 μM rapamycin and prepare activation media (*see* Subheading 2.2, **item 3**). Remove media from cells and replace with activation media and in one well add DMSO (using a volume equivalent to the volume used to activation media). Set a timer for 10 min.

5. After 10 min, remove media and lyse cells by adding 200 μL 1× SDS sample buffer with 10 μL protease inhibitor cocktail (PIC), and 10 μL 0.1 M PMSF. Thoroughly scrape with a cell scraper, and pass lysate through a 27 gauge needle ~5 times using a 1 mL syringe.

6. Let sit on ice for 30 min and spin at $20,000 \times g$ at 4 °C. Collect the supernatant and heat at 95 °C for 10 min.

7. Run lysates on a 10 % SDS-PAGE gel until the dye front reaches the bottom of the gel. Run the gel at 120 V using 1× running buffer.

8. Transfer the proteins to a nitrocellulose membrane using a semidry transfer cell (or equivalent western blot transfer apparatus) for 45 min at 15 V using 1× transfer buffer.

9. Rinse once quickly with 1× TBST, then incubate in blocking buffer with gentle shaking for at least 1 h. Subsequently incubate with anti-MBP using gentle shaking according to the manufacturer's recommendations.

10. Wash three times with 1× TBST (5 min with gentle shaking), then incubate with 1:100,000 goat anti-mouse HRP secondary antibody for ~1 h.

11. Wash three times with 1× TBST, then incubate in ECL detection substrate (e.g., luminata forte) according to the manufacturer's recommendations.

12. Expose blot to X-ray film for desired amount of time and develop using an X-ray film developer.

13. The expected result is the appearance of a spliced band at ~43 kDa (Fig. 3a and *see* **Note 15**). No band should be present for cells treated with DMSO vehicle.

14. Once splicing conditions are validated, proceed to testing the splice-competency of the toxin–intein fusions as described in Subheadings 3.3 and 3.4.

3.3   Determination of ^{C}Tox Splicing

1. Aseptically prepare plasmids encoding construct ^{N}MBP and ^{C}Sar (or ^{C}Tox) (Table 1), as well as pcDNA3.1(+) according to Subheading 3.2, **step 1**.

2. Distribute HeLa cells according to Subheading 3.2, **step 2**.

3. Using a ratio of 1:1 plasmid DNA (500 ng each), co-transfect HeLa cells with the following plasmid combinations: two wells with ^{N}MBP and ^{C}Sar (or ^{C}Tox), one well with ^{N}MBP and pcDNA 3.1(+), and one well with ^{C}Sar (or ^{C}Tox) and pcDNA 3.1(+). Perform transfection (*see* Subheading 3.2, **step 3**).

4. On the day of the experiment, thaw aliquots of 10 μM rapamycin and prepare activation media (*see* Subheading 2.2, **item 3**). Remove media from cells. In one well transfected with ^{N}MBP and ^{C}Sar (or ^{C}Tox) add DMSO (using a volume equivalent to the volume used to activation media). To all other wells add activation media and set a timer for 10 min.

5. After 10 min, collect lysates according to Subheading 3.2, **steps 5** and **6**.

6. Perform western blot according to Subheading 3.2, **steps 7–12** and use anti-MBP antibody to confirm splicing (Fig. 3b). For cells transfected with ^{N}MBP and ^{C}Sar and incubated with rapamycin, the expected result is a band corresponding to spliced MBP-^{C}Sar product at ~56 kDa. For experiments performed with alternative ^{C}Tox constructs, the spliced product should have a predicted molecular weight of the sum of: ~45 kDa (MBP) and the molecular weight of ^{C}Tox. No band corresponding to the spliced product should be present in the

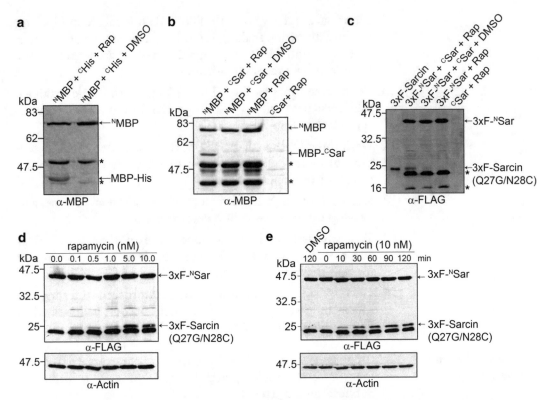

Fig. 3 Biochemical validation of splicing. (**a**) Western blot from HeLa cells treated as described in Subheading 3.2. The upper band represents the unspliced NMBP construct. The band at ~43 kDa indicated by the *arrow* corresponds to spliced MBP-His, and is only present following rapamycin treatment [34]. *Asterisks* indicate degradation products from the NMBP construct (*see* **Note 15**) [34]. (**b**) Western blot from cells treated as described in Subheading 3.3. The *upper band* represents the unspliced NMBP construct. The band at ~56 kDa indicated by the *arrow* corresponds to spliced MBP-CSar, and is only present following rapamycin treatment. *Asterisks* indicate degradation products from the NMBP construct. (**c**) Western blot from cells treated as described in Subheading 3.5. The band indicated by the *arrow* at ~23 kDa corresponds to spliced 3×FLAG-Sarcin. All other bands correspond to either unspliced 3×F-NSar (*arrow*) or degradation products (*asterisks*). (**d**) Western blot from cells treated as described in Subheading 3.6. The spliced 3×F-Sarcin product at ~23 kDa (*arrow*) increases in a rapamycin dose-dependent manner. Actin was used as a loading control. (**e**) Western blot from cells treated as described in Subheading 3.7, analyzed at the indicated time points. The spliced 3 × F-Sarcin product at ~23 kDa (*arrow*) increases in a time-dependent manner. DMSO was used as a control to ensure there was no background splicing. Actin was used as a loading control. Reproduced from [12] with permission from the Royal Society of Chemistry

lysates harvested from cells treated with DMSO vehicle and no bands should be present in the lysates from cells transfected with NMBP or CSar (or CTox) independently (*see* **Note 15**). If splicing is not observed, the splice junction is not permissive (*see* **Notes 9** and **16**).

3.4 Determination of ^{N}Tox Splicing

1. Aseptically prepare plasmids encoding construct ^{N}Sar (or ^{N}Tox) and ^{C}His, as well as pcDNA3.1(+) according to Subheading 3.2, **step 1**.

2. Distribute HeLa cells according to Subheading 3.2, **step 2**.

3. Using a ratio of 1:1 plasmid DNA (500 ng each), co-transfect HeLa cells with the following plasmid combinations: two wells with ^{N}Sar (or ^{N}Tox) and ^{C}His, one well with ^{N}Sar (or ^{N}Tox) and pcDNA 3.1(+), and one well with ^{C}His and pcDNA 3.1(+). Perform transfection (*see* Subheading 3.2, **step 3**).

4. On the day of the experiment, thaw aliquots of 10 μM rapamycin and prepare activation media (*see* Subheading 2.2, **item 3**). Remove media from cells. In one well transfected with ^{N}Sar (or ^{N}Tox) and ^{C}His add DMSO (using a volume equivalent to the volume used to activation media). To all other wells add activation media and set a timer for 10 min.

5. After 10 min, collect lysates according to Subheading 3.2, **steps 5 and 6**.

6. Perform western blot according to Subheading 3.2, **steps 7–12** and use anti-His antibody to confirm splicing. For experiments performed with alternative ^{N}Tox constructs, the spliced product should have a predicted molecular weight of the sum of the ^{N}Tox fragment and the polyHis tag (~1 kDa). If splicing is not observed, the splice junction is not permissive (*see* **Notes 9** and **16**). Once it is confirmed that both ^{N}Tox and ^{C}Tox intein fusions are splice-competent, in vitro toxin splicing can be performed.

3.5 Rapamycin-Mediated Activation of Toxin Splicing In Vitro

1. Aseptically prepare plasmids encoding constructs 3×F-^{N}Sar (or ^{N}Tox) and ^{C}Sar (or ^{C}Tox), as well as pcDNA3.1(+) according to Subheading 3.2, **step 1** (*see* **Note 14**).

2. Distribute HeLa cells according to Subheading 3.2, **step 2**.

3. Using a ratio of 1:1 plasmid DNA (500 ng each), co-transfect HeLa cells with the following plasmid combinations: two wells with 3×F-NSar (or ^{N}Tox) and ^{C}Sar (or ^{C}Tox), one well with 3×F-^{N}Sar (or ^{N}Tox) and pcDNA 3.1(+), and one well with ^{C}Sar (or ^{C}Tox) and pcDNA 3.1(+). Perform transfection (*see* Subheading 3.2, **step 3**).

4. On the day of the experiment, thaw aliquots of 10 μM rapamycin and prepare activation media (*see* Subheading 2.2, **item 3**). Remove media from cells. In one well transfected with 3×F-^{N}Sar (or ^{N}Tox) and ^{C}Sar (or ^{C}Tox) add DMSO using an equivalent volume to rapamycin. To all other wells add activation media and set a timer for 10 min.

5. After 10 min, collect lysates according to Subheading 3.2, **steps 5 and 6**.

6. Run western blot according to Subheading 3.2, **steps 7–12** and use anti-FLAG antibody to confirm splicing (Fig. 3c) (*see* **Note 17**). For experiments performed with alternative NTox and CTox constructs, the spliced product should have a predicted molecular weight of the sum of the full length toxin and the epitope tag(s) (e.g., plus ~3 kDa if 3×FLAG, as in 3×-FNSar). The NTox and CTox constructs may independently exhibit splicing (as in Subheadings 3.3 and 3.4), but fail to splice a full length toxin when used in combination. In this scenario, where the combination of extein sequences prevents splicing, alternative split sites should be explored to obtain splicing-permissive and compatible intein–extein junctions (*see* **Notes 9** and **16**).

3.6 Determination of Optimal Rapamycin Concentration for In Vitro Splicing

1. Use plasmids encoding construct 3×F-NSar (or NTox) and CSar (or CTox) prepared in sterile TE buffer (*see* **Note 14**).

2. Distribute HeLa cells according to Subheading 3.2, **step 2**.

3. Using a ratio of 1:1 (500 ng each), co-transfect HeLa cells with 3×F-NSar (or NTox) and CSar (or CTox) (6 total wells). Perform transfection (*see* Subheading 3.2, **step 3**).

4. On the day of the experiment, thaw aliquots of 10 μM rapamycin and prepare activation media with rapamycin at final concentrations of 0.1, 0.5, 1.0, 5.0, and 10 nM rapamycin. Remove media from cells and activation media and set a timer for 2 h (*see* **Note 17**). In a single well, add DMSO vehicle using a volume equivalent to the volume used to make 10 nM rapamycin.

5. After 2 h, collect lysates according to Subheading 3.2, **steps 5** and **6**.

6. Run western blot according to Subheading 3.2, **steps 7–12** and use anti-FLAG antibody to confirm splicing (Fig. 3d). Additionally, probe the blot with anti-Actin as a loading control.

7. Expect to observe a band corresponding to the spliced product of ~23 kDa. No band should be visible for cells treated with DMSO vehicle only. A dose-dependent increase in toxin splicing should be observed for the range of rapamycin concentrations tested.

3.7 Determination of Time Required for Rapamycin-Induced Toxin Splicing

1. Use plasmids encoding construct 3×F-NSar (or NTox) and CSar (or CTox) prepared in sterile TE buffer (*see* **Note 14**).

2. Distribute HeLa cells according to Subheading 3.2, **step 2** in two 6-well plates.

3. Using a ratio of 1:1 (500 ng each), co-transfect HeLa cells with 3×F-NSar (or NTox) and CSar (or CTox). Perform transfection (*see* Subheading 3.2, **step 3**).

4. On the day of the experiment, thaw aliquots of 10 μM rapamycin and prepare activation media with rapamycin at the optimal concentration determined in Subheading 3.6. In a single well, replace the overnight growth media with media containing DMSO vehicle. In all other wells, replace the overnight growth media with activation media. Collect cell lysates for the rapamycin-treated cells at the following time intervals: 1, 10, 30, 60, 90, and 120 min (post rapamycin addition). While waiting for the time points, store the collected cell lysates at −80 °C until ready for western blotting.

5. Run western blot according to Subheading 3.2, **steps 7–12** and use anti-FLAG antibody to confirm splicing. Expect to observe a band corresponding to the spliced sarcin product of ~23 kDa (Fig. 3e), or the predicted molecular weight of an alternative toxin. No band should be visible for cells treated with DMSO vehicle only (*see* **Note 15**). A time-dependent increase in toxin splicing should be observed.

3.8 Determination of Optimal Permissive Temperature for Toxin Splicing

1. Distribute HeLa cells as described in Subheading 3.2, **step 2**.

2. Using a ratio of 1:1 (500 ng each), co-transfect HeLa cells with 3×F-^{N}Sar and ^{C}Sar constructs (*see* Subheading 3.2, **step 3**).

3. Add rapamycin at a concentration of 10 nM, and incubate at 30 °C or 37 °C, 5 % CO_2 for up to 1 h.

4. Harvest and collect lysates for each temperature condition according to Subheading 3.2, **steps 5** and **6**. Perform western blot analysis according to Subheading 3.2, **steps 7–12** and use anti-FLAG antibody to detect splicing of full length toxin (*see* **Note 1**).

3.9 Detection of Apoptosis Using Flow Cytometry

1. Distribute and transfect cells as described in Subheading 3.5, **steps 2** and **3**. Instead of 3×F-^{N}Sar, use the untagged ^{N}Sar construct (*see* **Note 10**).

2. Wash cells once with PBS, add 1× Trypsin-EDTA and incubate at 37 °C, 5 % CO_2 for 2–5 min until cells begin to lift off the dish (*see* **Note 18**).

3. Neutralize trypsin with DMEM, 10 % FBS immediately after cells begin to lift off the dish and transfer to a 15 mL conical tube.

4. Process cells using the FITC Annexin-V apoptosis detection kit according to the manufacturer's instructions (*see* **Note 19**).

4 Notes

1. Temperature is a factor that may affect your application. While the VMA intein splices at 37 °C, it exhibits superior splicing at temperatures closer to 25 °C [22, 24]. Incubation temperature

for splicing experiments can be lowered to 30 °C (at least for HeLa cells) or lower (cell type-dependent). One must also consider the optimal functional temperature of the toxin used in given CTS system. For example, performing CTS in zebrafish (which grow near 28 °C), may be near optimal for VMA splicing, but the spliced toxin may only be functional at 37 °C.

2. To aid in evaluating the splice competency of new CTS constructs, it is strongly recommended to include epitope tags at the N-terminus of the VMA^N construct (e.g., a FLAG tag) and at the C-terminus of the VMA^C construct (e.g., a polyHis or myc tag). Inclusion of these tags permits diagnostic western blotting of cell lysates (e.g., with anti-FLAG and anti-myc antibodies) to determine the extent, if any, that splicing occurs upon addition of rapamycin.

3. As an alternative, a nontoxic rapamycin analog (in vitro or in mice) may be used: AP21967 or A/C heterodimerizer (Clontech).

4. Many commercially available transfection reagents may serve as alternatives, such as the Lipofectamine series of transfection reagents (ThermoFisher Scientific), and are effective at transfecting HeLa cells. Transfection conditions should be optimized for any transfection reagent, using a gene encoding a fluorescent reporter (e.g., GFP or mCherry) to evaluate efficiency.

5. While we used an inverted fluorescent microscope to live-image GFP splicing in cells and analyzing apoptosis, other types of microscopy with the ability to detect fluorescence, such as confocal, are equally suited for this purpose.

6. We used a BD FACSCalibur flow cytometer for the purposes of detecting GFP fluorescence and analyzing apoptosis, but any commercially available flow cytometers are fully capable of performing these tasks.

7. FlowJo is a data analysis software for dealing with flow cytometry data that has performed well in our hands. Alternative flow cytometry analysis software platforms are available as well (e.g., FCS Express).

8. For activation media, add rapamycin to the final desired concentration just prior to adding media to cells. Thaw the pre-made 10 μM solution of rapamycin using your fingers.

9. Special consideration should be given to the fragmentation site of the protein toxin. Several different split sites should tested when possible. The residues at the split site will become the +1 and –1 extein residues. Where possible, a split site should be chosen such that cysteine becomes the +1 extein position. This cysteine must not participate in any disulfide bridging. If cysteine is not chosen as the +1 position, it will have to be artificially installed at that site and will remain as a 'scar' in the spliced

product. In this scenario, a full length version of the toxin harboring this +1 cysteine must also be generated and tested for its cytotoxicity to cells. This may be evaluated by cloning the full length toxin (with +1 cysteine) into pcDNA3.1(+), transfecting HeLa cells, and quantitating cytotoxicity relative to a pcDNA3.1(+) vehicle control transfection. For the −1 extein position, care should be taken to avoid the following residues (when using VMA intein): valine, leucine, isoleucine, asparagine, and proline [20]. Preferred residues at the −1 extein position include: glycine, alanine, threonine, lysine, arginine, histidine, and methionine. If one of these preferred residues is not conveniently located at the split site, and the splicing is not observed, then a preferred residue may be artificially installed at this site to confer splicing competency (more in **Note 13**). Glycine is a good first choice as a substitute since it is the naturally occurring −1 extein residue for the VMA intein. Given each toxin sequence is different, it is possible additional extein residues (e.g., at the −2, −3, or +2, +3 positions) may need to be mutated to achieve splicing [38]. In addition to sequence determinants, split sites should be selected near the protein surface and where possible between clearly folded structural elements. For example, we unsuccessfully attempted to split the ricin A chain in the protein core between residues 170 and 171; an I170G mutation rendered a splice competent N-extein (albeit poorly), but the C-extein failed to splice despite a naturally occurring cysteine at position 171 (unpublished data). Finally, though not an absolute requirement, split sites should be chosen so as to avoid disrupting disulfide bridges that connect the C-terminal and N-terminal extein fragments.

10. If epitope-tagged CTS constructs are to be used for cell ablation experiments, the toxicity of the tagged toxin must be evaluated. Depending on the nature of the tag and the mode of action of the toxin, toxicity may be ablated (e.g., 3× FLAG-Sarcin [12]). Toxicity can be evaluated by cloning the full-length epitope-tagged toxin into a mammalian expression plasmid (e.g., pcDNA 3.1(+)) and transfecting HeLa cells with increasing doses. A dose-dependent cytotoxic response should be observed with increased cell killing with increasing amounts of toxin-encoding plasmid transfected [43]. If toxicity is ablated by inclusion of the tag, simply generate untagged versions of the CTS constructs for cell ablation experiments.

11. As a convenient control for protein splicing, GFP splicing may be included in all experiments to confirm conditions are permissive for splicing. Here, HeLa cells are transfected according to Subheading 3.2, **step 3** with ^{N}GFP and ^{C}GFP. The following day, overnight growth media is replaced with activation media and incubated for up to 12 hrs. The media can then be exchanged with HEPES-buffered HBSS and imaged for GFP

fluorescence using an inverted fluorescence microscope. After addition of rapamycin, development of fluorescence will be evident as early as four hours.

12. To quantify splicing efficiency of split fluorophores using flow cytometry, transfect cells with ^{N}GFP and ^{C}GFP (Subheading 3.2. **step 3**), treat cells with rapamycin (*see* **Note 11**), trypsinize cells (Subheading 3.9, **step 2**), and neutralize the trypsin with the addition of DMEM, 10% FBS. Transfer cells to a 15 mL conical tube, centrifuge for 5 min at $350 \times g$. Wash with 1 mL PBS, and centrifuge again for 5 min at $350 \times g$. Resuspend samples in 0.5 mL 0.5% BSA in PBS. Filter samples through a 70 μm cell strainer and transfer to 5 mL FACS tubes and acquire at least 1.0×10^4 events/sample on the flow cytometer. Ensure cells are in the middle of a forward scatter vs side scatter plot, and set a gate on the major population (Fig. 4a). Adjust the voltage of the FL1 channel using the untransfected sample to put the population between 10^0 and 10^1 on a log scale (Fig. 4b). Set a gate on any fluorescent intensity above background (as determined by the negative controls) on the FL1 channel, to calculate the percentage of cells that underwent successful intein splicing (Fig. 4c). As an additional control, run samples transfected with the ^{N}GFP and ^{C}GFP halves on their own, as well as samples transfected with both ^{N}GFP and ^{C}GFP but treated with DMSO, to determine any background fluorescence associated with these constructs. If using a different fluorophore than GFP, ensure it is compatible with the flow cytometer.

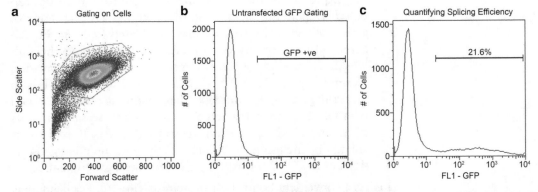

Fig. 4 Gating for GFP fluorescence using flow cytometry. (**a**) On a forward scatter vs. side scatter plot, make a gate (*pink circle*) around the major population of cells. (**b**) Using a histogram for GFP fluorescence intensity (FL1 channel for BDFACSCalibur), set the negative population to be approximately between 10^0 and 10^1 on a logarithmic scale. Draw a gate for GFP fluorescence intensity above the background fluorescence seen in the untransfected sample. (**c**) Running experimental samples will allow you to determine the percentage of cells that successfully underwent splicing to create functional fluorophores

13. While it is recommended to use [N]MBP and [C]His constructs [34] for pilot experiments and determining splice competency of new [N]Tox and [C]Tox constructs, it is also possible to use 3× F-[N]Sar and [C]Sar constructs in their place, respectively. For example, to determine the splice competency of new [C]Tox constructs, one may perform a splicing trial using the 3× F-[N]Sar construct. By performing an anti-FLAG western blot, one can detect the spliced product 3× F-[N]Sar-[C]Tox. Similarly, for new tagged [N]Tox constructs, splicing reactions with the [C]Sar construct will produce a protein corresponding to the mass of [N]Tox plus ~13.5 kDa ([C]Sar).

14. Several commonly used large scale DNA preparation kits may be used. Common choices for Maxipreps include kits provided by Qiagen and ThermoFisher Scientific.

15. CTS splicing constructs are prone to degradation and these products will be observed by western blot (see asterisks in Fig. 3a–c and [34]). For western blots probed with antibodies against the N-terminal epitope tag of the VMA[N] construct, expect to visualize the full-length VMA[N] precursor, the spliced product (where appropriate), and a few breakdown products of characteristic size for each construct. Depending on the size of the final spliced product, care should be taken to ensure resolution of any spliced products.

16. If splicing of the product is not observed and the precursor is present at the correct molecular weight, the extein context of the toxin (likely the −1 position) is likely not sufficient for splicing. It is recommended to (1) try an alternative split site and reevaluate splicing competency or (2) use site-directed mutagenesis to install a preferred residue (see **Note 9**) at the −1 position. If the latter option is pursued, the new −1 extein context will leave a splicing scar and thus a full-length version of the toxin harboring this mutation must also be generated and tested for its cytotoxicity to cells. This may be evaluated by cloning the toxin (with a −1 extein mutation) into pcDNA3.1(+), transfecting HeLa cells and quantitating cytotoxicity relative to a pcDNA3.1(+) vehicle control transfection. In the described case of α-sarcin, asparagine-28 was mutated to glycine, and the corresponding full-length toxin was found to retain cytotoxicity [12].

17. The splicing described for constructs 3×F-[N]Sar and [C]Sar produces an inactive toxin by virtue of the N-terminal 3×FLAG epitope. The 3×FLAG epitope conveniently blocks sarcin activity [43], thereby enabling detection of the spliced product by western blot [12]. Where possible, it is recommended to include an electrophoretic standard for the spliced product (e.g., a catalytically inactive and full-length epitope-tagged toxin). Removal of the tag (in the [N]Sar construct) de-attenuates the toxin, such that splicing between constructs [N]Sar and [C]Sar

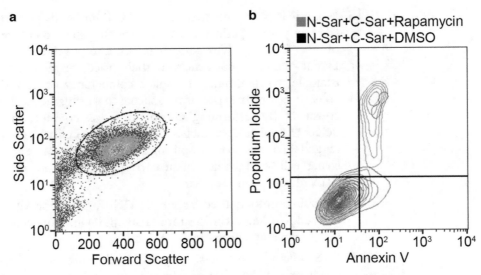

Fig. 5 Detecting apoptosis following CTS. (**a**) On a forward scatter vs. side scatter plot, make a gate (*black circle*) around the major population of cells. (**b**) Plot the events from the gate made in (**a**) using annexin V vs propidium iodide. Shown is a contour plot comparing the levels of apoptosis following expression of both halves of the toxin and treatment with either control DMSO (*black*) or rapamycin (*red*). Treatment with rapamycin induces CTS and apoptosis, as seen with the increased annexin V/propidium iodide positive population (*upper right quadrant*)

renders an active toxin. Inclusion of an epitope tag may not attenuate different toxins applied to the CTS system and, therefore, care should be taken to harvest cell lysates at appropriate times for western blots performed for biochemical characterization (i.e., for rapamycin dosing, timing of splicing, and temperature trials).

18. When cells are treated with trypsin prior to detection of apoptosis (Subheading 3.9), carefully monitor cells, and only use gentle agitation or swirling to lift cells off the dish. It is important not to over trypsinize, as this will disrupt the integrity of the lipid membrane and lead to increased staining with Annexin-V.

19. After cells have been lifted off the plate during the apoptosis assay, carry out all steps on ice or at 4 °C unless otherwise indicated to minimize cell death. When setting up the flow cytometer, ensure cells are in the middle of a forward scatter vs side scatter plot, and set a gate on the major population of cells (Fig. 5a). Plot the events from the gated cells using annexin V versus propidium iodide (Fig. 5b). Ensure to include the appropriate compensation control samples (unstained, annexin V alone, propidium iodide alone) to determine how to separate unstained events from apoptotic events (not shown). Annexin V staining will label cells undergoing early apoptosis and propidium iodide will label cells undergoing late apoptosis/necrosis.

Acknowledgements

This work was supported by an NSERC Discovery Grant to PLH. The ^{N}MBP and ^{C}His constructs described were the kind gift of Dr. Tom Muir.

References

1. Araki K, Araki M, Yamamura K (2006) Negative selection with the Diphtheria toxin A fragment gene improves frequency of Cre-mediated cassette exchange in ES cells. J Biochem 140(6):793–798

2. Fraser B, DuVal MG, Wang H, Allison WT (2013) Regeneration of cone photoreceptors when cell ablation is primarily restricted to a particular cone subtype. PLoS One 8(1):e55410

3. Schuldiner M, Itskovitz-Eldor J, Benvenisty N (2003) Selective ablation of human embryonic stem cells expressing a "suicide" gene. Stem Cells 21(3):257–265

4. Tanoue S, Krishnan P, Krishnan B, Dryer SE, Hardin PE (2004) Circadian clocks in antennal neurons are necessary and sufficient for olfaction rhythms in Drosophila. Curr Biol 14(8):638–649

5. Cui W, Gusterson B, Clark AJ (1999) Nitroreductase-mediated cell ablation is very rapid and mediated by a p53-independent apoptotic pathway. Gene Ther 6(5):764–770

6. Denny WA (2003) Prodrugs for gene-directed enzyme-prodrug therapy (suicide gene therapy). J Biomed Biotechnol 2003(1):48–70

7. Denny WA (2004) Tumor-activated prodrugs-a new approach to cancer therapy. Cancer Invest 22(4):604–619

8. Kirn D, Niculescu-Duvaz I, Hallden G, Springer CJ (2002) The emerging fields of suicide gene therapy and virotherapy. Trends Mol Med 8(4 Suppl):S68–S73

9. Malecki M (2012) Frontiers in suicide gene therapy of cancer. J Genet Syndr Gene Ther 2012(3):e114

10. Ciceri F, Bonini C, Gallo-Stampino C, Bordignon C (2005) Modulation of GvHD by suicide-gene transduced donor T lymphocytes: clinical applications in mismatched transplantation. Cytotherapy 7(2):144–149

11. Qasim W, Gaspar HB, Thrasher AJ (2005) T cell suicide gene therapy to aid haematopoietic stem cell transplantation. Curr Gene Ther 5(1):121–132

12. Alford SC, O'Sullivan C, Obst J, Christie J, Howard PL (2014) Conditional protein splicing of alpha-sarcin in live cells. Mol Biosyst 10(4):831–837

13. Eiklid K, Olsnes S, Pihl A (1980) Entry of lethal doses of abrin, ricin and modeccin into the cytosol of HeLa cells. Exp Cell Res 126(2):321–326

14. Jennings JC, Olson BH, Roga V, Junek AJ, Schuurmans DM (1965) Alpha sarcin, a new antitumor agent. II. Fermentation and antitumor spectrum. Appl Microbiol 13:322–326

15. Endo Y, Huber PW, Wool IG (1983) The ribonuclease activity of the cytotoxin alpha-sarcin. The characteristics of the enzymatic activity of alpha-sarcin with ribosomes and ribonucleic acids as substrates. J Biol Chem 258(4):2662–2667

16. Shah NH, Muir TW (2014) Inteins: nature's gift to protein chemists. Chem Sci 5(1):446–461

17. Elleuche S, Poggeler S (2010) Inteins, valuable genetic elements in molecular biology and biotechnology. Appl Microbiol Biotechnol 87(2):479–489

18. Eryilmaz E, Shah NH, Muir TW, Cowburn D (2014) Structural and dynamical features of inteins and implications on protein splicing. J Biol Chem 289(21):14506–14511

19. Amitai G, Callahan BP, Stanger MJ, Belfort G, Belfort M (2009) Modulation of intein activity by its neighboring extein substrates. Proc Natl Acad Sci U S A 106(27):11005–11010

20. Xu MQ, Perler FB (1996) The mechanism of protein splicing and its modulation by mutation. EMBO J 15(19):5146–5153

21. Appleby-Tagoe JH, Thiel IV, Wang Y, Wang Y, Mootz HD, Liu XQ (2011) Highly efficient and more general cis- and trans-splicing inteins through sequential directed evolution. J Biol Chem 286(39):34440–34447

22. Chong S, Williams KS, Wotkowicz C, Xu MQ (1998) Modulation of protein splicing of the Saccharomyces cerevisiae vacuolar membrane ATPase intein. J Biol Chem 273(17):10567–10577

23. Binschik J, Zettler J, Mootz HD (2011) Photocontrol of protein activity mediated by the cleavage reaction of a split intein. Angew Chem Int Ed Engl 50(14):3249–3252

24. Tyszkiewicz AB, Muir TW (2008) Activation of protein splicing with light in yeast. Nat Methods 5(4):303–305

25. Buskirk AR, Ong YC, Gartner ZJ, Liu DR (2004) Directed evolution of ligand dependence: small-molecule-activated protein splicing. Proc Natl Acad Sci U S A 101(29):10505–10510

26. Peck SH, Chen I, Liu DR (2011) Directed evolution of a small-molecule-triggered intein with improved splicing properties in mammalian cells. Chem Biol 18(5):619–630

27. Yuen CM, Rodda SJ, Vokes SA, McMahon AP, Liu DR (2006) Control of transcription factor activity and osteoblast differentiation in mammalian cells using an evolved small-molecule-dependent intein. J Am Chem Soc 128(27):8939–8946

28. Berrade L, Kwon Y, Camarero JA (2010) Photomodulation of protein trans-splicing through backbone photocaging of the DnaE split intein. Chembiochem 11(10):1368–1372

29. Wong CC, Traynor D, Basse N, Kay RR, Warren AJ (2011) Defective ribosome assembly in Shwachman-Diamond syndrome. Blood 118(16):4305–4312

30. Wood DW, Wu W, Belfort G, Derbyshire V, Belfort M (1999) A genetic system yields self-cleaving inteins for bioseparations. Nat Biotechnol 17(9):889–892

31. Zeidler MP, Tan C, Bellaiche Y, Cherry S, Hader S, Gayko U, Perrimon N (2004) Temperature-sensitive control of protein activity by conditionally splicing inteins. Nat Biotechnol 22(7):871–876

32. Wu H, Hu Z, Liu XQ (1998) Protein trans-splicing by a split intein encoded in a split DnaE gene of Synechocystis sp. PCC6803. Proc Natl Acad Sci U S A 95(16):9226–9231

33. Mootz HD, Blum ES, Tyszkiewicz AB, Muir TW (2003) Conditional protein splicing: a new tool to control protein structure and function in vitro and in vivo. J Am Chem Soc 125(35):10561–10569

34. Mootz HD, Muir TW (2002) Protein splicing triggered by a small molecule. J Am Chem Soc 124(31):9044–9045

35. Shi J, Muir TW (2005) Development of a tandem protein trans-splicing system based on native and engineered split inteins. J Am Chem Soc 127(17):6198–6206

36. Zettler J, Schütz V, Mootz HD (2009) The naturally split Npu DnaE intein exhibits an extraordinarily high rate in the protein trans-splicing reaction. FEBS Lett 583(5):909–914

37. Kanno A, Ozawa T, Umezawa Y (2006) Intein-mediated reporter gene assay for detecting protein-protein interactions in living mammalian cells. Anal Chem 78(2):556–560

38. Ozawa T, Nogami S, Sato M, Ohya Y, Umezawa Y (2000) A fluorescent indicator for detecting protein-protein interactions in vivo based on protein splicing. Anal Chem 72(21):5151–5157

39. Ozawa T, Takeuchi TM, Kaihara A, Sato M, Umezawa Y (2001) Protein splicing-based reconstitution of split green fluorescent protein for monitoring protein-protein interactions in bacteria: improved sensitivity and reduced screening time. Anal Chem 73(24):5866–5874

40. Ozawa T, Kaihara A, Sato M, Tachihara K, Umezawa Y (2001) Split luciferase as an optical probe for detecting protein-protein interactions in mammalian cells based on protein splicing. Anal Chem 73(11):2516–2521

41. Sonntag T, Mootz HD (2011) An intein-cassette integration approach used for the generation of a split TEV protease activated by conditional protein splicing. Mol Biosyst 7(6):2031–2039

42. Narayanan S, Surendranath K, Bora N, Surolia A, Karande AA (2005) Ribosome inactivating proteins and apoptosis. FEBS Lett 579(6):1324–1331

43. Alford SC, Pearson JD, Carette A, Ingham RJ, Howard PL (2009) Alpha-sarcin catalytic activity is not required for cytotoxicity. BMC Biochem 10:9

Chapter 14

Photocontrol of the Src Kinase in Mammalian Cells with a Photocaged Intein

Wei Ren and Hui-Wang Ai

Abstract

Recently developed methods that can photochemically control protein activities and functions in live cells have opened up new opportunities for studying signaling networks at the cellular and subcellular levels. Our laboratory has reported a genetically encoded photoactivatable intein, which allows the direct photo-control of primary sequences of proteins, and consequently, their activities and functions in live mammalian cells. Herein, we provide details on experimental design and the utilization of this photocaged intein to photoactivate the Src tyrosine kinase in human embryonic kidney (HEK) 293T cells. The described procedures may be adopted to photocontrol other proteins in other types of mammalian cells.

Key words Photocaged intein, Photochemical genetics, Genetic code expansion, Unnatural amino acid, Protein splicing, Photoactivation

1 Introduction

Methods that can selectively perturb or control biological processes, such as genetic manipulation, RNA interference (RNAi), and small-molecule-based chemical genetics, have greatly accelerated our understanding of complex biological networks [1, 2]. In the past few years, optogenetics, which uses light to achieve biological control, has gained increasing attention [3–5]. Light-based optogenetic perturbation often results in fast responses and excellent spatial and/or temporal precision. As a promising strategy, genetic code expansion, which is based on the engineering and expression of orthogonal aminoacyl-tRNA synthetases (aaRS) and orthogonal aminoacyl tRNA, has enabled site-specific incorporation of photo-caged lysine, tyrosine, serine and cysteine residues into proteins of interest in bacterial, yeast and mammalian cells [6–11]. These pho-tocaged unnatural amino acids have been incorporated into critical protein residues for optical control of enzyme activities [12–14], ion channels [15], gene expression and silencing [16], and protein translocation [17, 18]. We have recently introduced a photocaged

Henning D. Mootz (ed.), *Split Inteins: Methods and Protocols*, Methods in Molecular Biology, vol. 1495,
DOI 10.1007/978-1-4939-6451-2_14, © Springer Science+Business Media New York 2017

cysteine amino acid into a highly efficient *Nostoc punctiforme* (*Npu*) DnaE intein, which was further inserted into other proteins to disrupt their functions [19]. The *Npu* DnaE intein is a naturally occurring split intein, but in our study, the two fragments were fused with a floppy peptide sequence to generate a one-piece, *cis*-splicing intein. A light-triggered photochemical reaction could reactivate the photocaged intein and induce protein splicing to regenerate proteins with their native sequences [19]. This method is highly attractive, because the modulation only requires the information on the amino acid sequences of proteins of interest. An understanding of their detailed biochemical properties or 3D structures is not a necessity, so this method may be utilized to investigate the biological functions of new or underexplored protein targets.

Here we detail how to use this photocaged intein to control the enzyme activity of a human Src tyrosine kinase in human embryonic kidney (HEK) 293T cells. This process involves the use of molecular biology to create chimeric gene cassettes encoding the protein of interest and an *Npu* DnaE intein, in which the codon of a critical Cys1 residue is mutated to an amber (TAG) codon for the genetic insertion of a photocaged cysteine. The resultant plasmid, along with an amber suppression plasmid encoding an orthogonal aaRS/tRNA pair for the photocaged cysteine, is used to co-transfect mammalian cells. Cells are next cultured in the medium supplemented with the photocaged cysteine amino acid, and the full-length chimeric protein is biologically synthesized. Next, a light source can be utilized to photoactivate the intein, and subsequently, the protein of interest. Furthermore, downstream studies can be performed to examine biological consequences, which are directly or indirectly caused by the activation of this particular protein.

It is important to select right residue sites in the protein of interest for insertion of the photocaged *Npu* DnaE intein. Since a downstream C-extein Cys + 1 residue is required for protein splicing, so it is preferable to insert the intein before a cysteine residue. If no cysteine exists, one has to select other residues to make point mutations. For example, a serine-to-cysteine mutant is often tolerated without drastically affecting protein activities [20]. Moreover, because additional N- and C- terminal extein sequences might affect the kinetics of protein splicing [21] and intein insertions at certain residue sites may not affect protein activities, it is also preferable to select several sites for construction of genetic chimeras. The derived multiple mutants may be screened for protein expression levels, post-photoactivation splicing kinetics, and activities/functions before and after splicing. For example, the catalytic domain of the human tyrosine kinase Src has 8 cysteine residues and 12 serine residues [22], and we randomly selected three sites for insertion of the photocaged intein. We sandwiched the intein between Gly276 and Cys277, or Val399 and Cys400 of Src (F1 and F2 in Fig. 1a). We also built the third construct (F3 in Fig. 1a), in which the intein was placed downstream of Met341, in addition

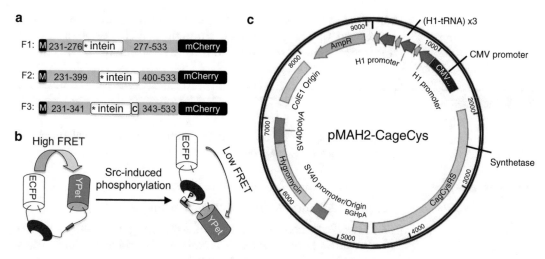

Fig. 1 Major genetic components used in this study. (**a**) Primary structures of the chimeric intein/Src proteins. The *grey portion* of the bars represents the sequence of the human Src kinase between the indicated residues. The *asterisk* indicates the Cys1 residue for UAA incorporation; "M" is methionine, as the translational start site; and "C" is cysteine, used to replace residue 342 of Src. (**b**) A FRET-based Src tyrosine kinase reporter. It is composed of a substrate peptide and a phosphotyrosine-binding domain between a FRET donor ECFP and a FRET acceptor YPet. The conformation change induced by Src-dependent phosphorylation causes the FRET change. (**c**) The plasmid map for pMAH2-CageCys, showing gene elements to express the corresponding aminoacyl tRNA (H1-tRNA) and aminoacyl-tRNA synthetase

to a Ser342Cys mutation. A red fluorescent mCherry was fused at the C-terminus for evaluation of amber suppression efficiency. Our studies showed that all three constructs produced inactive full-length proteins in HEK 293T cells, which could be further photo-activated to restore their Src kinase activities [19]. We also utilized a fluorescent kinase reporter based on Förster resonance energy transfer (FRET) between ECFP and YPet (Fig. 1b), to monitor the Src kinase activity before and after photoactivation. This kinase reporter was previously established by Wang et al. [23], and Src can reduce the ratio of the sensitized YPet fluorescence to the direct ECFP donor emission.

Our amber suppression plasmid, pMAH2-CageCys (Fig. 1c), contains gene cassettes to express an engineered aaRS and its corresponding tRNA. We used it to co-transfect HEK 293T cells along with the other plasmid harboring the chimeric intein/Src gene. The chimeric gene can be simply inserted into any common mammalian expression plasmid, such as pcDNA3, or used to replace the aaRS gene in pMAH2-CageCys to afford a plasmid that also co-expresses the suppression tRNA. These aforementioned plasmids are now deposited to the non-profit plasmid repository Addgene (http://www.addgene.org/Huiwang_Ai/). Since the molecular biology procedures used to create these plasmids are routine, this method paper will not focus on plasmid construction. Instead, we will use Construct F1 as the example to present experimental procedures for cell culture and transfection, expression of the photocaged chimeric protein, photoactivation, fluorescence measurements and imaging, and image processing.

2 Materials

Prepare all solutions using MilliQ water with electrical resistivity equal or above 18 MΩ·cm.

2.1 Plasmids

1. pMAH2-CageCys (Addgene Plasmid #71404).
2. pcDNA3-Intein/Src (Addgene Plasmid #71406).
3. KRas-Src kinase reporter (Prof. Peter Yingxiao Wang, UCSD). Prepare plasmid DNA with commercial Maxiprep kits. Adjust their concentrations to 1 μg/μL.

2.2 Chemicals for Synthesis of S-(1-(4,5-dimethoxy-2-nitrophenyl)ethyl) Cysteine (CageCys Amino Acid)

1. 4′,5′-dimethoxy-2′-nitroacetophenone.
2. Sodium borohydride ($NaBH_4$).
3. Ethanol (EtOH).
4. Phosphorus tribromide (PBr_3).
5. Dichloromethane (CH_2Cl_2).
6. L-cysteine (Cys).
7. Triethylamine (TEA).
8. Hydrochloric acid (HCl).
9. Tetrahydrofuran (THF).
10. Anhydrous sodium sulfate (Na_2SO_4).
11. Sodium bicarbonate ($NaHCO_3$).

2.3 Other Materials

1. HEK 293T cells (e.g., ATCC, Cat. # CRL-3216).
2. Dulbecco's Modified Eagle's Medium (DMEM).
3. Dulbecco's Phosphate Buffered Saline, no calcium, no magnesium (DPBS).
4. Dulbecco's Phosphate Buffered Saline, with calcium and magnesium (DPBS++).
5. Fetal bovine serum (FBS).
6. 35-mm cell culture dishes.
7. Linear polyethylenimine solution, 25 kDa (PEI): Mix 5 mL of 0.1 M HCl with 10 mg of linear PEI. After dissolving the solid, add 500 μL of 1 M NaOH to neutralize the solution. Add MilliQ water to a volume of 10 mL to make a final concentration of 1 μg/μL. Filter the solution through a 0.22-μm filter and store it at −20 °C.

2.4 Equipment and Software

1. UVA light source (365 nm radiation, e.g., Black Ray Lamp, Model XX-20BLB, VWR).

2. An epi-fluorescence microscope equipped with a DAPI filter cube (ex. 377 nm/50 nm), a CFP filter cube (ex. 436 nm/20 nm; em. 480 nm/40 nm), a CFP-YFP FRET filter cube (ex. 436 nm/20 nm; em. 535 nm/50 nm), and an RFP filter cube (ex. 560 nm/40 nm; em. 635 nm/60 nm).

3. A fluorescence plate reader or a fluorescence spectrometer.

4. Fiji software [24, 25]. Use the link http://fiji.sc/Downloads and follow the instructions to download and install Fiji, a pre-bundled ImageJ package. Check your Fiji installation for plug-ins, including TurboReg, MultiStackReg, Ratio Plus, and NucMed. If anything is missing, you may downloaded it from the plug-ins page in the ImageJ website (http://rsbweb.nih.gov/ij/plugins/).

---

3 Methods

Unless otherwise specified, carry out all procedures at room temperature.

3.1 Synthesis of S-(1-(4,5-dimethoxy-2-nitrophenyl)ethyl) Cysteine (CageCys)

1. Dissolve 4′,5′-dimethoxy-2′-nitroacetophenone (900 mg, 4 mmol) in THF/EtOH (1:1,15 mL) at room temperature, followed by intermittent addition of $NaBH_4$ (152 mg, 4 mmol) over 20 min.

2. After stirring the reaction mixture for another 3 h, add dilute HCl (1 M, 4 mL) to neutralize excess $NaBH_4$.

3. Remove the solvent by using rotary evaporator, and subsequently add H_2O (10 mL) to the residue.

4. Use CH_2Cl_2 (10 mL) to extract the mixture three times. Combine the organic layer from each of these extractions, and dry it over anhydrous Na_2SO_4. Remove the solvent by using rotary evaporator to afford yellow solid, which can be used directly without further purification.

5. Dissolve the yellow solid in CH_2Cl_2 (20 mL) cooled in ice bath. Add PBr_3 (475 μL, 5 mmol) dropwise.

6. Stir the mixture for another 3 h before adding saturated $NaHCO_3$ aqueous solution (15 mL).

7. Separate and keep the organic layer. Wash the organic layer twice with H_2O (10 mL), and dry it over anhydrous Na_2SO_4. Remove the solvent by using rotary evaporator to afford crude yellow oil. The yellow oil can be further purified by using silica chromatography (EtOAc/Hexane 1:4).

8. Dissolve L-cysteine (0.36 g, 3 mmol) in H_2O (5 mL) and TEA (405 μL, 2.8 mmol). Cool the mixture in ice/water bath.

9. Dissolve the yellow oil (2.79 mmol) from **step 7** in methanol (5 mL), and add it into the mixture in **step 8** in ice/water bath dropwise over 15 min.

10. Stir the mixture overnight. Yellow precipitation is expected to form.

11. Remove the solvent by using rotary evaporator. Dissolve the solid in saturated $NaHCO_3$ aqueous solution (10 mL). Wash twice with CH_2Cl_2 (10 mL each), and dispose the organic phase.

12. Cool the aqueous phase on ice. Acidify the mixture with HCl (1 M) to pH 5.5. Leave the mixture on ice for additional 15 min to form a yellow precipitate.

13. Filter the mixture under reduced pressure. Dry the yellow solid, which is CageCys·HCl.

14. Mix 5 mL 1 M NaOH with 183 mg CageCys·HCl to make a stock solution containing 100 mM CagCys amino acid. Filter the solution through a 0.22-μm filter and store it at –80 °C.

3.2 Cell Culture and Transfection

1. Culture HEK 293T cells in DMEM supplemented with 10% (v/v) FBS in the humidified incubator at 37 °C with 5% CO_2. In Day 1, seed 0.8×10^5 cells into 35-mm cell culture dishes.

2. In Day 2, prepare a transfection medium by mixing 2 μg pMAH2-CageCys, 1.5 μg pcDNA3-Intein/Src, 0.5 μg KRas-Src reporter plasmid, 10 μL PEI (1 μg/μL), and 500 μL fresh FBS-free DMEM (*see* **Note 1**). Incubate the mixture at room temperature for 20 min.

3. Remove FBS-containing DMEM from a 35-mm cell culture dish. Incubate the cells in the transfection medium for 1 h in the humidified incubator at 37 °C with 5% CO_2.

4. Remove the transfection medium, and add fresh DMEM supplemented with 10% (v/v) FBS and 1 mM CageCys amino acid (*see* **Note 2**).

5. Return the cell culture dish to the humidified incubator at 37 °C with 5% CO_2.

6. In Day 3, replace the medium with fresh DMEM supplemented with 10% (v/v) FBS and 1 mM CageCys amino acid. Culture the cells for another 2–3 days in the humidified incubator at 37 °C with 5% CO_2.

7. Transfer the culture dish with transfected HEK293T cells to the stage of an inverted epi-fluorescence microscope. Examine the red fluorescence of mCherry, which is an indicator for the expression levels of the full-length protein in individual cells.

3.3 Photoactivation and the Subsequent Imaging on an Epi-fluorescence Microscope

The following procedures can be utilized to photoactivate and image a small area of HEK 293T cells using a common epi-fluorescence microscope.

1. Remove the cell culture medium. Wash cells with DPBS++, and leave cells in DPBS++ for subsequent studies.

2. Identify chimeric-protein-expressing red fluorescent cells under the fluorescence microscope. After removing all neutral density filters in the light path to gain the highest intensity, irradiate the area using a DAPI excitation filter for 2 min.

3. Set up the instrument to acquire fluorescence images using the CFP channel and the CFP-YFP FRET channel. A time-lapse series may be acquired to monitor changes over 1 h.

3.4 Image Processing

The following procedures are adapted from a previous protocol [26] and can be utilized to analyze FRET ratios and render pseudocolor images.

1. Open the raw CFP and FRET images in the Fiji software (*see* **Note 3**).

2. Select a region of interest (ROI) by specifying a region using the 'Edit → Selection → Specify' command from the Fiji menu. Switch to the other channel and specify the same ROI. Crop both images using the command 'Image → Crop'.

3. Select 'Process → Subtract background' from the menu with all boxes in the dialog box unchecked. Set the rolling ball radius between 50 and 200 pixels. The radius of the rolling ball should be bigger than that of the largest object in the image. Perform this transformation for both channels.

4. There may be a small pixel shift between the raw CFP and FRET images. Select 'Plugins → Registration → MultiStackReg' to align two images or two time-lapse stacks. In the dialog box, select the CFP channel as the reference, and align the FRET channel to it. Select 'Rigid Body' as the transformation method.

5. Convert images or image stacks to 32-bit by selecting 'Image → Type → 32 bit'. Perform this transformation for both channels.

6. Select 'Process → Smooth' to reduce the image noise. Perform this transformation for both channels.

7. This step is applied only to the FRET channel, which converts the background pixels to 'not a number' (NaN). Select 'Image → Adjust → Threshold' from the menu. Set the following parameters: 'Default', 'B&W', and 'Dark Background'. Press 'Apply'. An icon should appear to show 'Set Background Pixels to NaN'. Press 'OK'.

8. Select 'Plugins → Ratio Plus' from the menu to generate a ratiometric image. Choose the FRET image as 'Image1' and the

CFP image as 'Image2' and press 'OK'. Set background and clipping values to 0, and 'multiplication factor' to 1.

9. Select 'Plugins → NucMed → Lookup tables' from the menu. Select 'Blue_Green_Red' from the list.

10. Select 'Image → Adjust → Brightness/Contrast' from the menu. Click 'Set' in the dialog box to precisely define the range. Save the image as a TIFF file.p.

11. Select 'Analyze → Tools → Calibration Bar' from the menu to derive a calibration bar showing the relationship between the ratios and false colors.

12. To get a number for the FRET ratio of a particular area, circle the area and select 'Analyze → Measure' to get quantitative information.

13. For presentation purposes, select 'Image → Type → RGB color' from the menu to convert it into RGB. Save the color image as a different TIFF file (*see* **Note 4**).

3.5 Photoactivation of Bulk Samples for Subsequent Characterizations

The following procedures can be utilized to photoactivate a large number of HEK 293T cells for downstream characterizations.

1. Remove the cell culture medium. Wash cells with DPBS++. Leave cells in 35-mm culture dishes containing 200 μL DPBS++.

2. Illuminate cells in culture dishes sitting on ice with the UVA light source (365 nm radiation) for 5 min.

3. Collect cells. Resuspend cells in DPBS and use a fluorescence plate reader or a fluorescence spectrometer to monitor the fluorescence of the Src kinase reporter.

4. Cells may also be lysed for downstream applications, such as Western Blot, protein purification, and mass spectrometry characterizations.

4 Notes

1. Other commercial transfection reagents may be utilized to replace PEI and gain better transfection efficiency. When difficult-to-transfect cell lines are utilized, using commercial transfection reagents may be necessary. Moreover, a viral delivery system may be utilized to gain the highest transduction efficiency [27].

2. The CageCys amino acid is photosensitive. Light exposure should be minimized. It is a good practice to wrap cell culture dishes with aluminum foil.

3. Commercial softwares can be utilized for image processing. When Fiji (or ImageJ) is being used, it is important to save intermediate steps, because this free software does not do it automatically.

4. After converting the ratio image to 'RGB', the quantitative information is lost. The original 32-bit ratio image has to be saved, if a continuing access of the quantitative information is needed.

Acknowledgement

This work was supported by the Start-up Fund to H.A. from the University of California, Riverside.

References

1. Alaimo PJ, Shogren-Knaak MA, Shokat KM (2001) Chemical genetic approaches for the elucidation of signaling pathways. Curr Opin Chem Biol 5(4):360–367

2. Dykxhoorn DM, Novina CD, Sharp PA (2003) Killing the messenger: short RNAs that silence gene expression. Nat Rev Mol Cell Biol 4(6):457–467

3. Deisseroth K (2011) Optogenetics. Nat Methods 8(1):26–29

4. Pastrana E (2011) Optogenetics: controlling cell function with light. Nat Methods 8(1):24–25

5. Miesenböck G (2011) Optogenetic control of cells and circuits. Annu Rev Cell Dev Biol 27:731–758

6. Wu N, Deiters A, Cropp TA, King D, Schultz PG (2004) A genetically encoded photocaged amino acid. J Am Chem Soc 126(44):14306–14307

7. Chen PR, Groff D, Guo J, Ou W, Cellitti S, Geierstanger BH, Schultz PG (2009) A facile system for encoding unnatural amino acids in mammalian cells. Angew Chem Int Ed Engl 48(22):4052–4055

8. Lemke EA, Summerer D, Geierstanger BH, Brittain SM, Schultz PG (2007) Control of protein phosphorylation with a genetically encoded photocaged amino acid. Nat Chem Biol 3(12):769–772

9. Liu CC, Schultz PG (2010) Adding new chemistries to the genetic code. Annu Rev Biochem 79:413–444

10. Deiters A, Groff D, Ryu Y, Xie J, Schultz PG (2006) A genetically encoded photocaged tyrosine. Angew Chem Int Ed Engl 45(17):2728–2731

11. Ai HW (2012) Biochemical analysis with the expanded genetic lexicon. Anal Bioanal Chem 403(8):2089–2102

12. Zhao J, Lin S, Huang Y, Zhao J, Chen PR (2013) Mechanism-based design of a photo-activatable firefly luciferase. J Am Chem Soc 135(20):7410–7413

13. Gautier A, Deiters A, Chin JW (2011) Light-activated kinases enable temporal dissection of signaling networks in living cells. J Am Chem Soc 133(7):2124–2127

14. Groff D, Wang F, Jockusch S, Turro NJ, Schultz PG (2010) A new strategy to photoactivate green fluorescent protein. Angew Chem Int Ed Engl 49(42):7677–7679

15. Kang JY, Kawaguchi D, Coin I, Xiang Z, O'Leary DD, Slesinger PA, Wang L (2013) In vivo expression of a light-activatable potassium channel using unnatural amino acids. Neuron 80(2):358–370

16. Hemphill J, Chou C, Chin JW, Deiters A (2013) Genetically encoded light-activated transcription for spatiotemporal control of gene expression and gene silencing in mammalian cells. J Am Chem Soc 135(36):13433–13439

17. Gautier A, Nguyen DP, Lusic H, An W, Deiters A, Chin JW (2010) Genetically encoded photocontrol of protein localization in mammalian cells. J Am Chem Soc 132(12):4086–4088

18. Baker AS, Deiters A (2014) Optical control of protein function through unnatural amino acid mutagenesis and other optogenetic approaches. ACS Chem Biol 9(7):1398–1407

19. Ren W, Ji A, Ai HW (2015) Light activation of protein splicing with a photocaged fast intein. J Am Chem Soc 137(6):2155–2158

20. Wang X, Pineau C, Gu S, Guschinskaya N, Pickersgill RW, Shevchik VE (2012) Cysteine scanning mutagenesis and disulfide mapping analysis of arrangement of GspC and GspD protomers within the type 2 secretion system. J Biol Chem 287(23):19082–19093

21. Cheriyan M, Pedamallu CS, Tori K, Perler F (2013) Faster protein splicing with the Nostoc punctiforme DnaE intein using non-native extein residues. J Biol Chem 288(9):6202–6211

22. Johannessen CM, Boehm JS, Kim SY, Thomas SR, Wardwell L, Johnson LA, Emery CM, Stransky N, Cogdill AP, Barretina J, Caponigro G, Hieronymus H, Murray RR, Salehi-Ashtiani K, Hill DE, Vidal M, Zhao JJ, Yang X, Alkan O, Kim S, Harris JL, Wilson CJ, Myer VE, Finan PM, Root DE, Roberts TM, Golub T, Flaherty KT, Dummer R, Weber BL, Sellers WR, Schlegel R, Wargo JA, Hahn WC, Garraway LA (2010) COT drives resistance to RAF inhibition through MAP kinase pathway reactivation. Nature 468(7326):968–972

23. Seong J, Lu S, Ouyang M, Huang H, Zhang J, Frame MC, Wang Y (2009) Visualization of Src activity at different compartments of the plasma membrane by FRET imaging. Chem Biol 16(1):48–57

24. Schneider CA, Rasband WS, Eliceiri KW (2012) NIH Image to ImageJ: 25 years of image analysis. Nat Methods 9(7):671–675

25. Schindelin J, Arganda-Carreras I, Frise E, Kaynig V, Longair M, Pietzsch T, Preibisch S, Rueden C, Saalfeld S, Schmid B, Tinevez JY, White DJ, Hartenstein V, Eliceiri K, Tomancak P, Cardona A (2012) Fiji: an open-source platform for biological-image analysis. Nat Methods 9(7):676–682

26. Kardash E, Bandemer J, Raz E (2011) Imaging protein activity in live embryos using fluorescence resonance energy transfer biosensors. Nat Protoc 6(12):1835–1846

27. Chatterjee A, Xiao H, Bollong M, Ai HW, Schultz PG (2013) Efficient viral delivery system for unnatural amino acid mutagenesis in mammalian cells. Proc Natl Acad Sci U S A 110(29):11803–11808

Chapter 15

LOV2-Controlled Photoactivation of Protein *Trans*-Splicing

Anam Qudrat, Abdullah Mosabbir, and Kevin Truong

Abstract

Protein *trans*-splicing is a posttranslational modification that joins two protein fragments together via a peptide a bond in a process that does not require exogenous cofactors. Towards achieving cellular control, synthetically engineered systems have used a variety of stimuli such as small molecules and light. Recently, split inteins have been engineered to be photoactive by the LOV2 domain (named LOVInC). Herein, we discuss (1) designing of LOV2-activated target proteins (e.g., inteins), (2) selecting feasible splice sites for the extein, and (3) imaging cells that express LOVInC-based target exteins.

Key words Protein *trans*-splicing, Photoactivation, Fluorescence, Light-oxygen-voltage domain, Inteins

1 Introduction

Protein *trans*-splicing is a posttranslational modification facilitated by two parts of a split intein (termed InN and InC) that bind and are excised in the process that joins the two parts of the extein (termed ExN and ExC) with a peptide bond. Notably, this *trans*-splicing modification via split intein does not require exogenous cofactors. The DnaE split intein was initially characterized from the cyanobacterium species *Synechocystis* sp. Strain PCC6803 (Ssp) and found to undergo protein *trans* splicing [1]. Later work exploited this naturally occurring *trans*-splicing activity of DnaE split intein to generate functional engineered systems that can control protein purification, protein cyclization, site-specific protein labeling, site-specific proteolysis and biosensors to detect protein–protein interactions [2]. Other work instead used the DnaE intein homolog from *Nostoc punctiforme* cyanobacterium as it was shown to have a *trans*-splicing efficiency greater than 98 % [3].

The Light-Oxygen-Voltage-sensing (LOV) domains are naturally occurring sensory proteins found in many organisms that respond to changes in light availability. Light serves as a ubiquitous signal to induce a variety of responses: stress response and cellular

Henning D. Mootz (ed.), *Split Inteins: Methods and Protocols*, Methods in Molecular Biology, vol. 1495,
DOI 10.1007/978-1-4939-6451-2_15, © Springer Science+Business Media New York 2017

adhesion in bacteria; virulence regulation in fungi; phototropism and chloroplast relocation in plants [4]. LOV domains acquire photoactivity through light dependent interactions with a flavin-based cofactor (e.g., flavin mononucleotide (FMN) or flavin adenine dinucleotide (FAD)). Usually the LOV domain is followed by a linker sequence and an effector domain that include enzymes (e.g., histidine kinases, serine/threonine kinases), DNA binding structural motifs (e.g., helix–turn–helix proteins), or other signaling proteins [4]. For example, the LOV2 domain found in phototropin-1 from *Avena sativa* (i.e., the Oat plant) controls the activity of a serine/threonine protein kinase [5]. In absence of blue light (e.g., 440–480 nm), the LOV2 domain tightly binds to a downstream amphipathic Jα helix; in the presence of blue light, LOV2 releases binding to the Jα helix. This light-induced conformational change is reversible, allowing a distinct closed dark and open lit state of the LOV2-Jα domain [6, 7]. This photoactive mechanism of the LOV2 domain was used to engineer genetically encoded photoswitches that spatiotemporally control an array of cellular processes involving signaling cascades, DNA binding proteins, small GTPases, membrane-bound Ca^{2+} channels, as well as cell death [8–12].

Recently, a split intein was engineered with protein *trans*-splicing activity upon blue light stimulation using the *N. punctiforme* DnaE split intein and the *A. sativa* LOV2-Jα domain (named LOVInC) [13]. To identify the minimum required functional residues of the DnaE InC, the N-terminal β-strand was serially truncated. With the removal of each subsequent residue, the *trans*-splicing activity was tested in a cell-based assay measuring fluorescence translocation. The truncation of up to four residues (i.e., Ile-Ala-Thr-Arg) from the N- terminal β-strand of InC allowed near full retention of *trans*-splicing activity and any further truncation significantly abolished activity. Next, the LOV2-Jα domain was fused N-terminally to this truncation mutant of the InC, which, in this case, caused the InC to acquire photoactivity to its *trans*-splicing activity in the presence of InN. In theory, this procedure works to create photoactive target proteins because the target protein has been rendered to the verge of instability at its N-terminal end and the LOV-Jα domain photoactive conformational change exerts steric hindrance on its C-terminal end (i.e., point of fusion with the target protein). In target proteins where this design works in creating photoactive proteins (e.g., LOVInC), the LOV2-Jα allows the target protein (e.g., InC) to teeter between stability and instability in the presence and absence of light, respectively.

2   Materials

2.1 *Visualizing and Predicting Protein Structure*

1. Protein visualization software: e.g., Swiss PdbViewer [14], Visual Molecular Dynamics [15], or BioBlender [16].

2. Protein secondary structure prediction software: e.g., SPIDER2 [17], s2D [18], or RaptorX-SS8 [19].

2.2 Target Amplification

1. PCR kit for target sequence amplification.
2. Thermal cycler to run the Polymerase Chain Reaction (PCR).

2.3 Subcloning

1. pCfvtx expression vector.
2. pTriEx-mCherry-PA-Rac1 (Addgene Plasmid #22027) as a source for the LOV2 gene.
3. Restriction enzymes and associated buffers.
4. Standard 1 % agarose gel.
5. Gel purification kit.
6. *E. coli* (DH5α strain).
7. DNA purification kit.

2.4 Cell Culture and Transfection

1. Culture containers: e.g., T25 flasks, 24-well multiwell plates or 96-well tissue culture plates.
2. Cell growth media: e.g., Dulbecco's Modified Eagle's Medium (DMEM) with 25 mM D-glucose, 1 mM sodium pyruvate, 4 mM L-glutamine and 30 % fetal bovine serum (FBS).
3. Trypsin-based reagent for trypsinization of cells, e.g., TrypLE with Phenol Red (Invitrogen).
4. Phosphate buffered saline (PBS).
5. Liposome-based transfection reagents.
6. CO_2 Incubator.

2.5 Imaging and Illumination

1. Widefield fluorescence microscope.
2. Excitation and emission filters: CFP (Ex: 438/24 nm; Em: 482/32 nm), YFP (Ex: 500/24 nm; Em: 542/27 nm), RFP (Ex: 580/20 nm; Em: 630/60 nm).
3. Image acquisition software: e.g., μManager [20], Advanced MetaMorph (Molecular Devices), IC Capture (The Imaging Source).
4. Waterproof strobe flashlight: e.g., P10 Tactical Strobe (Nitecore), M18 Striker (Olight).
5. 20 μm filter (e.g., PluriSelect), optional for single cell analysis.
6. Glass-bottom culture plates.

3 Methods

3.1 Designing LOV2-Activated Target Proteins (e.g., Inteins)

This section highlights the overall procedure while Subheading 3.3 details a specific example with a split intein.

1. Design primers (*see* **Note 1**) to amplify (*see* **Note 2**) N-terminal truncated sequences of the target gene (*see* **Notes 3** and **4**) (Fig. 1). In the overhangs of the primers, include the restriction sites (typically, *Spe*I and *Bgl*II or *Bam*HI and *Nhe*I) for inserting the PCR fragment into the pCfvtx expression vector [21].

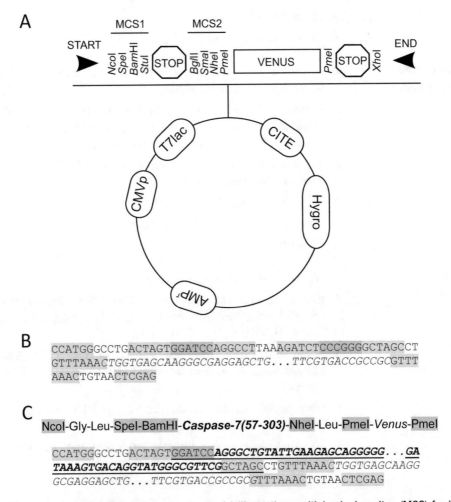

Fig. 1 Plasmid map of the pCfvtx expression vector (**a**) illustrating multiple cloning sites (MCS) for inserting target proteins. The DNA sequence (**b**) of the multiple cloning cite. Restriction sites are shaded grey and Venus is in *italics*. The DNA sequence (**c**) of the Caspase-7 fragment (amino acids 57–303) inserted into the pCfvtx expression vector. The Caspase-7 fragment is in *bold italics*. Underlined sequences are forward and reverse primers for PCR of Caspase-7 from human cDNA, respectively

2. Ligate the PCR fragment of the target gene sequence into the pCfvtx expression vector (*see* **Note 5**) at the selected restriction sites in **step 1**. The will create a fusion protein between the fragment and the Venus yellow fluorescent protein mutant.

3. Develop a cell-based screening assay (*see* **Note 6**) to test the function of each truncated sequence of the target gene. This assay will depend on the function of the target protein.

4. Transfect the cells, using liposome-based reagents following manufacturer's protocol, with each of the truncated target genes and test for function using the assay in **step 3**. Identify the truncation where the loss of function begins and use ±2 residues from this site to generate four testable fragments.

5. Design primers (*see* **Note 1**) to amplify the LOV-Jα fragment (L404 to L546) from pTriEx-mCherry-PA-Rac1 (Addgene Plasmid #22027). In the overhangs of the primers, include the restriction sites for *Nco*I and *Spe*I (*see* **Note 7**) for inserting the PCR fragment into each one of the four host vectors containing the testable fragments in **step 4**.

6. Perform the cell-based screening assay in the presence and absence of pulsed light (*see* **Notes 8** and **9**) overnight (*see* **Note 10**) using a low intensity CFP filter.

7. Select the LOV2-target protein constructs (*see* **Notes 11** and **12**) that demonstrate the greatest change in response to light (*see* **Note 13**).

3.2 Choosing Where to Split the Extein

1. For exteins with a solved protein structure (*see* **Note 14**), download the structure from the RCSB Protein Data Bank.

2. View the structure using Swiss PdbViewer or equivalent software.

3. Identify loops that separate the extein into two relatively stable parts (ExN and ExC) (*see* **Note 15**). Figure 2 shows illustrative examples of split sites in GFP and caspase-7.

4. Design primers (*see* **Note 1**) to amplify the ExN. In the overhangs of the primers, include the restriction sites *Nco*I and *Spe*I for inserting the PCR fragment into the InN expression vector [13].

5. Design primers (*see* **Note 1**) to amplify the ExC. In the overhangs of the primers, include the restriction sites *Nhe*I and *Xho*I for inserting the PCR fragment into the LOVInC expression vector [13].

6. Develop a cell-based screening assay to test the function of the extein (*see* **Note 16**). This assay will depend on the function of the extein (*see* **Note 17**).

7. Co-transfect cells, using liposome-based reagents following manufacturer's protocol, with the plasmids in **steps 4** and **5**.

A

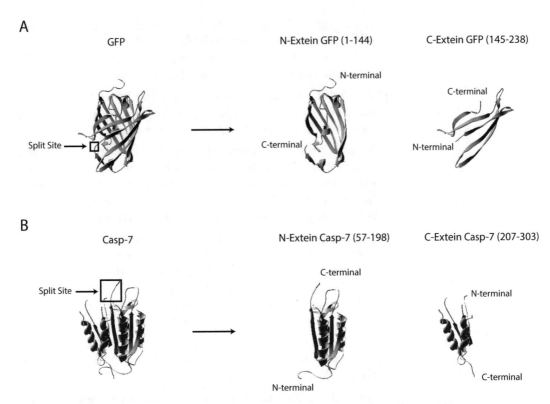

Fig. 2 Representative split sites for (**a**) Green Fluorescent Protein (GFP) and (B) Caspase-7 (Casp-7) generated using Swiss PdbViewer [14]. GFP was split at residues 144–145 as it corresponds to a loop that would separate the protein into parts that retain many bonds within their secondary structures. Casp-7 was split between residues 198–207 because this is the natural self-cleavage site that would separate Casp-7 into two well folded subunits p20 and p11

8. Perform the cell-based screening assay in **step 6** in the presence and absence of pulsed light (*see* **Notes 8** and **9**) overnight (*see* **Note 10**) using low intensity CFP filter. Figure 3 highlights this procedure schematically.

3.3 Using LOVInC-Extein

1. Culture cells of interest in DMEM supplemented with 10% FBS incubated at 37 °C and 5% CO_2 in T25 flasks.

2. Passage cells at 70% confluency using 0.05% TrypLE and seed into 24-well or 96-well plates at a dilution ratio of 1:20.

3. Transfect 0.5 μg each of the plasmids, using liposome-based reagents following manufacturer's protocol, encoding the ExN-InN and LOVInC-ExC (*see* **Note 18**) developed in the previous section into cells of interest (*see* **Note 19**). After an incubation period of 4 h, wash cells and replace growth media for overnight incubation at 37 °C.

4. Image cells (*see* **Notes 20** and **21**) after 24 h (*see* **Note 22**) in glass-bottom culture plates (*see* **Note 23**). If possible, when searching for cells, use red light to avoid pre-stimulation of the sample.

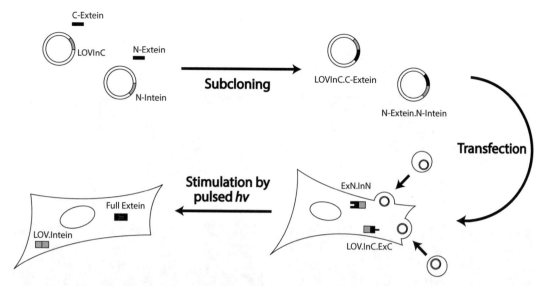

Fig. 3 Schematic overview of procedure. Subcloning is performed to insert extein genes into the plasmid at the appropriate location adjacent to LOVInC or N-intein (InN). Both plasmids are then co-transfected in target cells (e.g., HeLa, HEK 293) and stimulated by pulsed light to confirm extein activity

5. To activate protein *trans*-splicing, excite the LOVInC-ExC for 1 s at 30 s intervals using low intensity CFP filter. Complete protein *trans*-splicing should occur within 4 h.

6. Confirm that splicing occurred by a cell-based assay for the extein function developed in the previous section (*see* **Note 24**). Figure 4 illustrates a translocation assay used for LOVInC.

7. Further confirm splicing occurred by SDS-PAGE (*see* **Note 25**).

4   Notes

1. Primer design tips [22]:
   - Primer length should be 18–22 nucleotides.
   - The annealing temperatures (T_m) of both the forward and the reverse primers should range from 65 to 75 °C and be within 5 °C of each other.
   - Avoid GC or AT rich regions and try to achieve about 50 % GC content.

   Figure 1 illustrates examples of a set of primers for the Caspase-7 target gene. Sample forward and reverse primers with their characteristics are listed below in Table 1.

2. To amplify the target sequence, use an available Polymerase Chain Reaction (PCR) kit and follow manufacturer protocols.

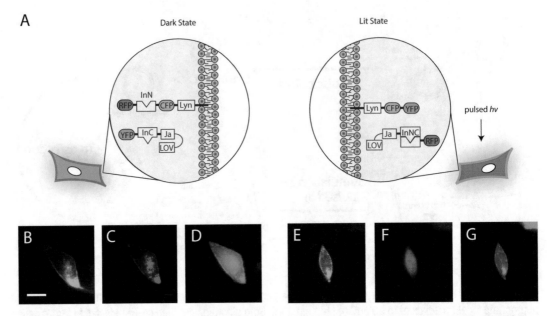

Fig. 4 Activation of a LOVInC system via light. (**a**) A representative diagram depicting the co-localization of fluorescent proteins before (*left*) and after (*right*) stimulation via light. Fluorescence images depict HeLa cells co-expressing plasma membrane tagged CFP-InN-RFP with LOVInC-Venus (**b–d**) before and (**e–g**) after photoactivation. Images are false colored: CFP, *cyan*; YFP, *green*; mRFP, *red*

Table 1
Sample primer sequences and relevant characteristics

	Sequence (5′–3′)	Overlap length	T_m	GC %
Forward	GCCGGATCCAGGGCTGTATTGAAGAGCAGGGGG	24	61	58
Reverse	GCCGCTAGCCGAACGCCCATACCTGTCACTTTATC	25	60	50

The restriction site is *underlined*

3. The series of truncated sequences should begin at the first N-terminal secondary structure (i.e., β-strand or α-helix) found in the target protein, assuming you have knowledge of the structure. Starting truncation in regions before the first N-terminal secondary structure usually does not destabilize the function of the target protein.

4. For split inteins, the truncation should be at the InC because the splicing event will remove the intein together with the LOV2 domain, leaving an extein with only scar residues (i.e., A-E-Y-C-F-N) at the splicing junction [13].

5. The pCfvtx expression vector [21] was generated to aid the fusion of multiple protein domains for driving high expression in both bacterial and mammalian systems. It features two

adjacent multiple cloning sites (MCS) followed by the Venus fluorescent protein. This aids in screening for correct insertion using the presence or absence of fluorescence. The plasmid is available from the author.

6. Fluorescent protein-based assays are convenient for screening because fluorescent proteins can be readily expressed and imaged in living cells. In split inteins, this can be a fluorescent protein translocation event wherein after splicing, fluorescence distribution changes from cytoplasmic to plasma membrane [13]. Aside from fluorescence, another assay can involve changes in the cell's morphological state. For example, using caspase-7 activation, cells appear characteristically round having undergone apoptosis [12].

7. The *Spe*I introduces the two residues (i.e., TS) at the linker between the LOV-Jα domain. These residues could potentially weaken the steric hindrance of the LOV-Jα domain on the target protein in the dark state. Another subcloning strategy is needed to avoid the inclusion of these residues.

8. Using continuous light will cause phototoxicity. Recommended intensity to be used is 25 mW/cm^2 [13].

9. The total duration of light stimulation depends on the activity as the fusion of LOV-Jα domain may have reduced the activity of the target protein. For split inteins, the pulse was administered for 1 s at 30s intervals for 2 h to overnight while for caspase, about 20 min continuous stimulation triggered morphological changes. The 30s interval is consistent with the time it takes the LOV domain to achieve relaxation after photoactivation [13].

10. A waterproof flashlight with strobe settings at the weakest light intensity worked well for our group. It does not matter that the light produced is white because it still emits blue light. For sources of light overnight in a CO$_2$ incubator (*see* Subheading 2.4, **item 4**).

11. In LOVInC, the dark C450M mutant of the LOV-Jα domain was used to better suppress *trans*-splicing activity in the dark state [13]. This dark mutant can irreversibly convert to the lit state after hours of light stimulation. While this ensured a tight control on leakage, it also reduced the speed of *trans*-splicing.

12. It is possible that the fusion of the LOV-Jα domain with the target protein does not create a photoactive target protein as some target proteins are simply not amenable to this approach of engineering light-sensitivity. However, from our experience, 4/7 of the target proteins that we have tested have been responsive.

13. Dynamic range of these engineered photoactive target proteins can be improved with directed evolution strategies [23].

14. In cases where the protein structure is not known, it will be difficult to choose a good splice site for the extein as you may

split at a secondary structure (i.e., α-helix or β-strand). From our experience, this results in extein fragments that aggregate or reformed exteins that no longer fold properly. For unknown protein structures, software packages exist that can predict secondary protein structure (*see* Subheading 2.1, **item 2**).

15. Do not make a split within a secondary structure. While splitting the protein will make it unstable, the task is to minimize this instability by conserving as many bonds as possible (hydrogen bond between secondary structures, ionic bonds found in salt bridges, covalent bonds between disulfide bridges, hydrophobic bonds, etc.). If the split in the extein is made such that there are many secondary structure parts that are not bonded, the fragments will most likely aggregate and decrease protein expression.

16. If the splice site selected does not reconstitute a functional extein, other splice sites should be attempted that have similar desirable characteristics. From our experience, every extein tested (i.e., 4/4) could be split and reformed to have near full functionality.

17. For a GFP extein, seeing a fluorescence signal under the YFP filter (Ex: 500/24 nm; Em: 542/27 nm) will confirm activity. For caspase-7 extein, cells appear characteristically round having undergone apoptosis [12].

18. When co-transfecting plasmids, if there is leak from the LOVInC-ExC, deviate away from equimolar concentrations for the transfection and favor ExN-InN to reduce leakage.

19. Transfect in low light in order to avoid pre-activation of LOVInC-ExC. You may want to wrap the culture plates with aluminum fold.

20. To image single cells, cells can be strained through a 20 μm filter (e.g., PluriSelect) before imaging.

21. Image in PBS because DMEM has phenol that has fluorescence properties.

22. Imaging should be done as soon as possible to ensure a low leak in the sample.

23. Glass-bottom plates will allow collecting of images with higher magnification (e.g., 40×, 60× and 100×) (*see* Subheading 2.4, **item 3**).

24. To improve photoactivation, you can use FMN at a concentration 10× the concentration found in cultured mammalian cells which ranges from 0.46 to 3.4 amol/cell [24].

25. If LOVInC-ExC is fluorescently tagged, you do not need to do a western blot because the fluorescent tagged protein will show up on the SDS-PAGE. Be careful not to boil the sample for more than 5 min before loading because it often abolishes the fluorescence of the fluorescent proteins.

Acknowledgments

This work was funded by grants from the Canadian Cancer Society Research Institute (#701936) and NSERC (#05322-14).

References

1. Wu H, Hu Z, Liu XQ (1998) Protein trans-splicing by a split intein encoded in a split DnaE gene of Synechocystis sp. PCC6803. Proc Natl Acad Sci U S A 95(16):9226–9231

2. Topilina NI, Mills KV (2014) Recent advances in in vivo applications of intein-mediated protein splicing. Mob DNA 5(1):5

3. Iwai H et al (2006) Highly efficient protein trans-splicing by a naturally split DnaE intein from Nostoc punctiforme. FEBS Lett 580(7):1853–1858

4. Herrou J, Crosson S (2011) Function, structure and mechanism of bacterial photosensory LOV proteins. Nat Rev Microbiol 9(10):713–723

5. Renicke C et al (2013) A LOV2 domain-based optogenetic tool to control protein degradation and cellular function. Chem Biol 20(4):619–626

6. Strickland D et al (2010) Rationally improving LOV domain-based photoswitches. Nat Methods 7(8):623–626

7. Jones MA et al (2007) Mutational analysis of phototropin 1 provides insights into the mechanism underlying LOV2 signal transmission. J Biol Chem 282(9):6405–6414

8. Rana A, Dolmetsch RE (2010) Using light to control signaling cascades in live neurons. Curr Opin Neurobiol 20(5):617–622

9. Strickland D, Moffat K, Sosnick TR (2008) Light-activated DNA binding in a designed allosteric protein. Proc Natl Acad Sci U S A 105(31):10709–10714

10. Wu YI et al (2009) A genetically encoded photoactivatable Rac controls the motility of living cells. Nature 461(7260):104–108

11. Pham E, Mills E, Truong K (2011) A synthetic photoactivated protein to generate local or global Ca(2+) signals. Chem Biol 18(7):880–890

12. Mills E et al (2012) Engineering a photoactivated caspase-7 for rapid induction of apoptosis. ACS Synth Biol 1(3):75–82

13. Wong S, Mosabbir AA, Truong K (2015) An engineered split intein for photoactivated protein trans-splicing. PLoS One 10(8):e0135965

14. Johansson MU et al (2012) Defining and searching for structural motifs using DeepView/Swiss-PdbViewer. BMC Bioinformatics 13:173

15. Humphrey W, Dalke A, Schulten K (1996) VMD: visual molecular dynamics. J Mol Graph 14(1):33–38, 27-8

16. Andrei RM et al (2012) Intuitive representation of surface properties of biomolecules using BioBlender. BMC Bioinformatics 13(Suppl 4):S16

17. Heffernan R et al (2015) Improving prediction of secondary structure, local backbone angles, and solvent accessible surface area of proteins by iterative deep learning. Sci Rep 5:11476

18. Sormanni P et al (2015) The s2D method: simultaneous sequence-based prediction of the statistical populations of ordered and disordered regions in proteins. J Mol Biol 427(4):982–996

19. Kallberg M et al (2012) Template-based protein structure modeling using the RaptorX web server. Nat Protoc 7(8):1511–1522

20. Edelstein AD et al (2015) Advanced methods of microscope control using muManager software. J Biol Methods 1(2):e10

21. Truong K, Khorchid A, Ikura M (2003) A fluorescent cassette-based strategy for engineering multiple domain fusion proteins. BMC Biotechnol 3:8

22. Feeney M, Murphy K, Lopilato J (2014) Designing PCR primers painlessly. J Microbiol Biol Educ 15(1):28–29

23. Packer MS, Liu DR (2015) Methods for the directed evolution of proteins. Nat Rev Genet 16(7):379–394

24. Huhner J et al (2015) Quantification of riboflavin, flavin mononucleotide, and flavin adenine dinucleotide in mammalian model cells by CE with LED-induced fluorescence detection. Electrophoresis 36(4):518–525

Chapter 16

A Cassette Approach for the Identification of Intein Insertion Sites

Tim Sonntag

Abstract

Over the past decade split inteins have established themselves as powerful tools for protein engineering, protein semisynthesis, and protein functional control approaches. Their key advantage lies in the protein *trans*-splicing (PTS) reaction that enables posttranslational protein assembly from two independent, even synthetic, peptide precursors. However, since most split intein applications deal with fragmentation and modification of proteins, various issues can arise, ranging from reduced stability to impairment of protein folding. In this chapter I address how the usage of DNA encoded intein cassettes can streamline and speed up the identification of functional split intein insertion sites in novel proteins of interest (POI).

Key words Protein splicing, Split intein, Conditional protein splicing, Protein semisynthesis, Protein expression, RF cloning, Homologous recombination, TEV protease, CREB

1 Introduction

1.1 Split Inteins as Tools for Protein Biochemists: PTS and CPS

Since their discovery more than 20 years ago inteins have evolved into a toolkit for protein engineers and chemical biologists [1]. Their applications range from protein purification and protein semisynthesis to segmental isotopic labeling for NMR spectroscopy. Furthermore, inteins can be utilized for generating cyclic peptides and to control protein function [2–7]. In most applications, split inteins orchestrate the modification of proteins and protein fragments. In contrast to *cis* inteins, the intein domain (~150 amino acids) is split into two independent intein fragments (Int^N and Int^C) (Fig. 1a), that have to encounter each other in solution and subsequently fold into an active complex. Naturally occurring split inteins can be found across various species in the Archaea and Bacteria domains as well as in viral genomes [8, 9]. Additionally, split inteins can be created artificially by separation of the intein domain fold [10, 11]. Similar to *cis* inteins, split inteins enable the protein splicing reaction, a consecutive series of nucleophilic attacks and rearrangements. This reaction, termed protein *trans*-splicing (PTS) for

Henning D. Mootz (ed.), *Split Inteins: Methods and Protocols*, Methods in Molecular Biology, vol. 1495,
DOI 10.1007/978-1-4939-6451-2_16, © Springer Science+Business Media New York 2017

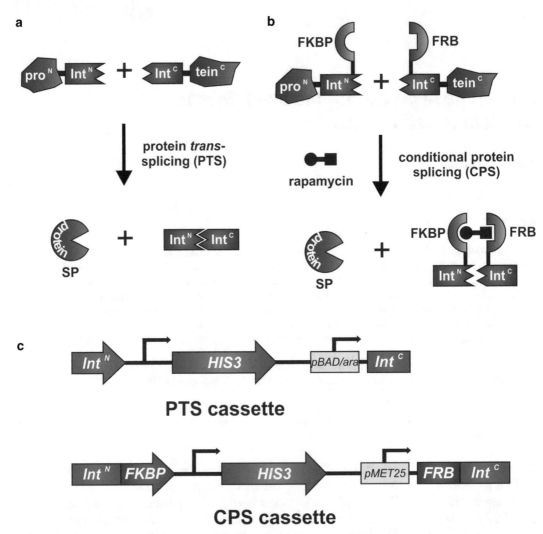

Fig. 1 Split inteins and corresponding intein cassettes. (**a**) Schematic representation of the protein *trans*-splicing (PTS) reaction. Intein activity is reconstituted upon association of the intein halves (Int^N and Int^C). During the PTS reaction the sequences flanking the intein, the exteins (pro^N and tein^C), are joined together generating the splice product (SP). (**b**) Schematic representation of the conditional protein splicing (CPS) reaction. The artificially separated *Sce*VMA intein fragments (Int^N and Int^C) have lost their inherent splicing capability. In CPS fusion with the FKBP/FRB heterodimerization system selectively reconstitutes intein activity: Upon addition of the bivalent small molecule rapamycin the FKBP-FRB complex forms, forcing the intein halves into proximity and hereby restoring activity. (**c**) Scheme of the described PTS and CPS cassettes. Both cassettes encode a pair of intein halves (Int^N and Int^C) enabling the protein *trans*-splicing reaction. The selection marker *HIS3* also functions as a transcriptional spacer and terminator. In the case of the PTS cassette, which contains the split NrdJ-1 intein, the *Int^C* gene is under the control of the arabinose inducible *pBAD* promoter from *E. coli* (*pBAD/ara*). The CPS cassette, which encodes the artificially split VMA intein, includes the FKBP/FRB domains. Expression control of the C-terminal sequences (*FRB-Int^C*) is achieved via the weak *S. cerevisiae MET25* promoter (*pMET25*)

split inteins, culminates in the ligation of the flanking regions of the intein—the exteins (pro^N and tein^C)—via a native peptide bond [12, 13]. The intein mediates the entire PTS reaction, which occurs spontaneously and without the need of an additional energy source. Hence, PTS can occur in vitro or inside virtually any host species. PTS typically takes place within the concentration range of nM (endogenous level inside cells) to low μM (in vitro) and conversion rates of above 90% are not unusual [4]. The half-life of the PTS reaction varies from hours to ~20 s, with the artificial split inteins being orders of magnitudes slower than the naturally occurring *Npu* DnaE or gp41-1 intein [4, 14–16].

Conditional protein splicing (CPS) is mediated by the artificially split VMA intein from *Saccharomyces cerevisiae* and can be classified as a specific subcategory of PTS (Fig. 1b). In contrast to the naturally occurring split inteins, the VMA fragments have no endogenous association ability and the PTS reaction cannot proceed [17, 18]. Therefore, it is possible to control splice product formation by dimerization approaches that force the VMA fragments into close proximity, reconstituting intein activity. Tested approaches are the small molecule mediated rapamycin-FKBP-FRB heterodimerization system, which is specifically known as CPS [17], light-controlled dimerization via the PhyB-PIF3 system [19], or selective coiled-coil domain association [20]. Furthermore, there are alternative ways to control the protein splicing reaction outside of the realm of the VMA intein, ranging from receptor–ligand binding domains to photo-control [21–23].

Previously three different types of PTS cassettes were generated [24]. Although the incorporated split inteins perform protein *trans*-splicing spontaneously, they harbor different amino acids at the C-terminal splice junction (+1 position; Cys, Ser, or Thr) [24]. The +1 position of an intein represents the only invariant remnant of the protein splicing process, residing inside the host protein. In case of the NrdJ-1 intein, which was originally identified in a metagenomic dataset, a serine is found at +1 [9]. Because of its superior reaction speed compared to the existing +1 serine *Ssp* DnaB intein [10, 14], the NrdJ-1 PTS cassette was generated. Besides showing additional beneficial characteristics (*see* **Note 1**), the whole Int^C part of the NrdJ-1 intein is cysteine free. Main applications of the NrdJ-1 intein comprise segmental isotopic labeling and site-specific cysteine based protein labeling.

1.2 Split Intein Insertion: Limitations

The general applicability of PTS/CPS to novel proteins of interest (POI) is limited by the necessity to identify functional split intein insertion sites. The inevitable feature of such a site is that the split intein is still able to perform all chemical reactions to link the exteins together. This can be problematic, because the intein is challenged by an unfamiliar extein sequence context when residing inside a nonnative POI. The potential causes for abolishment of splice product formation range from impairment of protein folding to abolishment of intein fragment association. Another common denominator for successful PTS and CPS is that all protein fragments and protein domains have to be stable upon fusion with the N- and C-terminal intein halves. A known byproduct of inserting a split intein into a new POI is an increase of protein splicing side reactions: N- and C-terminal cleavage. Herein, splicing intermediates are cleaved before the extein sequences could be joined together. The method of choice to suppress N- and C-terminal cleavage is the insertion of 3–5 amino acids directly flanking the N- or C-terminal catalytic residues of the split intein (most frequently a cysteine or serine). These amino acids are commonly selected based upon the native substrate of the intein, assuming increased sequence tolerance because of intein and host gene coevolution. Moreover, N- or C-terminal fusion proteins can increase solubility, stability, and folding of the intein halves. CPS is even more demanding than PTS, since the additional dimerization domains have to execute their function in the context of the fusion proteins. The more data about a POI is available—structural or biochemical—the easier it is to predict functional intein split sites. Although computational methods can offer further guidance [25], there will be cases where carefully selected sites fail for unknown reasons. In this chapter I describe a recently developed DNA cassette strategy that streamlines cloning and evaluation of novel intein insertion sites.

1.3 The PTS and CPS Intein Cassette Design

To this day the identification of functional intein insertion sites still resembles a trial and error process. Therefore, we developed CPS and PTS cassettes allowing the parallel screening of multiple and variable intein insertion sites [24, 26]. These DNA cassettes circumvent individual and time consuming intein fragment cloning steps by coding for both Int^N and Int^C (Fig. 1c). A shared component of the cassettes is the yeast HIS3 selection marker which also functions as a transcriptional spacer and terminator. Specific for the CPS plasmid is the incorporation of the FKBP-FRB heterodimerization domains. Finally, the cassettes encode their own promoter (*pBAD/ara* or *pMET25*), which generates a novel and individually addressable ORF upon cassette integration. By using variable primer sets for the initial intein cassette amplification, amino acid codons which reside next to the catalytic residues are quickly modified or inserted (*see* Subheading 3.1). Moreover, it is possible to perform the integration of the cassette into a gene of interest (GOI) via two strategies, homologous recombination (HR) in *S. cerevisiae* (Subheading 3.2), and in vitro restriction-free (RF) cloning (Subheading 3.3). Both approaches generate a bicistronic integration product that codes for all components to successfully perform PTS. Subheadings 3.4 and 3.5 describe the functional implementation on the protein level in *S. cerevisiae* and *E. coli*, respectively.

2 Materials

2.1 Plasmid Construction

1. TEV plasmid (His$_6$-GST-TEVprotease in p426GAL1), reporter plasmid (MBP-TEVsite-eGFP in p425TDH3), CREB plasmid (ST-GST-TEVsite-mCREB1-His$_6$ in pET22b), helper plasmid (AraC in pRSFDuet).

2. CPS and PTS cassette donor plasmids for PCR (split VMA and NrdJ-1 cassette), schematic representation *see* Fig. 1c (concerning the previously unpublished NrdJ-1 cassette *see* **Note 1**).

3. Restriction enzymes: *Hind*III-HF, *Xho*I, *Nde*I, *Dpn*I (including corresponding buffers).

4. Primers: *See* Table 1 (TEV protease) and Table 2 (CREB).

5. Phusion® High-Fidelity DNA polymerase and PfuUltra II Fusion HS DNA polymerase (including corresponding buffers).

2.2 Expression and CPS in S. cerevisiae

1. Competent *S. cerevisiae* ΔVMA strain (*MATα; his3Δ1; leu2Δ0; met15Δ0; ura3Δ0; tfp1Δ::kanMX4*).

2. Competent *S. cerevisiae* W303 strain (*MATα, ura3-1, ade2-1, trp1-1, his3-11, leu2-3*).

3. Competent *S. cerevisiae* W303 tor2-1 strain (*MATα, ura3-1, ade2-1, trp1-1, his3-11, leu2-3, tor2-1, fpr1Δ::HIS3MX6*).

4. YPD medium, 20 g/l Bacto™ peptone, 10 g/l yeast extract, 2% glucose.

5. SD medium (liquid and plates): –URA, –HIS, –URA –HIS, –URA –LEU, supplemented with either 2% glucose or 2% galactose and 1% raffinose.

6. Glucose stock solution, 40% dextrose.

7. GalRaf stock solution, 20% galactose, 10% raffinose.

8. Rapamycin stock solution, 1 mM rapamycin in DMSO.

2.3 Expression and PTS in E. coli

1. Competent *E. coli* BL21-Gold (DE3) cells.

2. Competent *E. coli* Top10 cells.

3. LB medium, 5 g/l NaCl, 10 g/l tryptone, 5 g/l yeast extract, pH 7.5, supplemented with ampicillin, 100 mg/l, or kanamycin, 50 mg/l, as indicated.

4. 100 g/l ampicillin stock solution.

5. 50 g/l kanamycin stock solution.

6. 400 mM isopropyl β-D-1-thiogalactopyranoside (IPTG) stock solution.

7. 20% L-arabinose stock solution.

Table 1
Primers for homologous recombination into TEV protease

Primer description	Primer sequence (5′–3′)[a]
TEV 16 FP	AAGCTTGTTTAAGGGACCACGTGATTACAACCCGATATCGtgtctttgccaagggtaccaa
TEV 16 RP	TTGTGTGCCCATCAGATTCATTCGTCAAATGACAAATGGTgcaattgtgcacgacaacctgg
TEV 97 FP	AATTATTCGCATGCCTAAAGGATTTCCCACCATTTCCTCAAAgtctttgccaagggtaccaa
TEV 97 RP	GACATATGGCGCTCTTCCCTTTGTGGCTCTCTAAATTTCAGgcaattgtgcacgacaacctgg
TEV 110 (I) FP	AAAGCTGAAATTTAGAGAGCCACAAAGGGAAGAGGCGCATAAtgctttgccaagggtaccaa
TEV 110 (A) FP	CAAAAGCTGAAATTTAGAGAGCCACAAAGGGAAGAGGCGCgcatgtctttgccaagggtaccaa
TEV 110 RP	TAGACATGCTCTTAGTTTGGAAGTTGGTTGTGTCACAAGACAaattgtgcacgacaacctgg
TEV 118 FP	AGGGAAGAGGCGCATATGTCTTGTGACAACCAACTTCCAAACTgtcgggtgctttgccaagggtaccaa
TEV 118 RP	AATGTGCAACTAGTGTCTGACACCATGCTAGACATGCTCTTttctccgcaattgtgcacgacaacctgg
TEV 130 FP	AAACTAAGAGCATGTCTAGCATGGTGTCAGACACTAGTtgctttgccaagggtaccaa
TEV 130 RP	ATGCTTCCAGAATATGCCATCAGATGAAGGGAATGTGCAattgtgcacgacaacctgg
TEV 135 FP	GTCTAGCATGGTGTCAGACACTAGTTGCACATTCCCTTCAtgctttgccaagggtaccaa
TEV 135 RP	CATCCTTGGTTTGAATCCAATGCTTCCAGAATATGCCATCgcaattgtgcacgacaacctgg
TEV 151 FP	TATTCTGGAAGCATTGGATTCAAACCAAGGATGGGCAGtgtctttgccaagggtaccaa
TEV 151 RP	ATGAACCCATCTCTAGTTGATACTAATGGACTGCCACAattgtgcacgacaacctgg
TEV 206 FP	AAATCAGGAGGCGCAGCAGTGGGTTAGTGTTGGGGATTAggttgctttgccaagggtaccaa
TEV 206 RP	TCACCATGAAAACTTTATGGCCCCCCACAATACTGAGTCgcaattgtgcacgacaacctgg

[a] *Uppercase letters* indicate sequence complementarity to the TEV protease plasmid and *lowercase letters* to the CPS cassette plasmid. Introduced/mutated codons are *underlined*

Table 2
Primers for restriction free cloning into CREB

Primer description	Primer sequence (5′–3′)[a]
CREB 161 (WT) FP	AGAAGAAAAATCTGAGGAAGAAACCtgtctggttggcagctctgag
CREB 161 (NPC) FP	AGAAGAAAAATCTGAGGAAGAAACC<u>aacccgtgc</u>tgtctggttggc
CREB 161 (WT) RP	TCACGGTGGTGATCGCCGGTGCGGAgttgtgcaccagaatgtcgttc
CREB 161 (EIV) RP	CGGTCACGGTGGTGATCGCCGGTGC<u>aacgatttc</u>agagttgtgcac

[a]*Uppercase letters* indicate sequence complementarity to the CREB plasmid and *lowercase letters* to the PTS cassette plasmid. Introduced flanking codons NPC/EIV are *underlined* and already encoded by the PTS plasmid

2.4 CPS and PTS Evaluation

1. 4× SDS loading dye, 500 mM Tris–HCl, 8 % (w/v) SDS, 40 % (v/v) glycerol, 20 % (v/v) β-mercaptoethanol, 0.08 % (w/v) bromophenol blue, pH 6.8.

2. NaOH/β-mercaptoethanol solution, 1.85 M NaOH, 7.5 % (v/v) β-mercaptoethanol. Generate a fresh solution by adding 7.5 % (v/v) β-mercaptoethanol to a 2 M NaOH stock solution.

3. 55 % trichloroacetic acid (TCA) solution. Dissolve TCA in ddH$_2$O to give a 55 % (w/v) solution.

4. HU sample buffer, 200 mM Tris–HCl, 8 M urea, 5 % (w/v) SDS, 1 mM EDTA, 1.5 % (w/v) DTT (w/v), 0.1 % (w/v) bromophenol blue, pH 6.8.

5. Strep-tag® II antibody, Penta·His antibody, GFP antibody, GST antibody, Anti-rabbit and Anti-mouse HRP conjugates.

3 Methods

Step-by-step procedures for integrating the split intein cassettes into a gene of interest (GOI) are given in the following protocol. The GOI examples are the tobacco etch virus (TEV) protease for the CPS cassette and the mammalian transcription factor cAMP response element-binding protein (CREB) for the PTS cassette (NrdJ-1 cassette).

3.1 Primer Design and Initial Intein Cassette Amplification

All split intein cassette projects start with the selection of a site for their integration. Depending on the type of application this region can be completely free to choose. However, if the aim is to segmentally label a POI domain with specific isotopes for the purpose of NMR spectroscopy, the selection is inherently confined (for more specific remarks about site selection *see* **Notes 2** and **3**).

General guidelines for the selection of integration sites are outlined in Fig. 2a:

1. The sites can be selected based on preexisting cysteine/serine/threonine residues in the primary sequence, which depends on the used CPS/PTS cassette. This selection is important because during the protein splicing and intein removal process, the nucleophile at position +1 is not removed from the POI. For the *Sce* VMA intein (CPS cassette), the invariant residue is a cysteine, and cysteine/serine residues were preferentially chosen as TEV protease integration sites (*see* Fig. 4b). For the PTS cassettes inteins covering all possible nucleophiles at the +1 position are available: the *Npu* DnaE has a cysteine, the Ssp DnaB & NrdJ-1 a serine, and the *Mxe* GyrA a threonine [14, 24].

2. Proteins often contain flexible linker regions that can be predicted by primary sequence analysis. These regions pose a primary target for the intein cassette integration because of reduced structural confinement.

3. In case structural data about the POI is available, it can guide the selection of integration sites. In particular loops or unresolved regions tend to be successful targets for integration.

As mentioned in Subheading 1.2, a common approach to enhance the intein performance is the addition of endogenous substrate amino acids to the catalytic intein residues. This approach can be easily incorporated in the intein cassette based integration strategy. In fact, all of the intein cassettes designed to date contain three additional endogenous extein codons, which reduce the primer size to amplify the respective intein cassettes.

The primer design to amplify the CPS cassette (Fig. 2b, primer list for TEV protease Table 1, and primer related *see* **Notes 4** and **5**) for subsequent homologous recombination (HR) in yeast is performed as follows:

1. Chose 40 nt complementary to the integration site of the CPS cassette into the TEV protease gene for the 5′ end of the upstream and downstream primers.

2. Use the middle region of the primers to include additional codons, for example codons for endogenous amino acids of the intein (*see* Table 1 and Fig. 4b).

3. At the 3′ end of the primers add ~20 nt complementary to the CPS cassette required for PCR amplification.

Design the primers to amplify the PTS cassette (Fig. 2c, primer list for CREB Table 2, and primer related *see* **Notes 4** and **5**) for subsequent restriction-free (RF) cloning as follows:

a   Selection of insertion sites

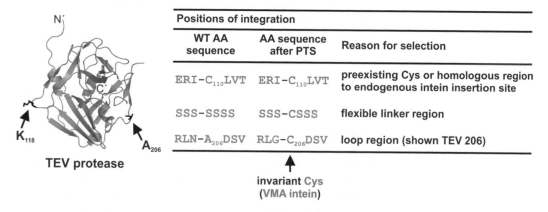

Positions of integration		
WT AA sequence	AA sequence after PTS	Reason for selection
ERI-C$_{110}$LVT	ERI-C$_{110}$LVT	preexisting Cys or homologous region to endogenous intein insertion site
SSS-SSSS	SSS-CSSS	flexible linker region
RLN-A$_{206}$DSV	RLG-C$_{206}$DSV	loop region (shown TEV 206)

invariant Cys
(VMA intein)

b   Primer design CPS cassette (homologous recombination)

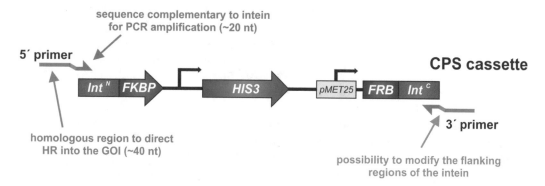

c   Primer design PTS cassette (RF cloning)

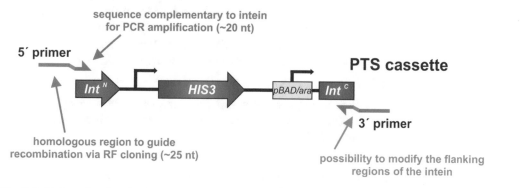

Fig. 2 Guidelines for choosing intein insertion sites and primer design for intein cassette integration. (**a**) Most common approaches to select for intein integration sites in novel proteins of interest (POI). Crystal structure of the TEV protease (PDB ID: 1Q31 [27]) with the TEV118 and TEV206 CPS integration positions highlighted. (**b**) The scheme depicts the different primer regions required for the amplification of the CPS cassette as well as the homologous recombination based integration into a gene of interest (GOI). The typical primer length is around 60 nt. (**c**) The scheme depicts the different primer regions required for the PTS cassette amplification as well as the restriction-free (RF) cloning based GOI integration. The typical primer length is around 45 nt

1. (alternatively to step 1) At the 5′ end of the upstream and downstream primers, include 25 nt complementary to the desired integration site of the PTS cassette into the CREB gene.

2. (alternatively to step 2) Additional codons can be incorporated in the middle region of the primers to vary the amino acids flanking the intein (*see* Table 2 and Fig. 5b).

3. (alternatively to step 3) At the 3′ end of the primers, add ~20 nt complementary to the PTS cassette to facilitate PCR amplification.

Following the primer design for multiple insertion sites and/or same site variants with and without additional codons, the primer pairs are used for the PCR amplification of the CPS and PTS intein cassettes. Despite the variable primer sizes a similar protocol can be used to amplify the intein cassettes with the Phusion® High-Fidelity DNA polymerase.

1. Perform a PCR protocol using Phusion® High-Fidelity DNA polymerase (98 °C 30 s, 30x[98 °C 10 s, 53 °C 30 s, 72 °C X s], 72 °C 8 min, 10 °C ∞). X is the elongation time and depends on the size of the cassette. Use 145 s for the CPS cassette 145 s (size 3207 bp) and 107 s for the NrdJ-1 PTS cassette (size 2377 bp).

3.2 CPS Cassette Integration via Homologous Recombination (HR)

One approach for integrating the intein cassette into the GOI is homologous recombination (HR) using *S. cerevisiae* (Fig. 3a). During the PCR amplification of the CPS cassette (Fig. 2b), the product is extended by 40 bp at both ends, which are complementary to the integration site inside the TEV protease.

1. To increase HR efficiency, the TEV plasmid (2 μ, URA3 marker) is previously incorporated in the ΔVMA strain (lacking the vacuolar ATPase including the VMA intein). The strain is propagated in –URA liquid SD media and LiAc transformed with the unpurified PCR mixture of the CPS cassette (*see* **Note 6**).

2. Select transformants on -URA -HIS SD plates.

3. Pick a single colony and grow it overnight in 5 ml -HIS SD medium. Colonies at this stage contain a heterogeneous mixture of plasmids.

4. The following day, extract the DNA, using a DNA/RNA phenol–chloroform extraction protocol (*see* **Note 6**).

5. W303 competent cells are grown in YPD medium and transformed with the isolated DNA/RNA mixtures from **step 4**.

6. Select transformants on -HIS SD plates.

7. Pick and grow a single colony (which now contains a homogenous plasmid content) overnight in 5 ml -HIS SD medium, followed by DNA/RNA phenol–chloroform extraction on the next day (*see* **Note 6**).

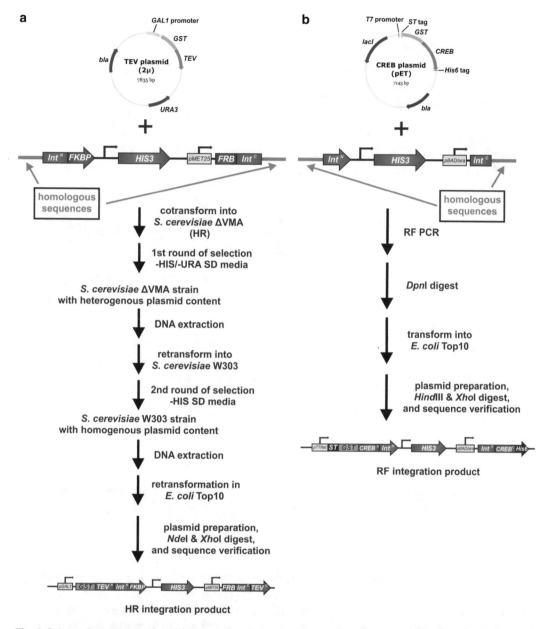

Fig. 3 Schematic protocols for split intein cassette integration. (**a**) Step-by-step procedure that is required to generate the homologous recombination (HR) based integration product in *S. cerevisiae*. (**b**) Step-by-step procedure that is required to generate the restriction-free (RF) cloning based integration product in vitro & in *E. coli*

8. Transformed chemically competent *E. coli* Top10 cells with the DNA/RNA solutions and select transformants on LB plates, 100 mg/l ampicillin.

9. For plasmid preparation, inoculate a single *E. coli* colony in 5 ml LB medium, 100 mg/l ampicillin, and growth overnight.

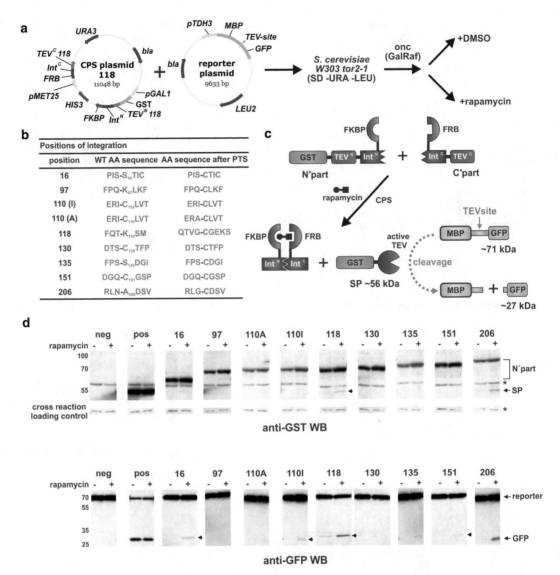

Fig. 4 Integration of the CPS cassette into the TEV protease. (**a**) Scheme of the co-expression and CPS induction in *S. cerevisiae*. The W303 *tor2-1* strain is grown in -URA –LEU SD medium, 2 % galactose, 1 % raffinose (the *URA3* marker is on the CPS plasmid 118 and the *LEU2* marker is on the reporter plasmid). 1 μM rapamycin is added and the cultures are incubated overnight (onc: overnight culture). (**b**) Table depicting the tested TEV protease integration positions. The table lists the primary amino acid sequence before and after the protein-*trans* splicing reaction. (**c**) Schematic representation of the CPS reaction. The protein splicing reaction of the N-terminal protein (GST-TEV^N-Int^N-FKBP) and C-terminal protein (FRB-Int^C-TEV^C) is induced by the small molecule rapamycin. The splice product (GST-TEV) is generated after dimerization of FKBP-FRB, active intein complex formation, and excision of the intein halves. TEV protease activity can be measured by cleavage of the MBP-TEVsite-GFP reporter. (**d**) GST western blot analysis of the whole lysate splicing reaction. The expression rate of the N-terminal fusion protein (N'part: GST-TEV^N-Int^N-FKBP) as well as the splice product formation (SP: GST-TEV) is detected. The GFP western blot detects MBP-TEVsite-GFP reporter cleavage (neg = negative control = reporter plasmid and empty plasmid; pos = positive control = reporter plasmid and GST-TEV)

10. Prepare plasmid solutions from the overnight culture and digest the DNA with *Nde*I & *Xho*I to verify successful integration into the TEV protease gene. After observation of the desired band pattern (size depends upon integration site, e.g., TEV118: 5.5 and 3.4 and 1.9 kbp). Further verify the integration product by DNA-sequencing of both ORFs that have been generated by CPS cassette integration.

3.3 Restriction-Free (RF) Cloning of the PTS Cassette

An alternative approach to generate intein cassette integration products is restriction-free (RF) cloning (Fig. 3b) [28]. RF cloning can be regarded as a modified site directed mutagenesis protocol, and in comparison to HR the integration is performed in vitro. In the final step RF cloning relies upon the DNA replication machinery of *E. coli* to generate the circularized integration product.

1. Digest the PCR product of the initial PTS cassette amplification using *Dpn*I (1 μl *Dpn*I in 50 μl PCR reaction volume, 37 °C 6–8 h). This step removes template DNA (*see* **Note 7**).

2. Isolate the linear PTS cassette DNA by agarose gel purification (*see* **Note 7**).

3. Perform RF cloning in a PCR reaction of 25 μl volume with 25 ng of the CREB plasmid as template DNA, 150 ng of the respective PTS PCR product and the PfuUltra II Fusion HS DNA polymerase (95 °C 2 min, 35x[95 °C 30 s, 55 °C 60 s, 68 °C 19 min], 68 °C 10 min, 10 °C ∞). Elongation time is calculated as 2 min/kbp, with the absolute integration product size being 9.5 kbp (*see* Fig. 5a, elongation time = 19 min).

4. Digest the PCR mixture with *Dpn*I (30 μl volume: 5 μl PCR reaction, 1 μl *Dpn*I, corresponding buffers and ddH$_2$O; 37 °C 6–8 h). This step removes template DNA.

5. Transform chemically competent *E. coli* Top10 cells with 1 μl of the reaction from the previous step and select colonies overnight on LB plates, 100 mg/l ampicillin.

6. Pick a single colony to inoculate 5 ml LB medium, 100 mg/l ampicillin, and grow overnight.

7. Prepare plasmid solutions from the overnight culture.

8. Digest the isolated plasmid with *Nde*I and *Hind*III-HF to verify the successful integration into the CREB plasmid. After observation of the desired band pattern (6.0 and 1.9 and 1.5 kbp), further verify the integration product by DNA-sequencing of both ORFs, which have been generated by the integration of the PTS cassette.

3.4 CPS of the TEV Protease in S. cerevisiae

The protocol given here represents the standard procedure to evaluate if CPS occurs in *S. cerevisiae*. To monitor TEV protease activity, a substrate protein is co-expressed from a co-transformed reporter plasmid (Fig. 4, *see* **Note 8** about initial GOI construct design).

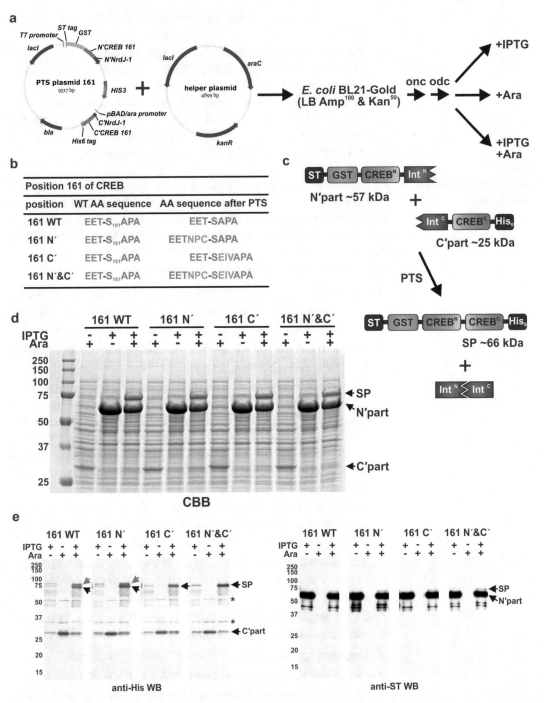

Fig. 5 Integration of the PTS cassette into the transcription factor CREB (**a**) Co-expression scheme for *E. coli*. A BL21-Gold strain harboring the PTS plasmid 161 (*bla*) and the helper plasmid (*kanR*) is propagated in LB medium, 100 mg/l ampicillin, 50 mg/l kanamycin. After overnight and overday cultivation (onc and odc), the promoters are individually induced and co-induced using 0.8 mM IPTG and 0.2% L-arabinose. (**b**) Table depicting variations at the PTS integration position 161. The table lists the primary amino acid sequence before and after the protein *trans*-splicing reaction. (**c**) Schematic representation of the PTS reaction. The splice product (ST-GST-CREB-His$_6$) formation occurs upon association of the intein fusion proteins (ST-GST-CREB^N-Int^N and Int^C-CREB^C-His$_6$) and excision of the intein halves. (**d**) Coomassie brilliant blue (CBB) stained SDS page and (**e**) Western blot analysis of whole lysate splicing reactions. The N-terminal (N'part: ST-GST-CREB^N-Int^N), the C-terminal protein

1. After successful generation of the integration products (depicted example CPS plasmid 118, Fig. 4a), transform *S. cerevisiae* W303 *tor2-1* strain with the CPS and the reporter plasmids using a LiAc protocol (*see* **Note 6**). Plate cells for selection on -URA -LEU SD plates.

2. Use a single colony to inoculate 5 ml -URA -LEU SD medium, 2% galactose 1% raffinose. For each individual CPS plasmid two cultures are inoculated (Fig. 4b: 9 integration variants, resulting in 18 cultures). Grow cultures overnight in 25 ml baffled Erlenmeyer flasks. Subsequently, add rapamycin to a final concentration of 1 μM to one culture and DMSO, 0.1% (v/v), as a mock control to the other. Hence, the expression of the N-terminal fusion protein (*pGAL1*, Fig. 4a) as well as the CPS reaction is induced at the same time (rapamycin-FKBP-FRB complex formation; Fig. 4c).

3. On the following day determine the optical density (OD_{600}) for all cultures and pellet a similar number of cells by centrifugation at $17{,}000 \times g$ for 2 min in a microfuge (corresponding to 1 ml of culture with an $OD_{600} = 1$).

4. Resuspend the pellet in cold 500 μl ddH_2O and keep the samples on ice.

5. Lyse the cells by addition of 75 μl NaOH/β-mercaptoethanol solution.

6. Incubate for 15 min on ice.

7. Add 75 μl of 55% TCA solution.

8. Incubate 10 min on ice.

9. Spin down the solution in a microfuge for 10 min at $17{,}000 \times g$ and 4 °C.

10. Thoroughly discard the supernatant and resuspend the pellet in 35 μl of HU sample buffer [29], followed by incubation at 65 °C for 10 min. The protein samples are centrifuged at $17{,}000 \times g$ for 5 min in a microfuge and then stored at −20 °C.

11. Assay the protein samples by Western blot using GST and GFP antibodies (Fig. 4d).

Western blot analysis showed the generation of full length TEV protease by CPS at positions 118 and 206 (splice product (SP), Fig. 4d). Furthermore, cleavage of the MBP-TEVsite-GFP reporter occurred in both cases. *See* **Note 9** about the uninduced SP formation (only DMSO treated cultures).

Fig. 5 (continued) (C′part: Int^C-$CREB^C$-His_6), and the SP (SP: ST-GST-CREB-His_6) can all be visualized. The His Western blot detects the C′part and the SP and the ST (strep tag II) Western blot the N′part and the SP. In samples 161 WT and 161 N′ of the His Western blot an additional band above the height of the SP was detected (*red arrow*)

This protocol describes the standard procedure to monitor PTS in *E. coli* via small scale expression and Western Blot analysis (Fig. 5, *see* **Note 8** about initial GOI construct design). The helper plasmid is essential for the PTS cassette method, because it codes for both the LacI (Lac repressor) and the AraC (arabinose operon regulatory protein) proteins. In combination with the two inducible ORFs on the PTS plasmid (T7 promoter and *pBAD/ara* promoter driven, Fig. 5a), LacI and AraC ensure selective expression.

1. After successful generation of the integration products (depicted example PTS plasmid 161), transform chemically competent *E. coli* BL21-Gold (DE3) (containing the helper plasmid) with the PTS plasmid and select on LB plates, 100 mg/l ampicillin, 50 mg/l kanamycin.

2. Pick a single colony harboring a PTS integration plasmid (four variants, *see* Fig. 5b) and inoculate 5 ml LB medium, 100 mg/l ampicillin, 50 mg/l kanamycin, and grow overnight at 37 °C.

3. On the following day, use each culture to inoculate fresh 6 ml LB medium, 100 mg/l ampicillin, 50 mg/l kanamycin, to give an $OD_{600} = 0.2$.

4. Grow the cultures at 37 °C to an OD_{600} of 0.5–0.7. Then split each culture split into three tubes of LB medium, 100 mg/l ampicillin, 50 mg/l kanamycin (each 1.5 ml volume): (a) supplemented with 0.8 mM IPTG, (b) supplemented with 0.2% L-arabinose, and (c) supplemented with 0.8 mM IPTG and 0.2% L-arabinose. These treatments induce the PTS plasmid expression, leading to the accumulation of the N- and C-terminal fusion proteins, and potentially formation of the splice product (Fig. 5c). *See* **Note 10** about remarks on the selective induction.

5. Grow cultures at 37 °C for 4 h.

6. Pellet similar amount of cells by centrifugation at $17,000 \times g$ for 2 min in a microfuge (corresponding to 1 ml of culture at $OD_{600} = 1$). Resuspend the pellet in 100 μl SDS loading dye ($2 \times$ SDS loading dye diluted 1:1 with ddH_2O).

7. Boil the samples at 95 °C for 10 min and centrifuge at $17,000 \times g$ for 5 min in a microfuge. Store samples at –20 °C for future use.

8. Analyze the supernatant of the protein samples by SDS PAGE, followed by both Coomassie Brilliant Blue (CBB) staining (Fig. 5d) and Western blot analysis using His and Strep-tag® II antibodies (Fig. 5e).

The analysis of the CBB staining and Western blot concluded that PTS took place at all variants of position 161. However, the His Western blot allowed to discriminate the splice product (SP) from a higher running species (red arrow, Fig. 5d). This species possibly represents a splice complex intermediate. The occurrence of this band was abolished by introducing the C-terminal codons of the NrdJ-1 intein (EIV, Fig. 5b).

4   Notes

1. The NrdJ-1 intein was used in the described PTS example, because in the authors hands it repeatedly demonstrated superior performance in terms of foreign extein context and suppression of N-terminal and C-terminal cleavage. This was determined for multiple insertions sites inside CREB, an *Arabidopsis thaliana* transcription factor, and a cyanobacterial kinase. In all cases the NrdJ-1 intein was directly compared to the *Npu* DnaE and IMPDH1 intein. The NrdJ-1 intein cassette plasmid is available from the author upon request.

2. As described in the main text, the examples about selection of split sites are considered general guidelines, and are not applicable for all POIs. For a given POI, an N-terminal fusion protein might be too unstable, making it impossible for PTS/CPS to occur. In these cases it is recommended to generally use a stabilizing and solubility increasing fusion protein in the initial screen for functional split sites. For TEV206, the SP was detected after integration into GST-TEV, but integration into His_{10}-TEV abolished SP formation and TEV activity. Also, the N-terminal fusion with FRB can function as a weak degron. Despite the fact that the addition of rapamycin can stabilize the whole FRB-Int^c protein, an additional N-terminal fusion might be necessary to improve CPS.

3. The TEV protease consists of a single catalytically active domain, limiting choices for intein insertion. However, if there are alternatives in larger and more complex POIs, the site of integration should be considered carefully. Assuming the aim of CPS is to functionally regulate a multi-domain kinase or protease, split intein integration inside or near catalytically active domains might be necessary to fully disrupt its function. In case the active domain is left intact, protein fragments can be generated, which retain residual activity or even become constitutively active.

4. The complementary sequences required for the integration via HR (40 bp) or RF-cloning (25 bp) might need extension in length depending upon the sequence context around the integration site. For example, homologous regions inside the same gene (e.g., domain repeats) or direct sequence duplications could increase the chance of miss-insertion.

5. It should be noted that the synthesis of primers in the range of 60 nt can introduce mutations or truncation products that can later interfere with the performance of the RF-cloning or HR

approaches. The author repeatedly encountered mutations in integration products at regions covered by the amplification primers. There is no need to order specially purified primers per se (e.g., HPLC purified). However, if integration products derived from independently re-amplified CPS/PTS cassettes carry identical mutations, primer resynthesis is recommended.

6. The *S. cerevisiae* protocols for LiAc transformation as well as DNA/RNA phenol–chloroform extraction are assumed to be standard lab protocols. They can be found in corresponding publications [30–32], for example.

7. The combination of *Dpn*I digest and agarose gel purification will improve RF cloning performance, because remnants of the circular PTS cassette plasmid as well as PCR reaction truncation products will be removed.

8. A critical step in both PTS and CPS is the design of the initial GOI construct. The aim of the first PTS/CPS experiment should be to characterize the protein *trans*-splicing reaction in full. To this end the N- and C-terminal POI derived fusion proteins should both be tagged to detect and quantify the SP, and potentially occurring N- and C-terminal cleavage products (*see* Subheading 1.2). Furthermore, it is highly recommended to incorporate a functional readout of POI activity in the experimental set-up, because even the observation of splice product does not necessarily result in folding and catalytic activity of the POI. The acquired information will facilitate and guide necessary future optimization steps (e.g., *see* **Notes 9** and **10**).

9. In case of CPS a potential side reaction is the occurrence of PTS without the addition of the inducer rapamycin. This property is known as "background splicing" and in the case of the TEV protease occurs at both positions 118 and 206 (SP in the GST Western Blot, Fig. 4d). Background splicing is likely influenced by the absolute and relative amount of the N- and C-terminal intein fragments. The *GAL1* promoter is a strong yeast promoter compared to the relatively weak *MET25* promoter (Fig. 4a). During prolonged expression high amounts of N-terminal precursor protein accumulate, which augments background splicing. In subsequent experiments background splicing could be diminished by changing to a weaker promoter and reducing the expression time [26].

10. The PTS expression protocol described here represents the most straightforward procedure to determine if an intein insertion site is functional or not. For other applications, such as segmental isotopic labeling, where tight control of gene expression is a prerequisite, additional optimization steps might be required. The general concept focuses on highly selective expression by "washing out" residual amounts of the

inducers (IPTG and arabinose) from the growth media as well as addition of promoter silencing agents (glucose, *pBAD* promoter). Detailed procedures are beyond the scope of this protocol, but can be found in subsequent publications [24, 33].

Acknowledgements

I thank the Mootz lab for continuous support and suggestions, in particular Joachim Zettler for his contribution to the PTS project. The author acknowledges funding from the Fonds der Chemischen Industrie, the National Institutes of Health (R01 GM074868 and R01 DK083834), and the Kieckhefer Foundation.

References

1. Gimble FS, Thorner J (1992) Homing of a DNA endonuclease gene by meiotic gene conversion in Saccharomyces cerevisiae. Nature 357(6376):301–306. doi:10.1038/357301a0

2. Lennard KR, Tavassoli A (2014) Peptides come round: using SICLOPPS libraries for early stage drug discovery. Chemistry 20(34):10608–10614. doi:10.1002/chem.201403117

3. Mootz HD (2009) Split inteins as versatile tools for protein semisynthesis. Chembiochem 10(16):2579–2589. doi:10.1002/cbic.200900370

4. Shah NH, Muir TW (2014) Inteins: nature's gift to protein chemists. Chem Sci 5(1):446–461. doi:10.1039/C3SC52951G

5. Topilina NI, Mills KV (2014) Recent advances in in vivo applications of intein-mediated protein splicing. Mob DNA 5(1):5. doi:10.1186/1759-8753-5-5

6. Volkmann G, Iwai H (2010) Protein trans-splicing and its use in structural biology: opportunities and limitations. Mol Biosyst 6(11):2110–2121. doi:10.1039/c0mb00034e

7. Wood DW, Camarero JA (2014) Intein applications: from protein purification and labeling to metabolic control methods. J Biol Chem 289(21):14512–14519. doi:10.1074/jbc.R114.552653

8. Perler FB (2002) InBase: the intein database. Nucleic Acids Res 30(1):383–384

9. Dassa B, London N, Stoddard BL, Schueler-Furman O, Pietrokovski S (2009) Fractured genes: a novel genomic arrangement involving new split inteins and a new homing endonuclease family. Nucleic Acids Res 37(8):2560–2573. doi:10.1093/nar/gkp095

10. Brenzel S, Kurpiers T, Mootz HD (2006) Engineering artificially split inteins for applications in protein chemistry: biochemical characterization of the split Ssp DnaB intein and comparison to the split Sce VMA intein. Biochemistry 45(6):1571–1578. doi:10.1021/bi051697+

11. Ludwig C, Pfeiff M, Linne U, Mootz HD (2006) Ligation of a synthetic peptide to the N terminus of a recombinant protein using semisynthetic protein trans-splicing. Angew Chem Int Ed Engl 45(31):5218–5221. doi:10.1002/anie.200600570

12. Mills KV, Johnson MA, Perler FB (2014) Protein splicing: how inteins escape from precursor proteins. J Biol Chem 289(21):14498–14505. doi:10.1074/jbc.R113.540310

13. Volkmann G, Mootz HD (2013) Recent progress in intein research: from mechanism to directed evolution and applications. Cell Mol Life Sci 70(7):1185–1206. doi:10.1007/s00018-012-1120-4

14. Carvajal-Vallejos P, Pallisse R, Mootz HD, Schmidt SR (2012) Unprecedented rates and efficiencies revealed for new natural split inteins from metagenomic sources. J Biol Chem 287(34):28686–28696. doi:10.1074/jbc.M112.372680

15. Shah NH, Dann GP, Vila-Perello M, Liu Z, Muir TW (2012) Ultrafast protein splicing is common among cyanobacterial split inteins: implications for protein engineering. J Am Chem Soc 134(28):11338–11341. doi:10.1021/ja303226x

16. Zettler J, Schutz V, Mootz HD (2009) The naturally split Npu DnaE intein exhibits an extraordinarily high rate in the protein trans-

splicing reaction. FEBS Lett 583(5):909–914. doi:10.1016/j.febslet.2009.02.003

17. Mootz HD, Blum ES, Tyszkiewicz AB, Muir TW (2003) Conditional protein splicing: a new tool to control protein structure and function in vitro and in vivo. J Am Chem Soc 125(35):10561–10569. doi:10.1021/ja0362813

18. Mootz HD, Blum ES, Muir TW (2004) Activation of an autoregulated protein kinase by conditional protein splicing. Angew Chem Int Ed Engl 43(39):5189–5192. doi:10.1002/anie.200460941

19. Tyszkiewicz AB, Muir TW (2008) Activation of protein splicing with light in yeast. Nat Methods 5(4):303–305. doi:10.1038/nmeth.1189

20. Selgrade DF, Lohmueller JJ, Lienert F, Silver PA (2013) Protein scaffold-activated protein trans-splicing in mammalian cells. J Am Chem Soc 135(20):7713–7719. doi:10.1021/ja401689b

21. Buskirk AR, Ong YC, Gartner ZJ, Liu DR (2004) Directed evolution of ligand dependence: small-molecule-activated protein splicing. Proc Natl Acad Sci U S A 101(29):10505–10510. doi:10.1073/pnas.0402762101

22. Berrade L, Kwon Y, Camarero JA (2010) Photomodulation of protein trans-splicing through backbone photocaging of the DnaE split intein. Chembiochem 11(10):1368–1372. doi:10.1002/cbic.201000157

23. Binschik J, Zettler J, Mootz HD (2011) Photocontrol of protein activity mediated by the cleavage reaction of a split intein. Angew Chem Int Ed Engl 50(14):3249–3252. doi:10.1002/anie.201007078

24. Zettler J, Eppmann S, Busche A, Dikovskaya D, Dotsch V, Mootz HD, Sonntag T (2013) SPLICEFINDER—a fast and easy screening method for active protein trans-splicing positions. PLoS One 8(9), e72925. doi:10.1371/journal.pone.0072925

25. Lee YT, Su TH, Lo WC, Lyu PC, Sue SC (2012) Circular permutation prediction reveals a viable backbone disconnection for split proteins: an approach in identifying a new functional split intein. PLoS One 7(8):e43820. doi:10.1371/journal.pone.0043820

26. Sonntag T, Mootz HD (2011) An intein-cassette integration approach used for the generation of a split TEV protease activated by conditional protein splicing. Mol Biosyst 7(6):2031–2039. doi:10.1039/c1mb05025g

27. Nunn CM, Jeeves M, Cliff MJ, Urquhart GT, George RR, Chao LH, Tscuchia Y, Djordjevic S (2005) Crystal structure of tobacco etch virus protease shows the protein C terminus bound within the active site. J Mol Biol 350(1):145–155. doi:10.1016/j.jmb.2005.04.013

28. van den Ent F, Lowe J (2006) RF cloning: a restriction-free method for inserting target genes into plasmids. J Biochem Biophys Methods 67(1):67–74. doi:10.1016/j.jbbm.2005.12.008

29. Silver PA, Chiang A, Sadler I (1988) Mutations that alter both localization and production of a yeast nuclear protein. Genes Dev 2(6):707–717

30. Gietz D, St Jean A, Woods RA, Schiestl RH (1992) Improved method for high efficiency transformation of intact yeast cells. Nucleic Acids Res 20(6):1425

31. Rose MD, Winston F, Hieter P (1990) Methods in yeast genetics: a laboratory course manual. Cold Spring Harbor, New York

32. Schiestl RH, Gietz RD (1989) High efficiency transformation of intact yeast cells using single stranded nucleic acids as a carrier. Curr Genet 16(5–6):339–346

33. Muona M, Aranko AS, Raulinaitis V, Iwai H (2010) Segmental isotopic labeling of multi-domain and fusion proteins by protein trans-splicing in vivo and in vitro. Nat Protoc 5(3):574–587. doi:10.1038/nprot.2009.240

Chapter 17

Computational Prediction of New Intein Split Sites

Yi-Zong Lee, Wei-Cheng Lo, and Shih-Che Sue

Abstract

Split inteins have emerged as a powerful tool in protein engineering. We describe a reliable in silico method to predict viable split sites for the design of new split inteins. A computational circular permutation (CP) prediction method facilitates the search for internal permissive sites to create artificial circular permutants. In this procedure, the original amino- and carboxyl-termini are connected and new termini are created. The identified new terminal sites are promising candidates for the generation of new split sites with the backbone opening being tolerated by the structural scaffold. Here we show how to integrate the online usage of the CP predictor, CPred, in the search of new split intein sites.

Key words Circular permutation, Split intein, Protein splicing, Protein ligation

1 Introduction

Many inteins were discovered in the single polypeptide form [1, 2]. However, in some cases they were found as split inteins with a backbone split, therefore dividing their sequences into N-terminal and C-terminal fragments [2–4]. When the two fragments are able to reconstitute into an entire intein fold, they could recover the catalytic activity of the canonical ligation reaction [1]. These two-piece split inteins fulfill the ligation process known as protein *trans*-splicing (PTS), which ligates the flanking exteins and excision of intervening split intein [5]. Thus, the split-intein system has emerged as a powerful protein-engineering tool for protein ligation, cyclization, segmental isotope labeling and posttranslational modification [5–7]. Many naturally or artificially identified split inteins have been reported [1, 2], while researchers are still looking for better split inteins with great enzyme efficiency and high assembling specificity between the two fragments [1, 8]. Another concerned issue is the length of the short fragment of split intein. If the fragment is short enough, chemical synthesis could be incorporated into the preparation of this short piece [8–11].

Henning D. Mootz (ed.), *Split Inteins: Methods and Protocols*, Methods in Molecular Biology, vol. 1495,
DOI 10.1007/978-1-4939-6451-2_17, © Springer Science+Business Media New York 2017

There is still no robotic method to identify the viable split sites. We have demonstrated the feasibility that valid circular permutation (CP) sites in proteins have the potential to act as split sites [12]. A developed CP predictor (CPred) has been used to search for internal permissive sites for creating new split proteins [12]. CP, a protein backbone rearrangement, could be visualized as the original amino- and carboxyl-termini of a protein are linked and new backbone opening is created at another position [13, 14]. The CP predictor facilitates the creation of circular permutants in which backbone opening imposes the least detrimental effects on protein folding. Many CPs have been identified in well-known protein families [15]. Studying these natural cases revealed that circular permutants usually retain native functions [16, 17], sometimes with altered enzymatic activity and functional diversity [18, 19]. Additionally, artificial CPs have been tested in different proteins [13, 14, 16, 20–22]. The derived concept reveals that if the CP site is not detrimental for folding, the circular permutants could retain their native structure [13, 14, 16, 17, 23], although the structural stability and folding mechanism might be changed [20–22].

To predict viable CPs is difficult because of the unavailability of a practical method. To facilitate the process, we developed a program, CPSARST (Circular Permutation Search Aided by Ramachandran Sequential Transformation), to search for CPs from protein structural databases [24]. We also performed a large-scale search and created the first CP database, CPDB [25]. Finally, we developed the CP predictor, CPred (Circular Permutation site predictor), to rank possible CP sites in a given protein structure and output a score for each residue [26].

To search for functional split inteins, we tested the CPred prediction in NpuDnaE intein and evaluated the structural features of the selected sites [12]. We converted the valid CP sites to the split sites to predict new backbone openings with little effect on protein structure and folding. If the two fragments of intein successfully assemble themselves, there will be great chance to obtain a viable split intein. Together with observations made in other split proteins [12], our results have demonstrated the feasibility of the strategy. The established protocol therefore offers an efficient and systematic way to rationally design a split intein. A current limitation of the CPred server is the requirement of a protein structure to make predictions. In the near future, a new version of CPred requiring only protein sequences will be released. By integrating the power of various sequence-based discriminatory features, it will be able to predict viable CP sites (and split sites) for proteins without determined structures. In this chapter, we describe how the current CP predictor can be used in split intein design.

2 Description of the Computational Method

2.1 The Concept of CPred Program

Theoretically, positions important for protein folding are improbable to be valid CP sites [13, 14, 16, 17, 23]. Creating new termini located in the protein core also are likely to be energetically disfavored. In agreement with these considerations, CP sites tend to occur at the positions with greater solvent accessibility [27], lower sequence conservation and lower "closeness," a measure describing the molecular contact with the surrounding residues [28, 29]. However, these properties only yielded marginal performance, indicating that there are still other discriminatory differences between viable and inviable CP sites. We improved the prediction by establishing a well-designed dataset (CPDB) and analyzing all available CP sites. The analyses were integrated into the development of CPred by properly reorganizing the dataset for training and testing, and meanwhile, employing four machine-learning methods [26]. In our implementation, the output of every machine learning method for each residue was designed to be a real number score between 0 and 1, termed as a "probability score" [30]. To integrate the prediction power of the various methods, we averaged the individual scores into an integrated score. The performances of these methods for predicting viable CP sites had been well tested and the detailed algorisms and parameter settings are described elsewhere [30]. Other structural features considered in CPred include the number of hydrogen bonds, centroid distance measurement [31], GNM-F (Gaussian Network Model-derived mean-square fluctuation), RMSF (root mean square fluctuation) and the distance from the buried core of protein. The CPred software is accessible at http://sarst.life.nthu.edu.tw/CPred [26].

2.2 Characterization of Residues at CP Sites

We summarized the amino acid propensities of CP sites in Fig. 1. There is a significant preference for Pro and Gly at viable CP sites. Hydrophilic residues, especially Asp and Asn, are preferred whereas bulky hydrophobic residues like Met, Leu, and Ile are disfavored (Fig. 1a). Hydrophilic and neutral residues also have higher occurrence whereas hydrophobic residues have lower occurrence at CP sites (Fig. 1b). Viable CP sites prefer negatively charged and polar uncharged residues and disfavor nonpolar aliphatic residues (Fig. 1c). Repeating these tests on another dataset confirmed that Cys and aromatic residues are disfavored for CP (Fig. 1e, g) and CP has little or no preference for positively charged residues (Fig. 1c, g). Additionally, the preference of viable CPs for positions with relatively low sequence conservation and high solvent accessibility [27, 28] implies that CPs favor loop conformation (Fig. 1d, h) [33]. Some of the mentioned propensities cannot be easily understood at present, but these might reflect biological significances, for example that Pro and Gly are frequently found in loops [34] and the observed preference for Asp and Asn over Glu, Gln, Arg and Lys is the reverse of the preference for the residues

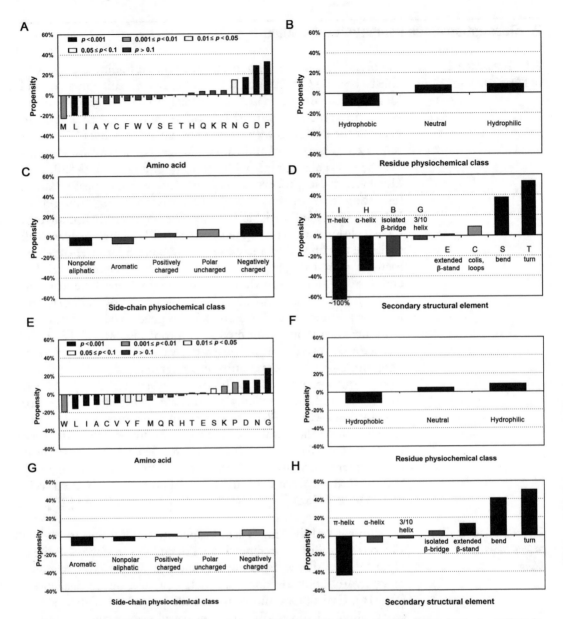

Fig. 1 Sequence and secondary structural propensities at viable CP sites. Each bar in these charts shows the relative occurrence of a pattern for the reference groups (nrCPDB-40, nrGIS-40) and viable CP site groups (nrCPsite$_{cpdb}$-40, nrCPsite$_{gis}$-40). The value derived from reference group was considered as the zero point in each experiment; therefore, a positive or a negative value indicates that the frequency of the pattern at CP sites was higher or lower than its background frequency. As shown in chart (**a**), the *dark* to *light gray colors* in the bar represent decreasing levels of *p*-values. Results shown in (**a–d**) were obtained from nrCPDB-40 and nrCPsite$_{cpdb}$-40. Patterns examined in this experiment include: (**a**) amino acids, (**b**) residue physiochemical types, (**c**) side-chain physiochemical types, and (**d**) secondary structural element determined by program DSSP [32]. The same test is repeated on another dataset, nrGIS-40 and nrCPsite$_{gis}$-40, and the similar analyses are shown in (**e–h**)

occurred in helices [35–37]. Thus, the statistics report that CP sites prefer the positions in loops and disfavor those in helices [30].

Important to the current work, the preference of CPs revealed that splitting the following positions in a protein would less affect the protein from folding into a stable structure: (1) hydrophilic/neutral residues, especially those with negatively charged or polar uncharged side chains, (2) residues in coils, loops, and turns, and (3) residues with large solvent-accessible surface areas, short distances to the protein surface, few hydrogen bonds, unpacked environments, and high flexibility. By contrast, CPs disfavor (1) hydrophobic residues, particularly those with bulky or aromatic side chains, (2) residues in helices and β-bridges, and (3) highly buried residues. To design a split intein is also a process of creating new termini. The same preference could be applied in the split-site search.

3   Method

1. Prepare structural coordinate of your target intein. If the target intein structure is available, you could either upload the structure coordinate (PDB file) or directly input the PDB 5-letter entry code (PDB identifier with chain identifier). X-ray or NMR structures both are readily submitted to CPred without any modification. For each prediction, CPred will output a value between 0 and 1 for each residue. The score closer to 1 corresponds to an increased likelihood. If the intein structure is unavailable, a structural homology modeling of the target intein is required before CPred prediction. Several structural modeling programs are suggested, such as Phyre2, SWISS-MODEL, and ROBETTA.

2. Consider the CP sites of score >0.7. CP sites with high CPred scores are competent to be backbone split sites. The CPred score = 0.7 is an empirical threshold form the currently known split proteins [12]. We also noticed that the reported functional split sites in inteins are mostly with CPred scores higher than 0.7. Meanwhile, there is a great correspondence between low CPred score (<0.5) and invalid backbone splitting. A position located in a region with high CPred scores is worthwhile to be tested as a split site. If there are residues with similar CPred scores in a region and thus confusing the split site selection, the amino acid propensities of CPs in Fig. 1a, e are well referable.

3. Prevent the sites involved in enzyme ligation. To maintain the PTS activity in the split intein, you should choose positions with at least a moderate separation from the active site. Although most functional active residues are located in the center of intein, there are some residues distributed on protein surface, involved in PTS activity.

4. Estimate the interface area between the two fragments. The PTS efficiency is highly related to the ability of fragments to

associate. The interface of split intein fragments controls the ability of fragment association. Therefore, the size of interface area could be positively correlated to the affinity that a larger interface most likely indicates better association. Software such as AREAIMOL (www.ccp4.ac.uk/html/areaimol.html) could be used to estimate the interfacial area [38].

5. Analyze the interface residue composition. The solubility of the two fragments of split intein also acts as a determined factor for PTS activity. The backbone opening might not be harmful on CP folding, but fragments in split intein might be insoluble if hydrophobic residues constitute the interface. The split site separates intein into two pieces and might expose many buried residues. To design a viable split intein, split sites resulting in relatively hydrophilic interfaces would have better choices than those creating hydrophobic interfaces.

4   Case Study

We analyzed six inteins (seven coordinates) with available 3D structures (Fig. 2). These commonly used inteins contain sequences with great variations in sequence identities, ranging from 15 to 60%. Meanwhile, they show similar 3D structures such that their backbone RMSD values are less than 2.5 Å. The output CPred profiles of these inteins are generally similar (Fig. 2). There are several consensus "hot spots" with scores >0.7 (representing great potential to be split sites), including sequence 10–15, 23–27, 34–41, 60–65, 76–80, 87–92, 96–104, and 122–125, where the number are referred to NpuDnaE intein sequence. Taking NpuDnaE as the model, most reported functional split sites were at or near the regions with local CPred score maximums (Fig. 3) [9, 12]. We calculated the corresponding interface areas by creating all possible split inteins (Fig. 3a). The interface area of each split intein is obtained by the equation of (surface area of N-terminal fragment + surface area of C-terminal fragment—surface area of one-piece intein)/2. The split sites distributed in the middle of the sequence contain the large interface areas and the values reach maximum at the sequence 75–100. The large interface area matches to the nM binding affinity of the naturally occurring split intein at split site 102 [12]. For the engineered split sites 36 and 123, the interface areas are generally half of the maximum, corresponding to the moderate association affinity [12]. Furthermore, for the engineered split sites 11 and 131, the association affinity became much weaker due to the small contact area. We converted the CPred scores on the sausage model and indicated the locations of functional split sites in Fig. 3b. By this procedure, the CP site 24

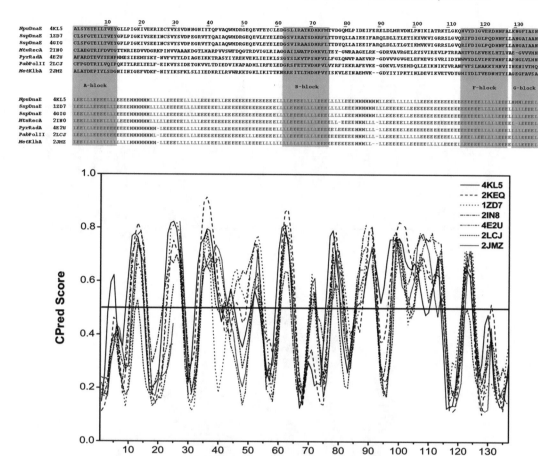

Fig. 2 Prediction of circular permutation probabilities of inteins by CPred predictor. Structure-based multiple sequences of different inteins are aligned using DALI-server: *Npu*DnaE from *Nostoc punctiforme* (4KL5 for crystal structure and 2KEQ for solution structure); *Ssp*DnaE from *Synechocystis sp.*; *Mtu*RecA from *Mycobacterium tuberculosis*; *Pyr*RadA from *Pyrococcus horikoshii*; *Pab*PolII from *Pyrococcus abyssi*; *Met*KlbA from *Methanococcus jannaschii*. The secondary structure of β-sheet is depicted in E, α-Helix in H and loop in L. The highly conserved blocks of A, B, F, and G, involved in the splicing mechanism, are indicated with *gray background*

was predicted. However, the site could not be successfully explored for the generation of new split NpuDnaE intein variant. Significant precipitation of the both fragments was observed, likely because of the hydrophobic interface (data not shown). This example shows that not all predicted sites will yield a useful split intein, despite successes at other positions [12].

266   Yi-Zong Lee et al.

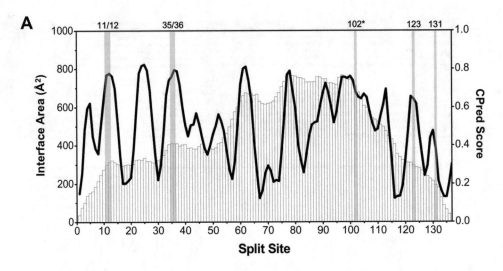

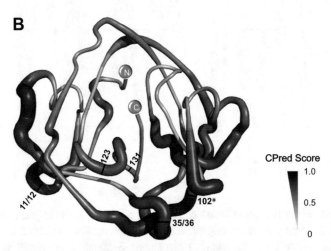

Fig. 3 The correlation of CP score, interface area and viable split sites. (**a**) *Npu*DnaE intein CP score predicted by CPred and the interface area of individual split inteins calculated by AREAIMOL. (**b**) CP scores mapping onto the structure by sausage model with a color gradient. The verified artificially and naturally occurring (*asterisk*) split sites are indicated

Acknowledgement

This work was supported by Ministry of Science and Technology (MOST), Taiwan (101-2311-B-009-006-MY2, 102-2113-M-007-014, and 103-2113-M-007-016) and National Tsing Hua University (104N2051E1).

References

1. Aranko AS, Wlodawer A, Iwai H (2014) Nature's recipe for splitting inteins. Protein Eng Des Sel 27:263–271

2. Shah NH, Muir TW (2014) Inteins: nature's gift to protein chemists. Chem Sci 5:446–461

3. Iwai H, Zuger S, Jin J, Tam PH (2006) Highly efficient protein trans-splicing by a naturally split DnaE intein from Nostoc punctiforme. FEBS Lett 580:1853–1858

4. Wu H, Hu Z, Liu XQ (1998) Protein trans-splicing by a split intein encoded in a split DnaE gene of Synechocystis sp. PCC6803. Proc Natl Acad Sci U S A 95:9226–9231

5. Muralidharan V, Muir TW (2006) Protein ligation: an enabling technology for the biophysical analysis of proteins. Nat Methods 3:429–438

6. Volkmann G, Iwai H (2010) Protein trans-splicing and its use in structural biology: opportunities and limitations. Mol Biosyst 6:2110–2121

7. Zuger S, Iwai H (2005) Intein-based biosynthetic incorporation of unlabeled protein tags into isotopically labeled proteins for NMR studies. Nat Biotechnol 23:736–740

8. Sun W, Yang J, Liu XQ (2004) Synthetic two-piece and three-piece split inteins for protein trans-splicing. J Biol Chem 279:35281–35286

9. Aranko AS, Zuger S, Buchinger E, Iwai H (2009) In vivo and in vitro protein ligation by naturally occurring and engineered split DnaE inteins. PLoS One 4:e5185

10. Ludwig C, Schwarzer D, Zettler J, Garbe D, Janning P, Czeslik C, Mootz HD (2009) Semisynthesis of proteins using split inteins. Methods Enzymol 462:77–96

11. Mootz HD (2009) Split inteins as versatile tools for protein semisynthesis. Chembiochem 10:2579–2589

12. Lee YT, Su TH, Lo WC, Lyu PC, Sue SC (2012) Circular permutation prediction reveals a viable backbone disconnection for split proteins: an approach enabling identification of a new functional two-piece intein for protein trans splicing. PLoS One 7:e43820

13. Tsai LC, Shyur LF, Lee SH, Lin SS, Yuan HS (2003) Crystal structure of a natural circularly permuted jellyroll protein: 1,3-1,4-beta-D-glucanase from Fibrobacter succinogenes. J Mol Biol 330:607–620

14. Ribeiro EA Jr, Ramos CH (2005) Circular permutation and deletion studies of myoglobin indicate that the correct position of its N-terminus is required for native stability and solubility but not for native-like heme binding and folding. Biochemistry 44:4699–4709

15. Lo WC, Lyu PC (2008) CPSARST: an efficient circular permutation search tool applied to the detection of novel protein structural relationships. Genome Biol 9:R11

16. Lindqvist Y, Schneider G (1997) Circular permutations of natural protein sequences: structural evidence. Curr Opin Struct Biol 7:422–427

17. Vogel C, Morea V (2006) Duplication, divergence and formation of novel protein topologies. Bioessays 28:973–978

18. Qian Z, Lutz S (2005) Improving the catalytic activity of Candida antarctica lipase B by circular permutation. J Am Chem Soc 127:13466–13467

19. Todd AE, Orengo CA, Thornton JM (2002) Plasticity of enzyme active sites. Trends Biochem Sci 27:419–426

20. Li L, Shakhnovich EI (2001) Different circular permutations produced different folding nuclei in proteins: a computational study. J Mol Biol 306:121–132

21. Chen J, Wang J, Wang W (2004) Transition states for folding of circular-permuted proteins. Proteins 57:153–171

22. Bulaj G, Koehn RE, Goldenberg DP (2004) Alteration of the disulfide-coupled folding pathway of BPTI by circular permutation. Protein Sci 13:1182–1196

23. Cunningham BA, Hemperly JJ, Hopp TP, Edelman GM (1979) Favin versus concanavalin A: circularly permuted amino acid sequences. Proc Natl Acad Sci U S A 76:3218–3222

24. Lo WC, Huang PJ, Chang CH, Lyu PC (2007) Protein structural similarity search by Ramachandran codes. BMC Bioinformatics 8:307

25. Lo WC, Lee CC, Lee CY, Lyu PC (2009) CPDB: a database of circular permutation in proteins. Nucleic Acids Res 37:D328–D332

26. Lo WC, Wang LF, Liu YY, Dai T, Hwang JK, Lyu PC (2012) CPred: a web server for predicting viable circular permutations in proteins. Nucleic Acids Res 40:W232–W237

27. Iwakura M, Nakamura T, Yamane C, Maki K (2000) Systematic circular permutation of an entire protein reveals essential folding elements. Nat Struct Biol 7:580–585

28. Paszkiewicz KH, Sternberg MJ, Lappe M (2006) Prediction of viable circular permutants using a graph theoretic approach. Bioinformatics 22:1353–1358

29. Amitai G, Shemesh A, Sitbon E, Shklar M, Netanely D, Venger I, Pietrokovski S (2004)

Network analysis of protein structures identifies functional residues. J Mol Biol 344:1135–1146

30. Lo WC, Dai T, Liu YY, Wang LF, Hwang JK, Lyu PC (2012) Deciphering the preference and predicting the viability of circular permutations in proteins. PLoS One 7:e31791

31. Shih CH, Huang SW, Yen SC, Lai YL, Yu SH, Hwang JK (2007) A simple way to compute protein dynamics without a mechanical model. Proteins 68:34–38

32. Kabsch W, Sander C (1983) Dictionary of protein secondary structure: pattern recognition of hydrogen-bonded and geometrical features. Biopolymers 22:2577–2637

33. Panchenko AR, Madej T (2005) Structural similarity of loops in protein families: toward the understanding of protein evolution. BMC Evol Biol 5:10

34. Crasto CJ, Feng J (2001) Sequence codes for extended conformation: a neighbor-dependent sequence analysis of loops in proteins. Proteins 42:399–413

35. Lyu PC, Liff MI, Marky LA, Kallenbach NR (1990) Side chain contributions to the stability of alpha-helical structure in peptides. Science 250:669–673

36. Chakrabartty A, Kortemme T, Baldwin RL (1994) Helix propensities of the amino acids measured in alanine-based peptides without helix-stabilizing side-chain interactions. Protein Sci 3:843–852

37. Moreau RJ, Schubert CR, Nasr KA, Torok M, Miller JS, Kennedy RJ, Kemp DS (2009) Context-independent, temperature-dependent helical propensities for amino acid residues. J Am Chem Soc 131:13107–13116

38. Lee B, Richards FM (1971) The interpretation of protein structures: estimation of static accessibility. J Mol Biol 55:379–400

INDEX

Henning D. Mootz (ed.), *Split Inteins: Methods and Protocols*, Methods in Molecular Biology, vol. 1495,
DOI 10.1007/978-1-4939-6451-2, © Springer Science+Business Media New York 2017

Printed in the United States
By Bookmasters